Clearer Skies Over China

Clearer Skies Over China

Reconciling Air Quality, Climate, and Economic Goals

edited by Chris P. Nielsen and Mun S. Ho

Dear Phil,

Thanks for the advice over the years.

Chris Nielsen

Ho Mun Sing

The MIT Press

Cambridge, Massachusetts

London, England

MIT Press books may be purchased at special quantity discounts for business or sales promotional use. For information, please email special_sales@mitpress.mit.edu or write to Special Sales Department, The MIT Press, 55 Hayward Street, Cambridge, MA 02142.

This book was set in 10/14 pt Sabon by Toppan Best-set Premedia Limited, Hong Kong. Printed and bound in the United States of America.

Clearer skies over China : reconciling air quality, climate, and economic goals / edited by Chris P. Nielsen and Mun S. Ho.
    pages   cm
Includes bibliographical references and index.
ISBN 978-0-262-01988-0 (hardcover : alkaline paper)
1. Air quality—China. 2. Air—Pollution—China. 3. Climatic changes—Government policy—China. 4. Environmental policy—China. 5. China—Economic policy—2000- 6. Economic development—China. 7. Economic development—Enviromental aspects—China.
I. Nielsen, Chris P., 1960– II. Ho, Mun S.
TD883.7.C6C59   2013
363.739′20951—dc23
2013006963

10 9 8 7 6 5 4 3 2 1

Contents

Preface and Acknowledgments   vii

Part I   Introduction, Review, and Summary   1

1   Atmospheric Environment in China: Introduction and
Research Review   3
Chris P. Nielsen and Mun S. Ho

2   Summary: Sulfur Mandates and Carbon Taxes for 2006–2010   59
Chris P. Nielsen, Mun S. Ho, Yu Zhao, Yuxuan Wang, Yu Lei, and Jing Cao

3   Summary: Carbon Taxes for 2013–2020   103
Chris P. Nielsen, Mun S. Ho, Jing Cao, Yu Lei, Yuxuan Wang, and Yu Zhao

Part II   Studies of the Assessment   159

4   Primary Air Pollutant Emissions of Coal-Fired Power Plants in China   161
Yu Zhao

5   Primary Air Pollutants and CO_2 Emissions from Cement Production
in China   203
Yu Lei, Qiang Zhang, Chris P. Nielsen, and Kebin He

6   An Anthropogenic Emission Inventory of Primary Air Pollutants in China
for 2005 and 2010   225
Yu Zhao, Wei Wei, and Yu Lei

7   Atmospheric Modeling of Pollutant Concentrations   263
Yuxuan Wang

8   Benefits to Human Health and Agricultural Productivity of Reduced
Air Pollution   291
Yu Lei

9   The Economics of Environmental Policies in China   329
Jing Cao, Mun S. Ho, and Dale W. Jorgenson

Part III   Appendixes   373

Appendix A   Economic-Environmental Model of China   375
Jing Cao, Mun S. Ho, and Dale W. Jorgenson

Appendix B   The Valuation of Health Damages   393
Yu Lei and Mun S. Ho

Appendix C   New Assumptions and Methods for the 2013–2020 Policy Cases   403
Yu Zhao, Yuxuan Wang, Yu Lei, and Chris P. Nielsen

Contributors   407
Index   409

Preface and Acknowledgments

The research of this book expands the inquiry into local and global air pollution in China of our 2007 edited volume, *Clearing the Air: The Health and Economic Damages of Air Pollution in China*, also from MIT Press. That study grew from a mandate in the mid-1990s of the Harvard University Center for the Environment (HUCE) to bring scholars together from across disciplines to jointly address environmental research topics. Building on that work, the current volume presents a new collaborative effort of researchers from various institutions in China and the United States to improve our knowledge of the complex web relating public policy, economic growth, energy use, local air quality, and the global atmosphere. This research was conducted over 2007–2012 and was orchestrated by the China Project of the School of Engineering and Applied Sciences (SEAS) at Harvard University, in collaboration with several schools of Tsinghua University.

The China Project was established by the HUCE and from the outset has been chaired by Michael B. McElroy of the Department of Earth and Planetary Sciences and SEAS and directed by one of us (Nielsen). The simulation of atmospheric transport and chemistry in this book was led by Yuxuan Wang of the Center for Earth System Science at Tsinghua University; Wang developed the China-specific, "nested" version of the GEOS-Chem atmospheric model in her Harvard Ph.D. dissertation, advised by McElroy, and in a series of subsequent scientific articles.[1] She now leads the China side of the Harvard-Tsinghua collaboration that collects data for use in the model at an atmospheric measurement station built by the team near Beijing in 2004, led on the Harvard side by J. William Munger.

The comprehensive emissions inventory used here was constructed at Harvard by Yu Zhao, with central contributions from Yu Lei, integrating and improving on the extensive work of their respective Ph.D. dissertations at what was then called the Department of Environmental Science and Engineering (DESE, now School of Environment) at Tsinghua. Zhao's Ph.D. research had been advised by Jiming Hao (who also led the air pollution component of *Clearing the Air*), and Lei's by Kebin He. Other researchers collaborated with Zhao or Lei at Tsinghua or Harvard, including Lei Duan, Shuxiao Wang, Qiang Zhang, and Wei Wei.

The health damage research of the book was conducted by Yu Lei at SEAS and builds on his work in a prior study with He and other colleagues at Tsinghua DESE, Xiaochuan Pan of the Peking University School of Public Health, and the Chinese and U.S. environmental protection agencies. Lei also drew strongly from chapters in *Clearing the Air* by colleagues at the Harvard School of Public Health, led by Jonathan I. Levy (now at Boston University), James K. Hammitt, and Ying Zhou (now at Emory University), all of whom also gave early advice. Lei completed the agricultural damage assessment at SEAS with our advice, the impressive resources of the Fung and Harvard-Yenching libraries, and research assistance from Samantha Go (who also assisted us on many other tasks for this book).

The economic model used to study the impact of policy on energy use and economic growth was constructed by Jing Cao, Dale W. Jorgenson, and one of us (Ho), and elements of the model formed parts of Cao's Ph.D. thesis at the Harvard Kennedy School of Government. This effort built on earlier work done with Richard Garbaccio and with substantial assistance from Shantong Li and Jianwu He from the China Development Research Center. Jorgenson also led the economics component in the previous *Clearing the Air*.

We owe deep gratitude to the friends and colleagues named above who led the work on individual chapters and to their collaborators. Coordinating such a large research team from such diverse fields into an integrated project was a complex task, taking a number of years and covering long distances. This task was possible because of the key participants' patience with the inevitable delays and their willingness to learn from, instruct, and accommodate the needs of other members of the team. Many project meetings and multiple rounds of reworking the research and revising the papers and book chapters took place.

Progress reports and drafts of this research were presented at seminars or workshops at both the Washington-based ChinaFAQs program and the Beijing office of the World Resources Institute, the Woodrow Wilson School at Princeton University, Tsinghua University School of Economics and Management, Nanjing University School of Environment, Peking University School of Physics, Sun Yat-Sen University (cohosted by the U.S. National Center for Atmospheric Research), the 2011 World Congress of the International Economic Association, and varied Harvard programs and units including the Center for Geographical Analysis, the Harvard Kennedy School, Harvard College, and, most gratifyingly, a 2010 symposium held by SEAS and HUCE in honor of Michael McElroy. We are indebted to the participants in all these events for their valuable critiques and comments.

Presenting such a complex economic, energy, and environmental assessment in a single book designed to bridge lay and expert readerships has been a challenging but rewarding undertaking for the editors and authors. We thank Clay Morgan and his colleagues at MIT Press for their constructive advice and commitment to the project. We owe special thanks to four anonymous reviewers for their detailed comments, which greatly improved the book.

The research of this volume was supported by a number of funding sources of the Harvard China Project, most of which also included research funding for Tsinghua University. The study was initiated under an award for collaborative research on China from the Harvard China Fund. This prepared the way for two successive grants for the core of the research from the China Sustainable Energy Program of the Energy Foundation. Harvard's Weatherhead Center for International Affairs provided a final major grant, to present the results in the scholarly form of this book. Supplementary support for the emissions research was provided from the Harvard Smeltzer Fund and by grants ATM-0635548 and ATM-1019134 from the U.S. National Science Foundation, and for the economics research by the Cheung Yan Fund of the Harvard Department of Economics. On behalf of the entire research team, we gratefully acknowledge the generous financial support of all of these sources.

Mun Ho thanks the Harvard Institute of Quantitative Social Science, Resources for the Future (Washington), the Environment for Development Ethiopia Center, the Ethiopian Development Research Institute, and the Tsinghua School of Economics and Management, all for hosting him at various points over the last several years while much of this work was conducted.

Chris Nielsen thanks Michael McElroy for his advice and patient support for the complex work of this initiative. He also thanks the Harvard School of Engineering and Applied Sciences, under Deans Cherry Murray and Venkatesh Narayanamurti, for its support of collaborative research that bridges both across disciplines and from the United States to China on our shared challenges to protect the atmospheric environment.

Chris P. Nielsen and Mun S. Ho

Note

1. Names of Chinese in this book are reversed to Western name order, with surnames last, to avoid erroneous citations of their work.

I

Introduction, Review, and Summary

1

Atmospheric Environment in China: Introduction and Research Review

Chris P. Nielsen and Mun S. Ho

1.1 Introduction

With each passing year, the future of the global environment becomes more affected by policy choices that China is making regarding its economy, use of energy, and atmospheric environment. Other nations further along the development path bear greater historical responsibility for the atmospheric loading of greenhouse gases (GHGs) that drive global climate change: the United States has emitted far more than China in cumulative terms, and many countries have larger per capita emissions. China's sheer size, rate of economic growth, and dependence on fossil fuels, however, have vaulted it far into the forefront in current national emissions of the dominant GHG, carbon dioxide (CO_2). One authoritative source estimates that of the 33.9 gigatons (Gt) of CO_2 emitted globally in 2011, China contributed 9.70, while the second-ranking United States emitted 5.42 (Olivier et al. 2012). There is little to suggest, moreover, that China's rising share of world CO_2 emissions will stabilize anytime soon, let alone decline.

International strategies to constrain the world's carbon trajectory must be equitable to all countries given their differing development stages. Many have also noted that China's emissions are due in part to its production of goods that are exported to other countries, and hence world consumption is ultimately responsible for a portion of its growing emissions. But regardless of how we assign the causes, and thus the burden of paying for mitigation, it is inescapable that China must play a central role if global GHG control strategies are to be effective.

At the same time, China's domestic air quality is severely and persistently degraded, especially in densely urbanized regions of the country. Even as concentrations of a few key "primary" pollutants—those emitted directly from sources, such as large particulates and sulfur dioxide—have been successfully reduced, China increasingly suffers the more complex air quality hazards of advanced economies. These are caused more by "secondary" pollutants, those formed chemically in the air from precursor gases, notably ozone and a large share of the particulate matter small enough to penetrate deeply into human lungs (i.e., less than 2.5 microns in diameter, termed $PM_{2.5}$). These conditions are exacerbated in part by exploding vehicle populations, which now clog China's heavily congested cities and have joined power plants and other coal-burning sources as the leading contributors to China's worst air pollution hazards.

However intractable China's atmospheric environmental challenges seem, its government deserves credit for trying energetically to respond over the last decade. China has pursued a raft of policies that target both fossil fuel use and atmospheric emissions, some major examples of which are useful to recount. In 2004, China enacted fuel economy standards for new cars that exceeded those of many much richer nations, including the United States, and it has been implementing phased vehicle emission standards modeled directly on those of the European Union (Oliver et al. 2009). The 2006–2010 11th Five-Year Plan (11th FYP) set targets for reducing the energy intensity of the economy by 20% and reducing emissions of sulfur dioxide by 10%, which were then followed by implementing mandates backed by newly vigorous enforcement mechanisms. In 2007, major energy plans launched breakneck development of hydroelectric, wind, nuclear, and other nonfossil electric power generation, driven by ambitious capacity targets for the year 2020 (NDRC 2007). Around this time, massive investments were made to relocate industries, upgrade pollution controls, and otherwise manage emissions to improve air quality in Beijing for the 2008 Summer Olympic Games, also as a test case for other Chinese cities. In late 2009, China unilaterally committed to reduce the carbon intensity of its economy 40–45% by 2020, compared to the 2005 level. The 12th Five-Year Plan has recommitted to these general aims, with targets for 2015 in carbon intensity (–17% compared to 2010), energy intensity (–16%), emissions of sulfur dioxide (SO_2, –8%) and nitrogen oxides (NO_X, –10%), and in the nonfossil share of total primary energy supply (11.4%).

While broad and ambitious, this program has produced only mixed results. Many of the individual elements have been successful judged by the narrow targets that motivated them, and indeed some of these successes may be historic given the scale of their effects and compared to earlier efforts that failed.[1] The 11th FYP successfully reversed a rise in energy intensity in the early 2000s and nearly achieved the 20% reduction target. China quickly leaped ahead of schedule on almost all the nonfossil power capacity goals, becoming a world leader in wind power generation in the process, and prompting most of the 2020 targets to be raised even higher. Environmental authorities proved many doubters wrong when the world's Olympic athletes competed under mainly clear Beijing skies. The 11th FYP SO_2 controls reversed a long growth trend and exceeded the 10% reduction target, the effects of which will be a primary subject of this book.

Judged by the overarching objectives of reduced carbon emissions and sustainably clean air, however, these efforts have been largely overwhelmed. Fossil fuel combustion drives most emissions of both global GHGs and local pollutants, and the energy demands of continued Chinese economic growth have outstripped the gains in energy efficiency and expansion of nonfossil energy supply. Consumption of coal and oil has continued to grow swiftly, driving greater emissions of CO_2 and pollutants not subject to high-profile national targets.

These successes and failures must be understood in a broader context. The government's environmental goals and efforts are embedded in the drive for national development. Few would question such economic objectives as raising living standards across a country where the average per capita income was only US$4940 in 2011, or US$8450 if adjusted for purchasing power (World Bank 2012). (The unadjusted per capita income of Hong Kong in 2011, for comparison, was US$36,010.) Carbon and air pollution control policies must be compatible with such national economic goals, as maintaining growth to benefit a billion lower-income Chinese is essential to the welfare not just of China, but also of the world. Otherwise, China's progress on improving the atmospheric environment will continue to be halting at best.

Conceiving such policies too narrowly in other respects may have also led to slower environmental progress than originally expected. The atmosphere is an exceedingly complex physical, chemical, and biological system, with dozens of relevant species and hundreds of reactions to consider. Air quality is influenced by meteorological, topographical, and land-use conditions that vary by location. High-profile, national prioritization of just a few pollutants for which tractable control options are well understood (such as primary particulates and SO_2) has neglected other important pollutants that are more difficult to control (NO_X until recently, for instance, or ammonia and volatile organic compounds currently) and complex secondary pollutant pathways (such as the photochemistry that produces smog). Narrowly targeted policies can lead to very real but similarly narrow environmental gains; achieving sustained improvement in overall air quality requires multipollutant strategies that are rooted in thorough understanding of physical and chemical atmospheric processes.

The high concentrations of $PM_{2.5}$ that afflicted China's northern cities in the first months of 2013 could prove to be a cautionary case in point. While many observers suggested that failed implementation of existing pollution controls were largely to blame,[2] careful scientific research is in fact needed to disentangle a wide range of possible causal factors. Such factors could also include, among others, anomalous meteorological conditions (such as persistent inversions and weakened average wind speeds) and the role of $PM_{2.5}$-precursors unrecognized to date in Chinese pollution control policies (such as ammonia). This inherent physical and chemical complexity also explains why air quality strategies will likely succeed only incrementally, after years of supporting research, policy development, and implementation. Unfortunately, quick fixes are highly unlikely.

Efforts to improve one dimension of the atmospheric environment, moreover, can also sometimes worsen another. Reducing SO_2 emissions by flue gas desulfurization (FGD) simultaneously raises energy use and CO_2 emissions because FGD systems themselves consume energy. Improving air quality in China's megacities by relocating industrial sources to adjoining rural areas or interior provinces with less stringent emission controls can simply relocate hazards to new areas (Zhao, Zhang, and Nielsen 2013*†). A transition to electric vehicles could reduce growth in oil consumption and vehicle emissions, but the effects on CO_2 emissions and regional air quality could be negative if the electricity is generated by the current coal-dominated capacity mix (Huo et al. 2010). Successful control of SO_2 reduces sulfate particles that harm public health, but also cancels their negative radiative forcing that may be moderating climate change in the short term (Solomon et al. 2007).[3]

With so many cross-cutting factors impinging on China's atmospheric environmental future, the most useful assessment of policy would have a very comprehensive scope. The urgency of reducing these environmental hazards has led to studies of the relationships between China's economic growth, energy use, local pollution, and carbon emissions. A review of such research, including our own prior book-length study, is presented later in this chapter. Given the difficulty of quantifying trade-offs across this confluence, however, no analysis to date is truly comprehensive.

This book tackles this challenge. It has two major goals. The first is to develop a rigorous framework for integrated analyses of national emission control policies

---

[†] A number of the studies cited in this chapter, including some by alumni lead authors subsequently collaborating from Tsinghua University, Nanjing University, or Peking University, were produced as part of the Harvard-led program conducting the research of this book. We note these studies with an asterisk (*) on the citation and in the reference list.

that recognizes the main elements of this complex web. Our approach links independent research in five main areas:

- Simulation of China's economy and energy use in response to emission control policies
- Plant-by-plant representations of key emitting industries, including coal-fired electric power and cement making
- Estimation of nationwide emissions of all major air pollutants and CO_2, based on independent scientific research and not official statistics
- Simulation of regional air quality resulting from emissions, including critical pollutants that are formed chemically in the air
- Estimation of the impacts of air quality on human health and crop productivity

We believe these steps, taken together, represent a major advance in research on the issues of interest. The core of this methodological advance is the integration of state-of-the-art economic and atmospheric models of China within a single framework, the linkage made possible only by recent progress in research on emissions. This type of integration has not been accomplished previously on China, and in fact has rarely been attempted even in research on the United States or other countries.[4] It brings the types of research tools and perspectives employed in the strongest current scholarship across relevant economic and scientific fields to policy assessment, raising its rigor and enhancing the credibility of its conclusions. For readers familiar with our prior book (Ho and Nielsen 2007), the major innovations and expanded scope of the current study compared with the older work are introduced briefly in section 1.6 and detailed in section 2.3.

The second goal of this book is to apply this framework to evaluation of two sets of national emission control policies, one concerning the recent past and the other the near future. The study of the past, summarized in chapter 2, examines the years 2006–2010 and considers two scenarios:

- *Scenario P1*, the measures actually implemented in China's 11th FYP to reduce emissions of SO_2 in the electric power sector
- *Scenario P2*, a carbon tax of 27 yuan per ton of CO_2 (100 yuan per ton of carbon) implemented hypothetically instead of the SO_2 controls, with the tax revenues transferred back to consumers

The future policy set concerns the years 2013–2020, as summarized in chapter 3, and broadly expands the investigation of carbon taxes to evaluate different tax rates and structures:

- *Scenario F1*, a carbon tax set at a static rate of 30 yuan per ton of CO_2 from 2013 to 2020, with revenues again transferred back to consumers
- *Scenario F2*, a carbon tax that begins in 2013 at only 10 yuan per ton of CO_2 but then grows steadily to roughly 50 yuan per ton in 2020 in order to achieve the same cumulative carbon reduction as F1, with the same use of revenues
- *Scenario F3*, a tax identical to F2 except that the level ramps up much faster to 100 yuan per ton in 2020, roughly equivalent to international prices of carbon in the late-2000s
- *Scenario F4*, a carbon tax that is identical to F2 except that it is instead revenue-neutral, with all revenues used to reduce preexisting taxes instead of being distributed to consumers

The "P" and "F" in the scenario labels denote past and future, and are used to distinguish the two sets of scenarios when discussing them together.

As summarized in chapters 2 and 3, we quantify the effects of each policy not only on their targeted emissions but also on their side effects on other emission types, on the resulting air quality, on associated damages to public health and crop productivity, and on GDP growth and other economic indicators. (Our investigations of carbon taxes consider a few cases in addition to the ones described previously, but judged only by economic criteria.)[5]

We choose the two 2006–2010 cases not to advise specific policy decisions—which is inapplicable, as they concern the past—but rather to evaluate and compare the effectiveness of two types of emission control approaches that China has implemented or might implement in the future: technology mandates and market-based controls. These cases also establish our interdisciplinary capacity to analyze diverse emission control policies.

In the case of technology mandates, the conclusions we draw are based not on hypothetical conditions but rather on what may have been one of the most effective environmental policies ever implemented in China: the SO_2 controls of the 11th FYP. When considering future policies, it is invaluable to know how well past ones have performed in practice. This book provides an exhaustive assessment of the 2006–2010 SO_2 controls, indicating that the policy was broadly successful at protecting the health of Chinese citizens at little net cost to economic growth (even if other pollution risks may have obscured its benefits since 2010).

For market-based emission controls, we cannot take the same approach, as no such policies have yet been attempted in China at more than experimental scales.

For our first such case, a carbon tax, we pick the same years as those of the SO_2 controls, 2006–2010, so that we can draw a useful comparison of the effects of the two policies. Making our first carbon tax analysis a hypothetical scenario of the past also has the advantage of being rooted in the observed conditions that actually played out over 2006–2010.

We are nevertheless ultimately interested in scenarios from the past to gain lessons for future policy choices. Investigating the future is inherently more uncertain, requiring stronger assumptions about how the economy, technologies, and other policies will evolve. Such analyses are strengthened by use of a framework already demonstrated in an empirically informed investigation of the past.

Our 2013–2020 policies are chosen to investigate the effectiveness of different carbon tax rates and structures, reflected in the cases described earlier. And, central to all of our tax scenarios, we want to explore the degree to which a carbon tax might function as a multipollutant control strategy, thereby bringing pressing domestic air quality goals into powerful alignment with global imperatives in climate protection.

1.2 Contents of This Book and How to Read It

Part I of this book, comprising the first three chapters, introduces the subjects and summarizes its assessments of policy in language and presentation that is easily accessible to nonspecialist readers.

The current chapter provides an overview of the topics of concern and the state of research that investigates them. It seeks to establish for all readers a strong foundation in the latest scientific advances relevant to policy choices and the ways in which they relate to official information and data. It serves not only as background on the trade-offs facing China in trying to control its atmospheric emissions in an economically efficient way, but also as an introduction and reference on the tools and data available to guide decisions about those trade-offs.

Chapters 2 and 3 summarize the integrated research of the rest of the book. Chapter 2 reports the costs and benefits for the emission control policy cases set in the past, the SO_2 controls and a hypothetical carbon tax in 2006–2010. Chapter 3 presents the results for the policies set in the future, prospective carbon taxes imposed over 2013–2020. These first three chapters thus also provide the context for understanding the specific research areas that together drive our policy conclusions.

Part II consists of chapters 4–9 and presents the underlying research component by component and in much richer detail. These chapters allow nonexpert readers to explore any elements that interest them, and they provide to other researchers the necessary basis to understand, critique, adapt, and improve upon our work. The research components are described in the following sequence:

- Chapters 4 and 5: Development of plant-by-plant inventories of emissions of two critical polluting industries, coal-fired electric power generation and cement manufacturing

- Chapter 6: Establishment of a complete national inventory of emissions of major air pollutants, CO_2, and other atmospheric species, linked to the output of all economic sectors and gridded spatially for atmospheric simulation

- Chapter 7: Application of an advanced model of atmospheric transport and chemistry in China that simulates how changes in emissions affect the concentrations and distributions of major air pollutants, including critical species formed chemically in the air from precursors

- Chapter 8: Development of an analytical system for estimating health and agricultural exposures and damages from air pollution, and thus the benefits of policies to control emissions

- Chapter 9: Application of an economic model of China designed to represent the critical interindustry linkages and to evaluate economic growth and energy use in response to emission control policies

To save readers time and effort, we present these research methods and analyses by focusing chapters 4–9 only on the 2006–2010 policy cases. The framework of analysis of the 2013–2020 carbon tax cases is the same, and it would be highly repetitive to present all the underlying research for each of the future scenarios as well as the past ones. Readers should keep in mind, however, that chapters 4–9 support the results summarized not only in chapter 2 but also in chapter 3. It is more efficient to present the relatively few adjustments to methods and assumptions needed in endnotes to chapter 3 labeled as "methodological notes" and in a technical appendix C.

For readers solely interested in practical policy concerns, chapters 2 and 3 will be of greatest interest. Most will also gain valuable background from the broad review in the current chapter. They may find their interest piqued by these chapters, moreover, and will then have an overarching understanding to help them explore the underlying research of the rest of the book, as their interests dictate.

Fellow researchers may be primarily interested in one or more of the detailed research investigations of chapters 4–9, though they will also want to begin with chapter 2 (and likely chapter 3) to first understand the integrated context. The reviews of literature in the latter half of this chapter are also likely to be of interest.

Taken together, the research of this book illustrates how the perspectives of different disciplines can be brought together to inform the complex and pressing challenges China faces to reduce its carbon emissions and air pollution damages without harming its economy. In fact from this broadened analytical scope emerges a compelling prospect—that meeting those challenges successfully might be more achievable than most observers currently believe.

1.3   Sources of Data

Before presenting an introductory review of economic and environmental trends in sections 1.4 and 1.5, it is important to consider the nature of the data that underlie all such studies, including our own and those of others.

This is not the technical tangent that it might first seem. Widespread skepticism about Chinese data undermines trust in research on these topics, and not just of lay observers, but also of government decision makers themselves. Yet China's ability to grapple successfully with the challenges of the economy-energy-environment confluence depends on the availability of objective information and respect for a research culture that values it.

We start by differentiating the two primary categories of data used as inputs in our analyses: official statistics produced by the government, and the results of independent scientific research. It may be tempting to assume that these are competing data types and that we need only to identify and put to use the "better" information. Both for understanding this book and for developing a sophisticated interpretation of Chinese conditions, it is critical that the reader gain a more nuanced understanding of different types of data and their relative strengths and weaknesses.

We address this topic in considerable detail, in part to explain how the economics of chapter 9 naturally relies on official statistics while the scientific chapters (4–8) capitalize on data generated by independent research. We additionally want to explain how inconsistencies in government and scientific data are inevitable, but can easily be misunderstood or misconstrued by those who view environmental problems foremost through a political perspective.

Although it may be clear to many readers, let us emphasize that all data have uncertainties, regardless of whether they are calculated or measured by instruments such as the smokestack devices described in chapter 4. Some data are less uncertain than others—in the language of statisticians, they have a narrower error bound or confidence interval—and researchers of course prefer more certain data if all else is equal. But in relatively few cases is all else equal, and seldom does a comparison of government and scientific data present such a stark choice. In many cases they measure different characteristics of a phenomenon, reflecting the different purposes of their collection: supporting regulatory objectives versus scientific inquiry. In the work of this book, the choice is more often deciding which data best fit the research needs, weighing off not only relative accuracy but also such factors as scope and spatial or temporal resolution.

This noted, it is a fact that Chinese government data have at times revealed serious errors or large internal inconsistencies. A now well-known example is an abrupt decline in coal use originally reported by government statistics for 1998–2001 (NBS 2005), which led to analysis and international press attention about China's remarkable sudden success at reducing its CO_2 emissions (Eckholm 2001; Streets et al. 2001). The supposed deep decline, however, improbably occurred at the same time that the economy and electric power generation swiftly expanded, according to the same official sources. This case is discussed in various studies, such as Sinton (2001), Sinton and Fridley (2003), and our own analysis (Ho and Nielsen 2007, chapter 1), which explained how the National Bureau of Statistics (NBS) in fact acknowledged data problems at the time by a ballooning error term that appears in the same coal balance tables for these years.

Despite such problems, there are compelling reasons why official statistics are invaluable for understanding Chinese conditions. First, Chinese authorities recognize that structural problems in the country's statistics impair their ability to manage the economy, and they have been reforming the statistical system in response. An important example is the institution by the State Council and NBS of a full economic census administered every four years, starting in 2004 (Y. J. Wang and Chandler 2011). The data are widely believed to be improving over time (X. Wang 2011; Y. J. Wang and Chandler 2011; IEA 2007, 264). Indeed the problematic coal data from 1998–2001 were subsequently revised to correct earlier errors.[6]

One should not underestimate the task of reforming a statistical system of an economy that is not only enormous, but is also growing and evolving as fast as China's. China is still transitioning from a data system originally designed with Soviet assistance to guide a planned economy, to one meeting the very different information needs of a largely market-based economy.[7] Errors in official data are indisputably unfortunate, but they are also, to some degree, inevitable.

Second, what makes official statistics essential for many analyses is their comprehensive national coverage. Alternative sources of data can rarely cover all of China, and never with a similarly encompassing breadth of economic, social, demographic, and geographic attributes.

This point is essential to studies like ours. When it comes to representing the economy—a large statistical challenge for any country because national economies are inherently so complex—our understanding must be based on data from China's NBS. Questions about those data can be, and are, raised by internal inconsistencies or checks against independent information. One of the biggest ongoing research challenges to the authors of chapter 9 is to work continuously with these statistics: tracking updated data sets, incorporating them in the economic model, cross-checking for inconsistencies, and devising reasonable solutions to data problems. But for constructing a representation of the national economy, there is simply no alternative to working in this way with official data.

As for trends in pollutant emissions and air quality, however, we report official statistics in the next section only briefly and for context, as these data are in fact not used in the research of this book. A chief purpose of this volume is to meld the full analytical power of current scientific research on atmospheric emissions and air quality with an economics-based framework for analyzing the costs of emission control policies. The most valuable data on emissions and air quality are thus those produced by scientists and published in peer-reviewed scientific journals. These include a variety of studies led by scientists in the same collaborative program that produced this book and its predecessor, some conducted concurrently with the integrated research presented here. It is these data from our own and other scientific research groups that mainly inform the assessments of chapters 4–8.

Relatively few analyses of energy and environmental policy in China capitalize on such scientific data and literature, relying instead on official statistics. This is perhaps unsurprising, as articles in scientific journals are often impenetrable to those outside of relevant fields, and linking policy and science research methods and data sets is far easier to imagine than to accomplish.

This scientific literature and data, however, have considerable and increasing value to redressing two of the most vital environmental challenges at hand in China: the continued rise in emissions of CO_2 and the growth of complex, secondary air pollutant forms as simpler, primary forms are successfully reduced. It is arguably a matter of fundamental national and international interest that the most advanced information available on these hazards becomes accessible both to nonexperts and to those advising emission control policies in China.

We aim to bridge this gap between scientific, economic, and policy perspectives by devoting substantial attention to them in this first chapter. We provide a thorough summary of the state of the research and available data in the overview of current trends that follows. Section 1.4 presents the perspective of official statistics on national economic growth, energy use, emissions, and urban air quality in China. It is followed in section 1.5 by a survey of independent scientific evidence on Chinese emissions and atmospheric environment. More detailed discussion of data relevant to research of chapters 4–9 is included in those chapters.

1.4 Official Perspectives

1.4.1 Economy and Energy Use

Our projections of future economic growth and energy use are based on the performance of the last 30 years, a period of remarkable growth and changes in policies. The high rate of growth of gross domestic product (GDP) is well known—officially 9.8% per year during 1978–2007, before the global financial crisis, and remaining above 9% during 2008–2010—but the more complicated changes in energy use deserve discussion.

Figure 1.1 gives the energy consumption in million tons of standard coal equivalents (Mtce) for the three main fossil fuels and other sources, but excluding

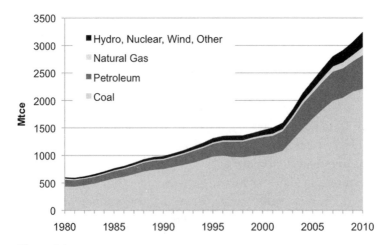

Figure 1.1
Total energy consumption in million tons of standard coal equivalents (Mtce) by source, 1980–2010. *Source:* NBS 2011.

Table 1.1

Historical growth in GDP, energy use, and emissions in China (percent)

	1985–2010	1999–2010
GDP	9.47	9.80
Total commercial energy (SCE)	5.81	7.62
Coal	5.48	7.28
Oil	6.10	6.49
Gas	8.07	14.79
Other primary energy	7.77	11.04
Electricity	9.27	11.11
SO$_2$ emissions	1.31	−0.55
TSP emissions	−4.97	−7.48

Source: Official statistics (NBS 2011).

noncommercial biofuels burned by rural households. Although the oil share rose to a peak of 22% in 2000–2002, it had fallen to 19% in 2010, while the coal share had fallen to 68%, compared to its 72% share in 1980 and its peak of 76% in 1985–1992. By 2010 the natural gas share had more than doubled since 1999, but still made up less than 5%. The "other" category, made up chiefly of hydroelectric power but also including nuclear, wind, and lesser sources of energy, had risen somewhat to 9%. Although recently declining slightly, the persistent dominance of the coal share hides a number of big changes in the economy and energy use.

The growth rates of GDP and energy use are given in table 1.1. GDP grew at 9.8% per year during 1999–2010, slightly faster than during the previous decade that included the 1989 crisis. During this time of rapid output growth, coal consumption rose at a 7.3% rate while oil use rose at 6.5%; this difference is due in large part to the growth of coal use in electricity generation being faster than the growth of oil use in transportation.

Figure 1.2 gives the total primary energy use per unit of GDP, in kilograms of coal equivalent (kgce) per yuan of GDP in 2000 prices.[8] This measure of energy intensity fell almost continuously at a rate of 5.1% per year until it bottomed out at 0.130 kgce/yuan in 2002, rose to 0.150 in 2005, and then began to fall again. Analysts have long debated the factors behind this rapid decline (Garbaccio et al. 1999; Ma and Stern 2007), and a number of papers address data anomalies around 1998–2002 as discussed in the prior section. The debate is by no means settled.

Cao and Ho (2009) survey the literature and describe the surprising behavior during 2002–2005, when the energy intensity at the industry level rose after

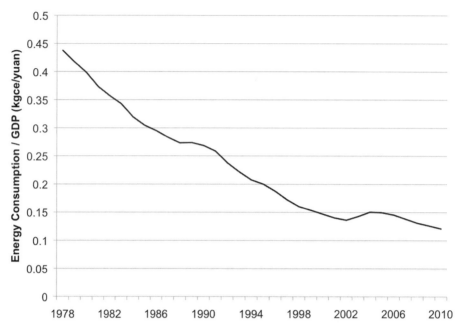

Figure 1.2
Energy consumption per unit GDP in kilogram standard coal equivalents (kgce) per yuan in constant 2000 yuan. *Source:* NBS 2011.

declining for two decades. For now we note the unusual characteristics of the economy during this period. First, the investment share of GDP rose to 43% in 2005 from an average of 37% during 1997–2003, associated with a boom in construction. The current account surplus (a broad measure that includes the trade surplus and other international payments) rose sharply, from an average of 2% during 1997–2004 to 11% in 2007, with a correspondingly sharp rise in the savings rate. These changes shifted the composition of output from consumption to investment goods, and from agriculture and consumer manufacturing to construction, heavy industry, and export-related manufacturing. This growth of heavy manufacturing (cement, iron and steel, motor vehicles) brought high growth in energy consumption (figure 1.1) and atmospheric emissions.

A similar shift of output to production of heavily emitting investment goods and construction is believed to have occurred in 2009–2010 as a result of government stimulus spending, although analogous data are not available at the time of writing.

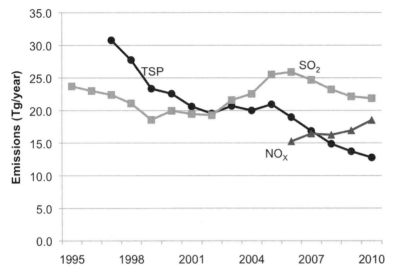

Figure 1.3
Official estimates of emissions of TSP, SO₂, and NOₓ in China in teragrams per year (Tg/year), 1995–2010. *Note:* TSP includes combustion and process emissions. *Sources:* 1996–1999 data are from Table 8B.2 of Fridley and Aden (2008), reproducing data from various years of the *China Environment Yearbook* and other State Environmental Protection Administration (SEPA) and National Bureau of Statistics (NBS) sources; 2000–2010 data are from the *China Statistical Yearbook on Environment* (NBS and MEP 2012) and the *China Environment Yearbook* (MEP 2009–2012a, 2012b).

1.4.2 Emissions

As noted earlier, official emission statistics are not used in the research of this book. They are discussed here as evidence of trends over a longer term than those reported in scientific studies, as well as for policy context. We first present official data on air pollutants, and then discuss recent developments in government inventories of CO₂ and GHG emissions.

Figure 1.3 presents officially estimated emissions of total suspended particulates (TSP), SO₂, and NOₓ, the only atmospheric pollutant emissions that the government routinely reports at the national level.

The pattern of SO₂ emissions shows a relatively steady decline between 1995 and 2002. Because the government calculation of emissions is based in part on energy data, this decline may reflect the fall in coal consumption originally reported in problematic official statistics for 1998–2001 that was discussed in section 1.3. Following this period, figure 1.3 shows estimated SO₂ rising again from 2002 to 2006, and then falling during the subsequent years of the 11th Five-Year Plan. This decline in the late 2000s occurred despite surging coal consumption shown in figure 1.1 for these years; whether and how China may have accomplished this reduction, and what its economic and environmental impacts may have been, is the subject of one of the first two policy cases analyzed in this book.

Figure 1.3 also shows emissions of TSP, a pollutant category that encompasses particles of all sizes and is a common early focus of environmental protection because it is the easiest form of airborne particulate matter (PM) to measure and control. (We will see that two different size fractions of TSP, $PM_{2.5}$ and particles less than 10 microns in diameter, or PM_{10}, are more closely associated with health impacts and gain emphasis over TSP as air quality protections advance.) The TSP values shown in figure 1.3 include particulates produced both by combustion of fuels and by mechanical or chemical means, termed "process" emissions.[9] Estimated emissions fell almost continuously throughout this period, with sharp declines from 1997 to 1999 and after 2004. Over the 1998–2008 period, when GDP was growing at 9.9% per year, TSP fell at 7.5% per year and SO_2 emissions had no change on net.

China began reporting emissions of NO_X only in 2006. The five years shown in the figure suggest modestly rising emissions. In section 1.5, however, we will see independent scientific evidence of continuous growth in emissions of NO_X since 1995, accelerating swiftly in the 2000s.

As for emissions of GHGs including CO_2, these are not included in the statistical yearbooks that report air pollution emissions, and, until recently, China had not developed a systematic institutional mechanism for estimating them. Doing so is a difficult statistical undertaking and, like other developing countries, China has not been obliged under the United Nations Framework Convention on Climate Change (UNFCCC) to submit GHG inventories beyond a first one, included in its "Initial National Communication" to the UNFCCC for the year 1994 (SDPC 2004).

This situation is now changing. China first joined many other developing countries in delivering a new inventory as part of a voluntary "Second National Communication" to the UNFCCC, which it submitted for the year 2005 in November 2012 (NDRC 2012). The inventory reported CO_2 emissions of 5554 Gt in 2005 on net—that is, taking account not only of sources but also of carbon uptake due to land use change and forestry activities. This total compares to a net 2666 Gt in 1994 reported in the initial communication (SDPC 2004), and thus represents a more than doubling over 11 years.

More importantly, in 2009 China pledged to begin reporting GHG inventories every two years according to the standardized methods used by wealthier ("Annex

I") countries under their annual reporting obligations to the UNFCCC (to measure their progress on quantitative commitments such as those of the Kyoto Protocol). China has been building a systematic management process for undertaking this commitment (Mintzer, Leonard, and Valencia 2010), although at the time of writing its biennial reporting had not yet begun.

Thus while official GHG inventories from China have been highly infrequent, and dated when produced, it has committed to improving its official emission accounting and reporting. With its recent 2005 inventory, it has taken a big step toward delivering on that commitment.

1.4.3   Atmospheric Concentrations

Emissions have resulted in high levels of pollution measured by the government in the major cities of China. Most policy-oriented analyses of China's air pollution rely on data from the government's extensive network of urban monitoring stations, chiefly operated by municipal Environmental Protection Bureaus (EPBs) with data compiled and reported by the Chinese National Environmental Monitoring Center (CNEMC).[10] Such data are collected, analyzed, and reported chiefly for regulatory purposes, as indicators of the progress of pollution control efforts, for guiding air pollution management, and for public communication.[11]

Encouragingly, concern about air quality and flush budgets for environmental monitoring have recently allowed some government institutions at central and provincial levels to construct new "supersite" monitoring stations. These are often located in rural districts near cities to allow measurement of a range of conditions affecting China's air quality problems.[12] These multistory stations are filled with arrays of state-of-the-art imported and domestic instrumentation. The scientific capacities of the teams operating the instruments and keeping them calibrated, not to mention interpreting the resulting data, are not always as advanced as the instruments, and developing strong data series will take time. Clearly, however, China's new supersites provide a basis for more scientifically informed air quality management in the future.

Figure 1.4 shows the average annual concentrations of three pollutant types in 31 major cities, as measured by the EPB urban networks. The city sample comprises the capitals of all current provincial-level jurisdictions. Official yearbooks report the average concentrations of more than 100 major Chinese cities; however, the sampled cities are not consistent from year to year. We instead draw a consistent sample of the noted 31 cities from the same data set, providing a more valid basis for detecting year-to-year trends.[13]

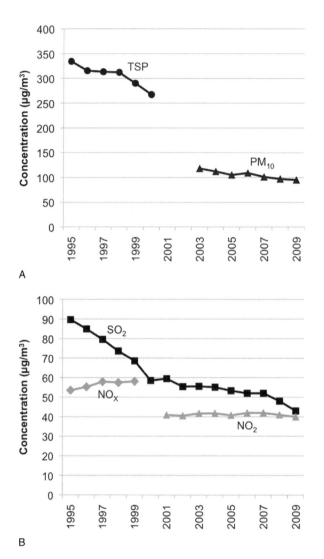

Figure 1.4
Official data on mean concentrations of pollutants for 31 major cities of China, 1995–2009, in micrograms per cubic meter ($\mu g/m^3$). (A) TSP and PM_{10}. (B) SO_2, NO_X, and NO_2. *Sources:* Data from 1995–2005 are selected from tables 8B.6–8B.8 of Fridley and Aden (2008), summarizing data from various years of the *China Environmental Yearbooks* and other MEP and NBS sources. Data from 2003–2009 are from the *China Statistical Yearbook on Environment* (NBS and SEPA 2004–2010). Gaps in the TSP-PM_{10} and NO_X-NO_2 transition years are due to missing data for some of the 31 cities in the sample.

In recent years, the concentrations of three pollutants have been monitored and reported as daily averages on websites, and as annual averages in published sources: SO_2, particulate matter (TSP until 2000, switching to PM_{10} thereafter), and nitrogen oxides (NO_X until 1999, NO_2 thereafter).[14] Ozone and carbon monoxide (CO) were added to China's national urban air measurement standards in 2011, and $PM_{2.5}$ was added for major cities and key regions in 2012. Some systems have also upgraded the reporting from daily to hourly averages (Shi 2012). Both $PM_{2.5}$ and ozone are central to the environmental impact assessments of this book, but with no official nationwide data yet available, we defer discussion of their trends to the scientific evidence discussed in section 1.5.

The average levels of urban TSP or PM_{10} concentrations have fallen continuously since 1995, roughly consistent with the trend in TSP emissions discussed earlier, and this reduction is an indicator that China's efforts to control TSP have made considerable progress. Particulates, notably PM_{10} and especially $PM_{2.5}$, are associated with the greatest total health effects of air pollution, as discussed in chapter 8. Note that average PM_{10} levels, however, have been declining only slowly since 2003. Moreover, at $95\,\mu g/m^3$ (micrograms per cubic meter) in 2009, the national annual average concentrations remain very high judged by the World Health Organization's sequence of "interim" annual average targets recommended to developing countries: 70, 50, and $30\,\mu g/m^3$, leading to a final guideline level of $20\,\mu g/m^3$ (WHO 2006).[15]

That officially reported urban average SO_2 levels have declined consistently since 1995, particularly prior to 2000, is encouraging. It is inconsistent, however, with the national SO_2 emission trend shown in figure 1.3. These two trends may be reconcilable by the fact that direct coal burning—for example, in building boilers—has been increasingly banned in the centers of many Chinese cities since the 1990s, even as coal consumption has surged in electric power and industrial plants. Such plants are increasingly located outside city centers and, in any case, disperse more of the SO_2 downwind from cities through tall smokestacks. While WHO guidelines do not advise an annual average for SO_2 concentrations, we note that the 2009 annual average of $43\,\mu g/m^3$ in the 31 selected Chinese cities far exceeds the WHO 24-hour mean guideline of $20\,\mu g/m^3$ (WHO 2006), a finding that suggests sizable health risk given that guideline pollution levels decrease as averaging times increase.

According to official statistics, monitored ambient NO_X or NO_2 concentrations in the cities shown in figure 1.4 have been remarkably steady since 1995, and indeed close to the WHO guideline annual average level for NO_2 of $40\,\mu g/m^3$. As with emissions, however, we will see that independent scientific research suggests that, on the national scale, NO_X and NO_2 constitute a major and growing air quality hazard for China. This concern is due less to its direct risks to human health as a primary urban pollutant, and more to its role as a central precursor to key secondary pollutants including nitrate aerosols (a form of $PM_{2.5}$) and ozone.

1.5   Scientific Research Perspectives

We now turn to describing how the scientific research literature approaches the estimation of emissions and air quality, contrasting it with the official methods just discussed.

1.5.1   Emissions

We focus first in this section on emissions of conventional air pollutants, followed by CO_2 emissions, and then offer a brief note on the uncertainties of estimating these emissions.

The scientific literature estimating Chinese emissions of air pollutants has expanded over the last decade, improving in terms of spatial resolution, temporal resolution, and species coverage. The motivation behind this research is the study of atmospheric transport and chemistry, from Chinese regional to global scales. The tools to model China's domestic air quality and transboundary export of air pollutants have also advanced in recent years, as discussed in chapter 7, providing a means to test the accuracy of emission inventories against field observations. For instance, chemical transport models can be used in inverse mode, in which instrumental measurements of pollution concentrations serve as inputs and the model is run "backward" to estimate the scale and location of emissions required to explain the observations. This testing, in turn, prompts refinements in emissions estimation.

Chapters 4–6 of this book report such emission inventories, describing in detail the combination of statistical data, stack measurements, and bottom-up calculations employed to estimate emissions. The term "bottom-up" refers to methods using information at the greatest level of detail available (in some industries this may comprise data for individual plants), in contrast to "top-down" procedures that start with national aggregates.[16] We limit our summary here to the key year-to-year trends indicated by this scientific literature and how they compare to the official trends just discussed.

First, however, we should help readers understand how emission estimates from independent researchers differ from those of the official sources, and why these two estimation approaches coexist. This discussion will also illustrate why the integrated assessments of this book could not be conducted until now; the resolution of emission inventories had to advance sufficiently in both sector and spatial dimensions to link a multisector economic model with a spatial atmospheric one.

A key distinction between official and scientific emission estimates is their scopes of coverage. As already noted, official national statistics cover emissions of only a few air pollutants covered by national regulatory objectives: TSP, SO_2, and NO_X. The scientific literature, by contrast, seeks to understand pollution and climate-forcing phenomena taking place in the actual physical atmosphere, with all its meteorological, chemical, and biological complexity. This science must therefore consider dozens of species involved in hundreds of chemical reactions and other effects.

The science similarly must extend beyond sources currently subject to regulatory control. These include noncombustion sources such as fugitive gases released by refineries. They include human activities seldom recognized as relevant to air quality, such as fertilizer use and the raising of livestock, both leading sources of ammonia. And they also must include natural sources, such as forest fires and pine forests [which emit volatile organic compounds (VOCs)].

The scientific research also demands higher spatial resolution than government estimates. Where the latter generally reflect only the reporting jurisdictions subject to regulatory oversight, the scientific inventories must be gridded spatially for use in modeling, sometimes at resolutions smaller than typical counties.

Scientists have to characterize sources more comprehensively in order to judge the effects of changes in anthropogenic emissions on concentrations. Most of these species react chemically in the atmosphere, and cannot be evaluated in isolation. Atmospheric simulations are not considered reliable until they perform well when tested against instrumental measurements, and they can fail such tests if the geographical distribution of emissions is too crude or if important pollutant types are left out.

Conventional Air Pollutants TSP/PM$_{10}$, SO$_2$, and NO$_X$/NO$_2$ Emission inventories for primary TSP and PM_{10} have not been developed as extensively as those for SO_2 and NO_X because they must account for a wider diversity of emission processes and are more difficult to construct. Figures 1.5A and 1.5B plot estimates of TSP and PM_{10} emissions from chapter 6 of this text and four other scientific studies.[17] Figure 1.5A also reproduces the official TSP emission estimates from figure 1.3. The independent research indicates a quite distinct trend from that of the official statistics,

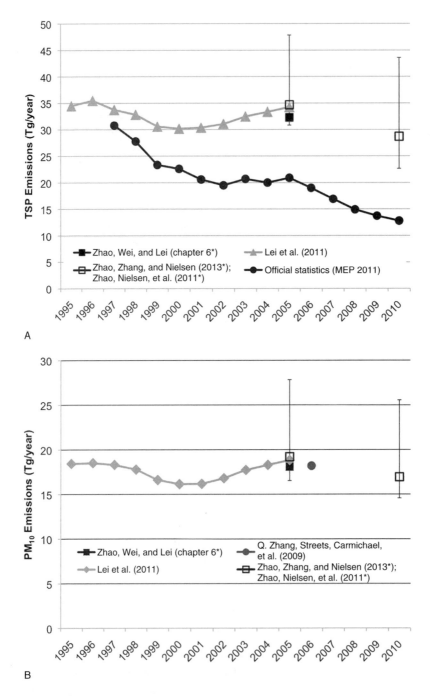

Figure 1.5
Scientific and official estimates of emissions in China, 1995–2010, in teragrams per year (Tg/year). (A) TSP. (B) PM$_{10}$. *Note:* Zhao, Zhang, and Nielsen (2013*) and Zhao, Nielsen, et al. (2011*) calculated 95% confidence intervals, shown as error bars.

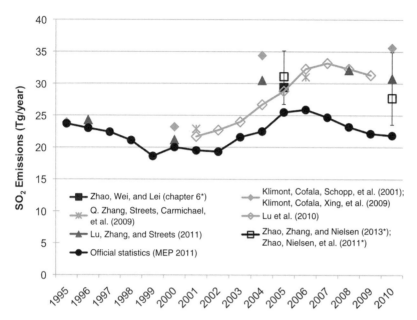

Figure 1.6
Scientific and official estimates of emissions of SO_2 in China, in teragrams per year, 1995–2010. *Note:* Zhao, Zhang, and Nielsen (2013*) and Zhao, Nielsen, et al. (2011*) calculated 95% confidence intervals, shown as error bars.

falling more slowly in the late 1990s and rising again instead of holding steady in the early 2000s. Chapter 6 will show, as does Zhao, Zhang, and Nielsen (2013*), a decline over the 2006–2010 11th FYP, in that respect tracking the official estimates. But the growing disparity between the two paths over the time period is considerable, and the independent estimates are much higher. This discrepancy is due in part to more complete coverage of unconventional sources in the independent science, but biases in official reporting protocols likely contribute to the differences.

Figure 1.6 shows a number of independent estimates of SO_2 emissions, along with the official ones reproduced from figure 1.3. The four most recent studies are quite consistent. The rates of growth of emissions in the first half of the 2000s of the independent and official estimates are similar, but the absolute levels of the former are at all times higher, and the disparity grows in the late 2000s. Lu et al. (2010, 2011) and Zhao, Zhang, and Nielsen (2013*) reflect, as will chapter 6, the abrupt reversal of emissions as a result SO_2 controls under the 11th FYP.

The independent estimates of emissions of NO_X (or just NO_2) mostly follow a similar path, as shown in figure 1.7, trending higher than the brief official record

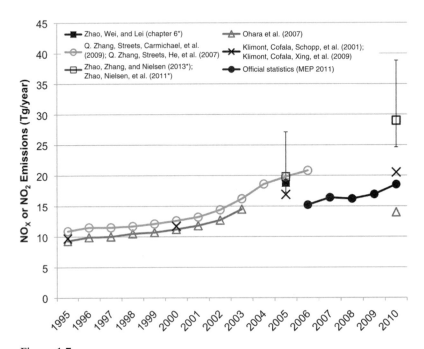

Figure 1.7
Scientific and official estimates of emissions of NO_X in China, in teragrams per year, 1995–2010. *Note:* Zhao, Zhang, and Nielsen (2013)* and Zhao, Nielsen, et al. (2011*) calculated 95% confidence intervals, shown as error bars.

for 2006–2010 (marked in solid black circles). Zhao, Zhang, and Nielsen (2013*) and related research in chapter 6 suggest a much swifter rise in NO_X during the 11th FYP period than earlier studies and official data indicate. In any case, this growth in emissions prompted inclusion of a NO_X reduction target in the 12th FYP (–10% in 2015 compared to 2010).

Carbon Dioxide   Most independent estimates of China's CO_2 emissions are calculated top-down, based on energy use and other data aggregated at the national level (U.S. EIA 2011; PBL 2010, from the Netherlands Environmental Assessment Agency; Boden, Marland, and Andres 2010, from the Carbon Dioxide Information Analysis Center).[18] Zhao, Nielsen, and McElroy (2012*) produced the first CO_2 emission inventory using a bottom-up approach that combined newly compiled emission factors with activity data developed in studies of China's conventional air pollutants, capitalizing on prior tests of the underlying data against instrumentally measured concentrations of those pollutants (Akimoto et al. 2006; Zhao, Nielsen, McElroy,

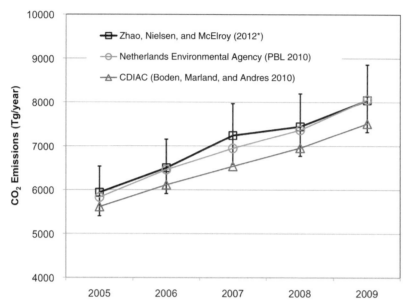

Figure 1.8
Scientific estimates of anthropogenic emissions of CO_2 in China, in teragrams per year, 2005–2009. *Note:* Zhao, Nielsen, and McElroy (2012*) calculated 95% confidence intervals, shown as error bars. *Source:* Zhao, Nielsen, and McElroy (2012*), adapted by C. P. Nielsen. Reprinted under terms of Creative Commons 3.0 Attribution license, at http://creativecommons.org/licenses/by/3.0/.

et al. 2012*). Including emissions from biomass and biofuel combustion omitted in other studies, they estimated that total Chinese CO_2 emissions rose from 7126 million metric tons (Mt) in 2005 to 9370 Mt in 2009. In order to compare with the Boden, Marland, and Andres (2010) and PBL (2010) estimates shown in figure 1.8, emissions from biomass and biofuel use are omitted and the result is plotted with the black squares. All three series indicate similar growth rates, but emission estimates by Zhao, Nielsen, and McElroy (2012*) and PBL (2010) are substantially higher than Boden, Marland, and Andres (2010).

The greater spatial resolution resulting from the bottom-up methods of Zhao, Nielsen, and McElroy (2012*) allows comparison to regional enhancements of CO_2 concentrations observed instrumentally in China, initiating direct tests of CO_2 estimation by empirical measurements. Y. X. Wang, Munger, et al. (2010*) analyzed a record of observed CO_2 concentrations at the Tsinghua-Harvard atmospheric station north of Beijing, using analytical techniques to differentiate the CO_2 of Chinese provenance from the much larger global background level, then comparing it to local carbon monoxide (CO) levels. The ratio of regionally derived CO_2 to CO concentrations observed in the air in 2006 (26.0) is indeed close to the ratio of emissions estimated for north China in the same year (27.6, reported in Zhao, Nielsen, McElroy, et al. 2012*). This is an encouraging if limited early indicator of the improving accuracy of emission inventories of CO_2 and CO. Y. X. Wang, Munger, et al. (2010*) is discussed further in section 1.5.2.

Uncertainties   While we must leave detailed discussion of uncertainties of national emission estimates to chapters 4–6, these may be much larger than expected and readers should have a sense of their scales. As reported in an emerging literature, improved input data and bottom-up methods now support formal statistical quantification of parameter uncertainties.[19] Zhao, Nielsen, et al. (2011*) and Zhao, Zhang, and Nielsen (2013*) used these methods to yield 95% confidence intervals (CIs) for 2005 and 2010 emission values, shown as vertical error bars around the open-square central estimates in figures 1.5–1.7; the 2010 CIs are (–21%, +52%) for TSP, (–14%, +51%) for PM_{10}, (–15%, +26%) for SO_2, and (–15%, +34%) for NO_X. These studies improve upon earlier uncertainty estimates using simpler approximating methods for SO_2 (Streets et al. 2003), NO_X (Q. Zhang et al. 2007), and several PM categories (Bond et al. 2004). Note that official emission estimation is based on simpler methods, implying larger uncertainties.

The same statistical techniques were applied to CO_2 emissions estimated for 2005, as shown in figure 1.8, and for 2010, yielding 95% CIs of (–9%, +11%) and (–10%, +9%), respectively (Zhao, Nielsen, and McElroy 2012*; Zhao, Zhang, and Nielsen 2013*). These values contrast with a ±15–20% uncertainty for the 2005 CO_2 estimate reported in Gregg et al. (2008), based not on statistical methods but on the scale of subsequent revisions of underlying energy data.[20] Note that the systematically lower Boden, Marland, and Andres (2010) estimates fall within, or very close to, the 95% confidence ranges in all years.

1.5.2   Atmospheric Concentrations

When research scientists need an aggregated measure of urban pollutant levels in China and lack other options, they sometimes analyze daily mean concentrations collected by municipal environmental protection bureaus. See, for example, Qu et al. (2010) and Lin et al. (2010*). For the most part, however, official monitoring data are of limited use to scientists studying atmospheric environment. These data are almost always reported in aggregated terms, both temporally (such as daily means) and spatially (such as the average of all stations in a city). The nature of the instrumentation and the methods of measurement are rarely specified, limiting the interpretation. And the locations of stations often are not clearly disclosed or are scientifically disadvantageous (for instance, too close to streets to be representative). In any case, official monitoring data are limited only to China's cities, which constrain their value to understanding the important regional dimensions of many of China's most difficult air quality problems.

It is important for critics of this system to keep in mind, however, that the monitoring networks and their reporting protocols were designed chiefly to meet regulatory objectives, not scientific ones. Nor does it follow that government should invest public resources so that monitoring stations meet the higher and more costly standards of experimental science.

There are three main sources of independent scientific observations of air pollution or GHG levels in China. They are made by instruments that are deployed (1) at the surface, at permanent stations or in short-term field campaigns; (2) in the air, on aircraft or balloons; and (3) in space, onboard satellites. Space-based observations are ideal for our immediate interest in year-to-year trends over major regions of China but cannot currently cover all species of interest, and so we introduce these observation techniques in turn.

A number of Chinese research institutions maintain permanent surface measurement sites, often on the roofs of buildings on urban campuses. These sites provide much of the available information on year-to-year trends analyzed in published research, though typically concerning only a subject city. It is difficult to generalize from such data to other cities, not to mention to the country at large.

It may seem counterintuitive, but there are scientific benefits to observing urban pollution outside of a target city itself, as downwind rural sites can measure a well-mixed urban plume that better represents the average air coming from the city, uninfluenced by specific sources in the immediate vicinity of the instruments. (Imagine, for instance, an instrument on an urban rooftop that is four stories high and 100 meters from a traffic-clogged street. The air measured at such a site will disproportionately sample vehicular exhaust, and may not be representative of air quality of the neighborhood, let alone of the city.)

The most common strategy in China to gain such data has been through short-term measurement campaigns, in which instruments are set up temporarily in a rural area near a city for weeks or months. These data are essential for many research purposes, but do not provide evidence of year-to-year trends.

Permanent scientific stations established outside cities can measure urban plumes and regional conditions over successive years, and the number of such stations is growing in China.[21] The government supersites described in section 1.4.3 offer excellent platforms for scientists to sharply expand such work, if allowed appropriate research independence. There are also a few permanent stations sited at isolated areas to sample clean, background air flowing into the densely populated areas of China.[22]

Balloon- (or "sonde-") and aircraft-based measurement campaigns are less common in China, but have been conducted domestically to analyze such issues as the transport and chemistry of acid precursors (Xue et al. 2010) and the climatology of ozone (Ding et al. 2008). There have also been aircraft campaigns off the coast of Asia or further east over the Pacific Ocean sponsored by scientific agencies of other countries to detect transpacific influence of pollutants with sources in China and East Asia.

Satellite-based instruments have the advantage of greatly expanding the spatial and temporal scopes of observation, which allows an independent record of concentrations averaged over time and space to emerge. A key limitation is that satellites are less able to characterize the vertical distributions of pollutants. They typically measure the "column density" over a footprint, essentially the total integrated occurrence of the pollutant from the altitude of the satellite to the surface.[23] A vertical column density (VCD) of a gas—sometimes called simply the column density— has to be interpreted using other tools to differentiate concentrations along that column, notably near the surface for pollutants, and this process introduces uncertainty. Nevertheless, for year-to-year trends over large regions of China, observations from space, when available, can provide the broadest empirical scientific evidence.

Nitrogen Oxides   The capacities of satellites to observe air pollution concentrations over China gained media attention after a high-profile study of NO_2 over 1996–2004 appeared in the journal *Nature* (Richter et al. 2005). (NO_X comprises NO_2 and nitric oxide, NO; for technical reasons the instruments measure only NO_2, but it is highly indicative of total NO_X.)[24] Richter et al. (2005) found a highly significant, and accelerating, increase of about 50% in the column density of NO_2 from 1996 to 2004 over central-east China including the North China Plain, the country's industrial and population heartland. The study attributed this to higher growth in NO_X emissions than that indicated by bottom-up inventories at the time, suggesting a larger NO_X hazard than previously appreciated.

Subsequent analyses of column densities observed from space similarly revealed rapid increases of levels of NO_2 in recent years compared to the bottom-up NO_X

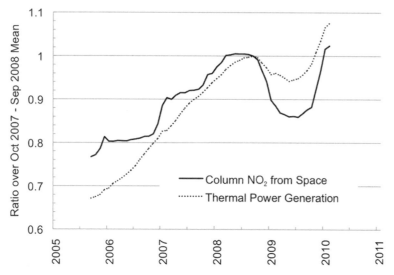

Figure 1.9
Trends in column NO_2 observed from space and official statistics on thermal electric power generation for central-east China, 2005–2010. *Note:* This figure shows the 12-month moving average of regional mean column NO_2 in China, relative to values averaged over the 12 months prior to the global financial crisis, October 2007–September 2008. The time series of column NO_2 is processed by the Royal Netherlands Meteorological Institute from observations by the OMI instrument onboard NASA's Aura satellite. *Source:* Lin and McElroy (2011*), adapted by J.-T. Lin. Reprinted under terms of Creative Commons 3.0 Attribution license, at http://creativecommons.org/licenses/by/3.0/.

emission estimates (e.g., Q. Zhang et al. 2007). Newer bottom-up inventories, as discussed previously, are growing closer to observations (Zhao, Zhang, and Nielsen 2013*).

Another analysis built on this prior NO_2 evidence and extended it into 2010 (Lin and McElroy 2011*). The domain is roughly the same as the earlier studies, the densely industrialized and populated central-east region.[25] Figure 1.9 shows the resulting record, reflected in 12-month moving averages of the regional mean column densities of NO_2 observed by satellite. The NO_2 VCD in the figure is expressed relative to the average of October 2007–September 2008 in order to accentuate the trends leading up to, and during, the global economic crisis. It reveals a remarkably sharp signal of the Chinese economic downturn that began in fall of 2008, a bottoming out in summer of 2009, and the subsequently steep industrial recovery from late–2009 into 2010 propelled by the Chinese government economic stimulus.

More pertinent for our purposes is the record in the more typical economic conditions of the four years prior to fall of 2008,[26] in which the annual mean VCD of NO_2 increased by 30% (and noting that two other satellite records analyzed by Lin and McElroy indicated similar values, 27% and 33%). The general upward trend continues that identified by Richter et al. (2005) for 1996–2004, and Q. Zhang et al. (2007). It is evidence that China's policy focus on SO_2 and primary PM control up to and including the 11th FYP allowed growth in NO_X emissions to accelerate. We will see this growth trend also reflected in observations of fine aerosols (i.e., particles) and ozone, since NO_X is a precursor of both of these secondary pollutant forms.

Particulate Matter: $PM_{2.5}$ and PM_{10}   Turning to fine and coarse particles, we use Lin et al. (2010*) to provide representative insights into the scientific challenges. The paper investigated year-to-year trends in a quantity termed aerosol optical depth (AOD), observed both by satellites and by ground-based instruments in China. AOD measures the extinction of solar radiation by the scattering and absorption by particles in the air, and is analogous to VCD of gases in that it similarly represents the pollutant load integrated over a column through the atmosphere.[27] While AOD measures the total aerosol load, the processing of satellite data described in Lin et al. measures AOD with stronger sensitivity to the $PM_{2.5}$ share of total aerosols, a distinction we will return to later.

The paper again focuses on the heavily populated and industrialized east-central region of China, similar to the domains emphasized in previously cited studies including Lin and McElroy (2011*). In fact Lin et al. (2010*) also examines column NO_2 in addition to AOD and, for current purposes, helpfully investigates the relationships between $PM_{2.5}$ and PM_{10} with column NO_2.

The two solid black lines in figure 1.10 show the AOD in recent years, the thicker one the average over the entire region and the thinner one the average over 37 cities within the region (to be further explained in the next paragraph). Both lines follow a growth path leading into the economic downturn similar to that seen for the NO_2 VCDs in figure 1.9, one of which is included in this figure as the heavy line marked by "+" symbols. However, the growth in particles in the atmosphere in the period before the economic downturn in 2008, as measured by the AOD, is more gradual than that for NO_2 VCD.[28]

The average PM_{10} levels measured by government monitoring stations at major cities within the region provide another useful source of concentration data. This is the fifth line in figure 1.10, shown in gray. The cities included are the same 37

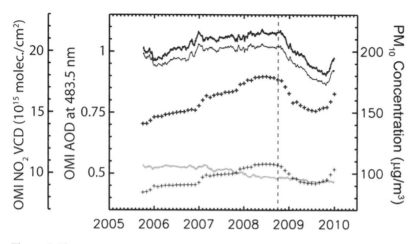

2005   2006   2007   2008   2009   2010

Figure 1.10
Trends in aerosol optical depth (AOD) and column NO_2 observed by satellite, and urban PM_{10} measured at the surface, over central-east China. The plot shows changes in the 12-month moving average of daily AOD, monthly column NO_2, and daily PM_{10} during October 2004–December 2009. The thick solid black line and thick "+" line denote mean AOD and column NO_2 over the region, respectively, from space. The thin solid black line, thin "+" line, and gray line denote for the same 37 cities the mean AOD and column NO_2 from space, and PM_{10} from surface stations, respectively. *Source:* Lin et al. (2010*). Reprinted with permission, Copyright 2010 American Chemical Society.

defining the lighter lines of AOD and column NO_2 in the figure, so these three trend lines are especially comparable. In contrast to the pre-downturn, upward trends of satellite-derived AOD and column NO_2 over the cities, surface PM_{10} is seen to decline steadily throughout the period. How can these trends be reconciled?

The explanation by Lin et al. (2010*) is instructive about the complexity of air quality problems and the challenges of control. First, recall that the AOD measurement method is more sensitive to the finer particles, $PM_{2.5}$. The upward trend in AOD is therefore more reflective of the fine particle fraction of the total load, whereas the surface stations monitor the combined fine and coarse fractions categorized as PM_{10}. Second, note that PM_{10} is mainly a primary pollutant form and therefore easier to target with control policies, which environmental authorities have gradually but successfully implemented over recent decades, particularly in urban areas. Much $PM_{2.5}$, in contrast, is formed by secondary chemistry from precursor gases, notably SO_2, NO_X, ammonia, and VOCs (Yang et al. 2011), which may come from sources far from the cities themselves. While SO_2 emissions have declined over the time period illustrated here, their effect on $PM_{2.5}$ concentrations may have been overwhelmed by concomitant increases in other $PM_{2.5}$ precursors: ammonia, VOCs, and especially NO_X, as discussed earlier. The observation that AOD increased, but at a rate slower than column NO_2 in the period prior to the economic downturn, also would follow because of the dampening effect of successful abatement of SO_2 and primary PM_{10} on the rate of growth of aerosol loading. However, the effect of successful reduction of SO_2 on $PM_{2.5}$ levels may be less consequential than expected in northern China, by freeing ammonia to react with abundant NO_X to form ammonium nitrate $PM_{2.5}$ (Y. X. Wang et al. 2013); in this case, successful SO_2 and sulfate $PM_{2.5}$ control may be negated by resulting growth in nitrate $PM_{2.5}$.

This complex analysis should serve as further caution to nonexpert readers who assume that the solutions to China's $PM_{2.5}$ pollution, including severe episodes like those in Beijing and other cities in the winter of 2012–2013, are fully understood and need only to be implemented successfully. The reader should keep in mind that $PM_{2.5}$ is defined only by the size and not the chemical composition of particles, which in fact result from many, often intersecting, primary and secondary chemical pathways. Redressing $PM_{2.5}$ is as much a challenge to expand scientific understanding of diverse and sometimes spatially distant causes, to weigh off widely ranging costs of mitigation of different source types, and to incrementally develop and implement effective multipollutant control strategies over time.

Sulfur Dioxide   Satellite-based observation of SO_2 VCDs is possible, but it has scientific constraints that make the results much more uncertain than those for NO_X and AOD, and scientists are very cautious about drawing firm conclusions about SO_2 observed from space. Nevertheless, available evidence is at least suggestive that China's implementation of the 11th FYP SO_2 control policy modeled in this book met with considerable success.

This was first indicated at a limited spatial scale, in a study of three areas in Inner Mongolia (Li et al. 2010). The researchers investigated summertime SO_2 (and NO_2) VCDs for 2005 through 2008 over nine locales containing power plants in the three regions. The nine plants represent varied conditions, including both rural and urban locations. Trends in the VCD of NO_2, a species not targeted by control measures, were used to indicate steady levels of electricity generation at the plants throughout the time period. Observed SO_2 was shown to rise (as much as 50%) over all nine plants from 2005 to 2007, and then to dramatically decline over all nine from 2007 to 2008. This is the time when FGD systems required under the 11th FYP were expected to become operational, thus appearing to instrumentally confirm from space the effectiveness of the SO_2 control policy in this region. This capacity to detect SO_2 (and NO_2) emissions of individual plants, when they are located relatively distant from other major emissions sources that could blur the atmospheric signal, is also notable for potential policy enforcement purposes (Li et al. 2010; Q. Zhang, Streets, and He 2009; S. W. Wang et al. 2010).

A subsequent study by contributors to this book (Zhao, Zhang, and Nielsen 2013*) included mapped annual mean SO_2 column densities over China for 2006–2009, shown in figure 1.11, to compare to new emission inventories. Noting again the strong concerns about potential errors and biases in SO_2 retrievals, the satellite evidence is nevertheless plainly suggestive of the already indicated temporal policy trend: an expansion of SO_2 emissions nationally to 2007, followed by sharp reductions in 2008 and 2009 as the 11th FYP SO_2 measures modeled in this book took effect.[29] Given data concerns, however, this evidence should be considered only tentative.

Ozone   Ozone in the lower atmosphere, the troposphere, is chiefly an anthropogenic secondary pollutant, produced photochemically in the air from NO_x, VOCs, and CO in the presence of sunlight. Tropospheric ozone levels are more difficult to extract from space-based measurements because they are dominated by ozone that occurs naturally at the higher altitudes of the stratosphere,[30] comprising 85–95% of the column total (Fishman, Wozniak, and Creilson 2003). Because of this difficulty there is little firm evidence yet from satellite records of year-to-year trends of ozone within China.

The lack of broader satellite-based evidence for China is unfortunate because the knowledge base on its ozone pollution has long been underdeveloped, despite the health and agricultural risks and its central place in the air quality strategies of more advanced economies. Only in 2011 was ozone added to China's national monitoring protocols. The limited national awareness and oversight result from China's focus to date on more immediate and tractable primary air pollution forms. The inherent complexity of ozone photochemistry, the role of unique meteorology and geography wherever it occurs, and the diversity of precursor source types (discussed in chapter 7) do not lend themselves to quick understanding, let alone control. (Note that ozone was first identified as an important urban air pollutant in the 1950s in southern California, which then became the center of ozone research and control policy in the world. Yet successful reduction of ozone concentrations in Los Angeles has been won in exceedingly small increments, over more than 50 years of coordinated effort.)

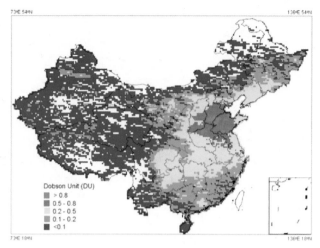

Dobson Unit (DU)
- > 0.8
- 0.5 - 0.8
- 0.2 - 0.5
- 0.1 - 0.2
- < 0.1

A

Dobson Unit (DU)
- >0.8
- 0.5-0.8
- 0.2-0.5
- 0.1-0.2
- <0.1

B

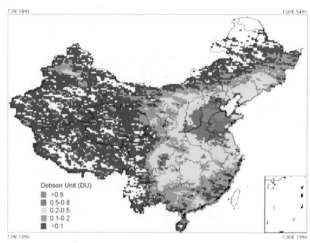

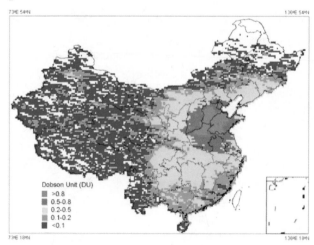

Dobson Unit (DU)
- >0.8
- 0.5-0.8
- 0.2-0.5
- 0.1-0.2
- <0.1

C

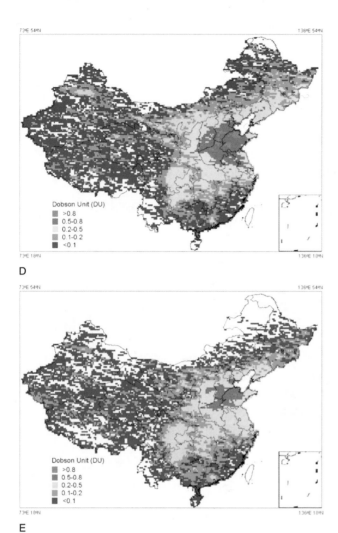

D

E

Figure 1.11
Annual mean SO$_2$ vertical columns of the boundary layer in China 2005–2009 from SCIA-MACHY satellite retrievals. (A) 2005. (B) 2006. (C) 2007. (D) 2008. (E) 2009. *Note:* A Dobson Unit is a measure of columnar density of an atmospheric trace gas, equivalent to a 10-μm-thick layer at standard temperature and pressure. *Source:* Zhao, Zhang, and Nielsen (2013*). Reprinted under terms of Creative Commons 3.0 Attribution license, at http://creativecommons.org/licenses/by/3.0/.

The small but growing atmospheric chemistry community in China has been building a literature analyzing ozone formation in the varied chemical, meteorological, and geographical conditions of particular Chinese cities, based on measurements at individual field sites. Most of these papers focus not on year-to-year trends but on other dimensions including variations by season or hour of day (e.g., T. Wang et al. 2006; Y. X. Wang, McElroy, et al. 2010*; Y. X. Wang et al. 2008*). One study of interest to policy-oriented readers is Y. X. Wang et al. (2009*), which found that aggressive emission controls and not just convenient weather were responsible for reduced ozone levels during the 2008 Beijing Olympics, as measured at the Tsinghua-Harvard atmospheric station north of the city.

For year-to-year trends, a pertinent study is T. Wang et al. (2009), which reports the first continuous record of surface ozone in the background air of south China, observed 1994–2007 at a remote coastal site at the southern tip of Hong Kong Island. Using techniques to screen out the local influences of Hong Kong itself, it then employed atmospheric "back-trajectories" to group observations by the provenance of the sampled air masses. The local climatology of the region is such that a high percentage of these air masses pass over the North China Plain (including Beijing, Tianjin, and many other cities) and/or the Yangtze River Delta (including Shanghai). These observations thus reflect the chemical influences of emissions of some of the most densely populated and industrialized parts of China. The study found that ozone increased by 0.94 parts per billion (ppb) per year from 1994–2000 to 2001–2007, or roughly 2% a year compared to the observed mean concentration of 48 ppb for these air masses. More geographically limited studies, of the Beijing region, were generally consistent, suggesting annual growth rates of summer ozone of 2–4% (Ding et al. 2008; Tang et al. 2009).

We caution here that the complexity of ozone photochemistry results in large regional, seasonal, and time-of-day (diurnal) variations, and the rates of growth cited here should be regarded as general indicators only. Still, the available evidence of 2–4% annual growth in concentrations in major regions already suffering from a high annual mean of 50–75 ppb (which implies much higher levels on highly polluted days)[31] strongly suggests a serious and accelerating hazard. Without expanded commitments to scientific research and informed policy strategies, China appears destined to wrestle with rising ozone risks for decades to come.

Carbon Dioxide CO_2 is a not a conventional air pollutant, and our interest in it concerns its effect on climate as a globally mixed greenhouse gas. Slight differences

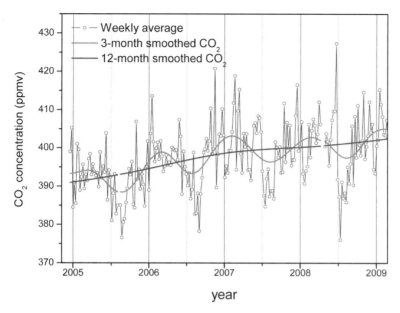

Figure 1.12
Record of CO_2 concentrations measured 100 km north of Beijing, 2005–2009. The sharply varying thin black line with circles represents weekly mean CO_2 at Tsinghua-Harvard Miyun atmospheric observatory from December 2004 to February 2009. The sinuous gray and steadily rising black lines are curves for the seasonal and annual variations, respectively. *Source:* Y. X. Wang, Munger, et al. (2010*). Reprinted under terms of Creative Commons 3.0 Attribution license, at http://creativecommons.org/licenses/by/3.0/.

in concentrations of CO_2 at national scale are largely irrelevant to its environmental impacts, which result instead from the rise in the global average CO_2 levels.

Those differences, however, are scientifically valuable, and will become more so over time. A year-to-year record of instrumental CO_2 measurements by Y. X. Wang, Munger, et al. (2010*) taken at our Tsinghua-Harvard observatory 100 km north of Beijing, as shown in figure 1.12, reflects higher concentrations of CO_2 in the region around Beijing than the global average level, and a rate of growth that is also higher (by about 1 part per million, or ppm, per year). This observed CO_2 enhancement is generally as expected based on the swift growth of Chinese emissions, taking account of local biological and meteorological influences. It also fits CO_2 observations at other stations upwind and downwind of the site, on the Tibetan Plateau and in Mongolia, Korea, and Japan.

These measurements are noted to spotlight that records of CO_2 concentrations in China are now being collected at a growing number of surface stations of scientific groups and government agencies. Such measurements will be invaluable to domestic and international scientific efforts to better understand the role of China in the global carbon cycle.

Carbon dioxide observation is now also extending into space. One dedicated satellite-based instrument is successfully measuring column CO_2, and another with improved spatial resolution is scheduled to launch in 2014.[32] The resolution of satellite-based CO_2 observations will inevitably increase over successive missions, providing improving evidence of emission trends independent of bottom-up estimates now at the heart of national policies and international diplomacy concerning GHGs and climate.

1.6   Review of Other Integrated Assessments of Atmospheric Environment in China

Before turning to our research framework and a summary of our first two policy assessments in the next chapter, we review similarly integrated analyses that use methods from the international literature and report results in peer-reviewed publications. These include some studies that estimate aggregate damages of air pollution in China, and others that evaluate the effects of emission control policies. The following review includes a brief summary of the prior analysis by our team reported in Ho and Nielsen (2007*) and updates a review of earlier literature that appeared in chapter 1 of that book. This update is limited to research on China with the same national scope as the current book, and it excludes the body of peer-reviewed research of the same issues at city and other subnational scales in China. Key parameters and results for the studies reviewed are summarized in table 1.2 for quick reference.

Ho and Nielsen (2007*)   Our prior book on the topic similarly reported development of an interdisciplinary assessment framework by a U.S.-China research team, applying it to estimate the costs and benefits of two green tax policies.

The structure of the economic component of this previous study was similar to that in this volume (chapter 9). The approach to evaluating emissions, air quality, and human exposure, however, was fundamentally different. Because of limited available data, Ho and Nielsen (2007*) used a methodological tool developed by health scientists termed "intake fraction," which can be interpreted as the proportion of total emissions of a pollutant from a given source type that ultimately ends up in human lungs. The concept and methods to apply intake fractions

Table 1.2

International collaborations on health damage of air pollution in China

Name (Year)	Geographical Domain	Sector Coverage	Air Pollutants	Types of Air Pollution Damages	Mortality PM Dose-Response Coefficient (with range, if applicable)	Method(s) and Central Value of Statistical Life (with range, if applicable)	Central Estimate of Aggregate Effects
Ho and Nielsen (2007*)	China	All, based on 33-sector CGE model and intake fractions derived from air dispersion modeling	PM_{10}, SO_2, CO_2	Mortality and morbidity from ambient PM_{10} and SO_2	1.95 (1.3, 2.6) acute mortalities per 1 million adults per $\mu g/m^3$ of PM_{10}	Range of China and income-adjusted U.S. WTP by contingent valuation = ¥370,000 (¥130,000, ¥950,000), in 1997	With caveats,[a] 93,000 acute or 561,000 chronic premature mortalities from ambient PM_{10}, plus acute mortality from SO_2, in 1997
World Bank–SEPA (2007)	China	All, based on ambient concentrations reported by government	PM_{10}, SO_2	Mortality and morbidity from urban ambient and rural indoor PM_{10}. Also acid rain and SO_2 on crops and materials	Log-linear relative risk function for chronic mortality from PM_{10}, with coefficient of $RR = 0.073$	China WTP by contingent valuation = ¥1,000,000; adjusted human capital method = ¥280,000 (urban)	350,000–400,000 premature mortalities from ambient urban PM_{10}, in 2003[b]
Aunan et al. (2007)	China, in two regions: Guang-dong and rest of China	All, 61-sector CGE model incorporating air pollution effects, based on simplified dispersion (PM_{10}, SO_2) and reduced-form CTM (O_3)	TSP, PM_{10}, SO_2, NO_x, VOCs, CO, O_3, CO_2	Mortality and morbidity from ambient PM_{10}, SO_2. Also ozone on crops	2.2 (0, 4.1) acute mortalities per 1 million adults per $\mu g/m^3$ of PM_{10}	Taiwan WTP scaled to China by income ratio, = ¥355,000 for all of China, ¥572,000 for Guangdong, in 1997	Not applicable; study quantifies monetized benefits and costs of several carbon tax levels

Table continued

Table 1.2 (continued)

Name (Year)	Geographical Domain	Sector Coverage	Air Pollutants	Types of Air Pollution Damages	Mortality PM Dose-Response Coefficient (with range, if applicable)	Method(s) and Central Value of Statistical Life (with range, if applicable)	Central Estimate of Aggregate Effects
Vennemo et al. (2009)	China, in three regions: Guangdong, Shanxi, rest of China	All, 61-sector CGE model incorporating air pollution effects, based on simplified dispersion (PM_{10}) and reduced-form CTM (O_3)	PM_{10}, O_3, CO_2	Mortality and morbidity from ambient PM_{10}. Also ozone on crops	Log-linear relative risk function for chronic mortality from PM_{10}, with coefficient of $RR = 0.073$	Adjusted human capital method = ¥900,000 for Guangdong, ¥270,000 for Shanxi, ¥460,000 for rest of China, in 2003	Not applicable; study quantifies monetized benefits and costs of three carbon policy structures
Saikawa et al. (2009)	China, global	All, based on ambient concentrations modeled from global emission inventory by CTM	Three $PM_{2.5}$ species: SO_4^{-2}, OC, and BC	Mortality from ambient $PM_{2.5}$. Also aerosol radiative forcing	0.4% increase chronic mortality per $\mu g/m^3$ of $PM_{2.5}$, for adults ≥30 years old	Not valued	470,000 premature mortalities from ambient $PM_{2.5}$, in 2000. Study also quantifies nonmonetized benefits (but not costs) of three emission scenarios
He et al. (2010)	China	All, based on bottom-up energy model and air quality model	PM_{10}, $PM_{2.5}$, CO_2	Mortality and morbidity from ambient $PM_{2.5}$	0.41%[c] increased chronic mortality per $\mu g/m^3$ of $PM_{2.5}$	Chongqing WTP by contingent valuation scaled to Beijing by GDP ratio and applied nationwide = US$166,775, in 1999[d]	Not applicable; study quantifies monetized benefits (but not costs) of two energy or emission control scenarios

| Matus et al. (2012) | China | PM_{10}, O_3 | All, PM_{10} based on ambient TSP concentrations reported by government, O_3 based on output of global model | Morbidity from ambient PM_{10} and O_3, mortality from ambient $PM_{2.5}$ | 0.25% chronic (0.02%, 0.48%, adults ≥30 years old) and 0.06% acute (0.04%, 0.08%) mortality per μg/m³ of PM_{10} | Not applicable, assesses instead the average value of years-of-lost-life per mortality by adjusting European WTP to China, = US$662, in 1997[e] | 2.7 million chronic and 0.2 million acute premature mortalities from exposure to ambient urban PM_{10}, in 2005 |

Notes: Base years: We are unfortunately unable to list all the base years of these assessments, required for careful comparisons across studies. The base years often differ by study component (e.g., emissions vs. valuation) and in many cases are unclear because they are not described clearly or at all in published reports. We caution that this table is introductory and researchers should consult the original literature.

General abbreviations: CTM is chemical tracer (atmospheric) model; CGE is computable general equilibrium (economic model); GDP is gross domestic product; GNP is gross national product; PPP is purchasing power parity; μg/m³ is micrograms per cubic meter; YOLL is year of life lost; WTP is willingness to pay; CV is contingent valuation; and BAU is business as usual.

Pollutant abbreviations: TSP is total suspended particulates; PM is particulate matter, and PM_{10} and $PM_{2.5}$ are categories of PM under diameters of 10 and 2.5 microns; SO_2 is sulfur dioxide; SO_4^{-2} is sulfate particles (of various chemical forms); NO_X is nitrogen oxides; NO_3^- is nitrate particles (of various forms); O_3 is ozone; OC is organic carbon; BC is black carbon; CO is carbon monoxide; Pb is lead; VOCs are volatile organic compounds; CO_2 is carbon dioxide; CH_4 is methane; and N_2O is nitrous oxide.

[a]Our prior study stressed use of damage estimation for comparing impacts of policies, cautioning about effects of large assumptions and uncertainties on point estimation of aggregate damages in all such research. Accordingly, we applied acute mortality epidemiology from China in a base case and high and low alternatives, and U.S. chronic mortality epidemiology in a separate case. The last is reported in monetized terms in table 9.11 of Ho and Nielsen (2007*). We also applied base case, high, and low valuations.

[b]This total is not reported in the text of World Bank–SEPA (2007), reportedly removed in a controversy that prevented formal publication of the study and led to its release only in draft form, without health damages, as a conference report. The estimate of premature mortalities comes from news reports at the time of the conference and controversy (e.g., Kahn and Yardley 2007).

[c]This is not directly reported in He et al. (2010), but is derived according to methods in a cited paper.

[d]This is not directly reported in He et al. (2010), but in a cited report.

[e]The unit of the low value "per case" of mortality in table 3 of the paper is not obvious. The interpretation here is based on discussion in section 5.3 of the paper.

were specifically developed for conditions for which data are too limited to support comprehensive simulation of air quality, as is the case in many developing countries.

Our team calculated intake fractions for nearly 800 emission sources sampled from five sectors that dominate emissions, applying two simple air dispersion models to plant-level data and population maps for exposure and health risk assessment. The intake fractions for the remaining sectors were then approximated from these results. The methods allowed differentiation of air pollution health risks by sector, taking into account not only emission levels but also other factors that influence exposure risk, such as average smokestack heights and proximity to population centers. The study found that a tax on fossil fuels proportional to the air pollution damages that they cause could reduce monetized total health damages by at least several times the loss of GDP, meaning the benefits would substantially exceed the costs.

Readers familiar with our prior book, the starting point for the current one, may want to know how the two books differ. The advances are detailed in section 2.3, but we outline the main innovations here. Sharply improved emission inventories allow us to introduce advanced atmospheric research methods while linking them to our structurally dissimilar economic and policy framing, replacing entirely the much rougher intake fraction methods of the previous book. The current study also expands the assessment of benefits beyond public health to crop productivity. And it widens the scope of policy assessment, adding technology mandates to market mechanisms, and evaluating an actual policy of the past as well as prospective policies of the future.

World Bank–SEPA (2007)   In 2007, the World Bank and the Chinese State Environmental Protection Administration (SEPA, now the Ministry of Environmental Protection) nearly completed a joint application of a newly developed integrated "Environmental Cost Model" (ECM) for China (World Bank–SEPA 2007; this is a draft conference report of a peer-reviewed study that was never finalized for formal publication because of political considerations). The study covered a range of pollution damages in both the water and air of China, the latter including not only impacts on health but also acid rain damages on agriculture and buildings. The ECM had a bottom-up framework, conducting air pollution assessment at the city level using annual average concentrations from official sources, aggregating them into provincial and national totals. It also assessed indoor air pollution damage in rural China, though it does not present those calculations in the conference report.

The costs in the study refer to monetized damages caused by pollution. (We note that this use of the term "costs" differs from its use in policy analyses of costs and benefits. In the framework of this book, the *reduction* of such damages constitutes a *benefit* of the policy, while the *costs* refer instead to the costs of control equipment and other direct and indirect economic costs of a policy. The World Bank–SEPA [2007] study did not assess costs in this sense, nor did it seek to assess the effects of emission control policies.) The study concluded that the monetized damage of premature mortality and morbidity from air pollution amounted to $157 billion yuan in 2003 (1.16% of GDP), and damages to crops and buildings from acid rain cost an additional 37 billion yuan annually.

Joint Economic Study (JES 2007)   This analysis of emission abatement policies was conducted as part of the official U.S.-China Strategic Economic Dialogue. The JES, coordinated by the U.S. Environmental Protection Agency (U.S. EPA) and China's SEPA, focused on the 11th FYP policies in the electric power sector. Using the CMAQ air quality model, the study estimated the change in pollution concentrations due to the FYP requirements for expanded use of FGD and the forced retirement of small, inefficient power plants. The Harvard economic model of China used in the current book was used to estimate the economic impacts of the policies.

The JES estimated that by 2010 the FGD policy alone would reduce annual SO_2 emissions by 5.4 million tons and concentrations of fine particles ($PM_{2.5}$) by an average of 5% nationally, with a benefit-to-cost ratio of 5-to-1. The plant shutdown policy was estimated not only to reduce coal consumption but also SO_2 emissions, by 2.1 million tons annually by 2010. As described in the next chapter, one of the policy cases of the current book is a much more data-intensive and analytically integrated version of this policy scenario. While JES (2007) is not a peer-reviewed publication, elements have been published in refereed journals (Cao, Garbaccio, and Ho 2009* and L. T. Wang et al. 2010a, 2010b, all three of which included coauthors contributing to the current book).

Aunan et al. (2007)   These authors estimated the effects of China taking on CO_2 targets, using a two-region computable general equilibrium (CGE) model of the economy (Guangdong and the rest of China). They included the effects not only of PM_{10} and SO_2 on human health but also of ozone on crop productivity. The study notably also differentiated relative welfare gains or losses between rural and urban areas. The air pollution levels were estimated using simplified reduced-form relationships between emissions and concentrations, including the photochemistry of ozone formation, that were incorporated directly into the economic model. Monetized agricultural and health benefits were judged to be of similar scales. The study estimated that a carbon tax that reduces CO_2 emissions by up to 17.5% may have negative costs; that is, the monetized health and agriculture productivity benefits would be higher than the economic costs.

Vennemo et al. (2009)   This study built on the modeling framework of Aunan et al. (2007) to assess the benefits and costs to China of three climate treaty designs: (1) a cap on CO_2/GDP intensity; (2) a cap on CO_2 emissions; and (3) a cap on CO_2 intensity of given sectors. Enhancements of Aunan et al. (2007) included more detailed distributional features, specifically adding a low-income, coal-producing province (Shanxi) as a third region and including 14 household income categories for each region. The study's distributional analyses are unique in the current literature—for instance, quantifying the larger negative impact of carbon controls on the economic welfare of provinces like Shanxi, of rural versus urban residents, and of poor versus rich households. While this is partly a function of assumed mechanisms for imposing an implied tax and recycling the revenues, potentially redressed by alternative mechanisms, the regressive nature of such policies for the rural poor was deemed more structural and challenging.

Saikawa et al. (2009)   This work is a study of the impacts of three species of $PM_{2.5}$—sulfate particles, organic carbon (OC), and black carbon (BC)—resulting from Chinese emissions in a base year of 2000 and in three scenarios for 2030. The specification of OC and BC, in contrast to other studies reviewed here, reflected the interest of this chiefly scientific research team in estimating not only the health impacts of Chinese emissions but also their effects on radiative forcing, which are negative for sulfate and OC and positive for BC. Saikawa et al. (2009) thus represents a significant innovation in quantifying a link between Chinese air pollution and agents of global climate change, through radiative effects of aerosols and not association with CO_2 (as in the current study). Regarding the health effects of Chinese emissions, the study estimated total damages, not only in China but also (considerably smaller) transboundary damages in other parts of Asia and North America. A simple policy evaluation evaluated and compared benefits but not costs of the three 2030 emission scenarios.

He et al. (2010)   These authors reported results of a collaborative study supported by the U.S. EPA to quantify the benefits (but not costs) of two speculative policy scenarios compared to a base case. One scenario included a range of energy-related policies to reduce carbon, and the other added additional policies to control pollutant emissions. These scenarios were evaluated bottom-up, using the Long-Range Energy Alternatives Planning (LEAP) model and emission inventories, the CMAQ air quality modeling system, and the Benefits Mapping and Analysis Program (BenMAP) for evaluating health effects. Chapter 8 of this book, by a coauthor of He et al. (2010), builds on the health benefit assessment component of that study.

Matus et al. (2012) This study investigated the historical effects on the Chinese economy of health damages of air pollution, through the lens of one of 16 regions in a CGE model of the world economy. The considerable value of Matus et al. (2012) is conceptual and methodological, making a strong case based on prior work on Europe for dynamic (not static) analysis of age-conditioned environmental health impacts on Chinese consumption and welfare over time. Strict accuracy, however, may not have been a primary aim of the study, which relied on highly aggregated data inputs and functions adapted from studies of the United States and Europe rather than of China. For instance, it represented PM_{10} concentrations in each year as a single national urban mean value derived from official TSP data; combined with other assumptions, this yielded a huge estimate of 2.9 million premature deaths in 2005 from acute and chronic PM_{10} exposures.

A Note on Indoor Air Pollution We detour momentarily here to acknowledge a body of research that does not pertain directly to the work of this book, but should not be overlooked by those interested in the total health burden of air pollution in China. J. F. Zhang and Smith (2007) provide an excellent review of the important literature on rural indoor air pollution and health in China. Such papers are rarely combined with research on outdoor air, in part because the research methods and policy options differ enormously. Our focus on outdoor air derives substantially from interest in links to CO_2 emissions, which are much larger in aggregate from large-scale combustion of commercial fossil fuels than the combustion of often noncommercial fuels (such as crop wastes) in rural households. This is not to diminish the importance of rural indoor air; as World Bank–SEPA (2007) and J. F. Zhang et al. (2010) indicate, the total health effects of indoor and outdoor air pollution are of similar magnitudes. Indoor air risks are also intertwined with critical challenges in rural poverty alleviation.

The research presented in this book differs from the preceding studies most fundamentally in that it incorporates advanced research methods across a wider range of disciplines related to both costs and benefits, from economics, to emissions, to atmospheric transport and chemistry, to health and agricultural impacts. It is also rooted in particularly detailed data sources, from national economic accounts to plant-by-plant emission inventories to independent atmospheric measurements used to test air quality simulations.

Aunan et al. (2007), Vennemo et al. (2009), and our prior study, Ho and Nielsen (2007*), were designed for policy assessment, considering both costs and benefits of possible emission control interventions. All three featured strong economic frameworks to evaluate costs, but relied on comparatively simple emission data and atmospheric tools to evaluate environmental benefits. World Bank–SEPA (2007), Saikawa et al. (2009), He et al. (2010), and Matus et al. (2012) were more limited in their scopes, in that they focused on pollution damages (or benefits of emission reductions) without attempting to evaluate costs. This limited their applicability to assessment of policy options.

These studies all have strengths, including important methodological innovations, some of which we exploit. Aunan et al. (2007) advanced estimation of ozone damages to crops. The regional and household differentiation in the economic framework of Vennemo et al. (2009) allowed compelling insights into distributional inequities of national emission control policies. Matus et al. (2012) introduced a valuable dynamic approach to assessing health damages of air pollution. World Bank–SEPA (2007) encompassed a much wider range of environmental damages than the current study, including the health impacts of China's rural indoor air and degraded water quality.[33] He et al. (2010) developed a health benefit tool that we adapted, and one of our cases expands directly on JES (2007). Saikawa et al. (2009) provided a valuable link between research on the health impacts of China's air quality and the burgeoning interest in the short-term radiative forcing of sulfate particles and black carbon.

1.7 Conclusion

This chapter's review of what is currently known and not known about the balance China has struck between economic growth and protection of its atmospheric environment is intended to rebut any overly convenient, black-and-white perceptions. In recent years there have been both deeply worrisome and remarkably encouraging developments: sharp rises in NO_X emissions, ozone levels, and, most critically, emissions of CO_2; at the same time, a booming economy combined with sustained declines in energy intensity and emissions of primary PM and now SO_2, substantially attributable to forceful policy design and implementation.

When evaluating China's record of grappling with these challenges, it should not be treated as somehow categorically distinct from that of other nations. Reconciling sustained 7–10 percent economic growth with protection of air quality and control of GHG emissions would tax even the wealthiest economies and the most environmentally committed societies. Attempting to do so while still validating data strategies and building up basic knowledge at the intersections of economy, energy use, emissions, atmospheric transport and chemistry, and environmental health and agriculture could certainly overwhelm any nation. China will inevitably experience both failures and successes in its efforts.

The study that follows in the rest of this book—policy cases for the recent past summarized in the next chapter and prospective ones for the future in chapter 3, with in-depth presentation in chapters 4–9—will offer further evidence of these challenging, intersecting realities facing China. But by building an integrated framework for understanding their trade-offs, it will also offer a basis for testing possible options in national emission control. And ultimately, readers may discover that the range of policies it evaluates suggests that comprehensive reconciliation of economic growth, protection of air quality, and control of carbon emissions may be more conceivable for China than they previously thought.

Notes

1. For example, the SO_2 target in the previous 10th FYP was 18 million tons, but the actual emissions for 2005 turned out to be 25 million tons. See Cao, Garbaccio, and Ho (2009*) for other examples.

2. Resistance of China's oil companies to mandated reductions in the sulfur content of vehicle fuels became a common theme in international news reports (e.g., Wee and Li 2013; Wong 2013), despite the very small share of SO_2 emissions attributable to transportation sources and their more critical roles in emissions of NO_X and volatile organic compounds.

3. Among other mechanisms, light-colored sulfate particles reflect incoming radiation back to space.

4. As described in section 1.6, the closest previous studies of China are structured around either equivalent economic models or equivalent atmospheric models, but not both. Advanced models of the world economy and global climate system have been joined in a number of "integrated assessment" research programs around the world (a leading example in the United States is led by MIT, described at http://globalchange.mit.edu/research/IGSM, while others are described in U.S. DOE [2009]). These focus more on climate issues than air quality ones, however, and their global scopes do not require or support the deep national perspective and data of the sorts developed in our models of China.

5. We could not conduct the full analysis for every tax policy case of interest because for each one of them, the analytical steps from emissions to air quality to impacts on health and

crops together constitute a very complex research undertaking, as chapters 4–9 will clearly demonstrate. Whereas different tax levels engender substantial changes in emissions and thus in air quality benefits, variations in other features of a tax often influence emissions only slightly, even as their effects on the economy remain of great interest. We therefore limit our comprehensive analyses to the cases in which we expect to gain the most insight about the emission and atmospheric implications, and employ only the economic framework in cases where the new insights are mainly economic.

6. We might note that economic, energy, and emission statistics are also routinely revised in Western countries, as more evidence is compiled and data are cross-checked. The revisions in China nevertheless have often been at larger scales, reflecting the developing nature of its statistical system.

7. World Bank (1992) provides a brief history of the economic statistics and the transition to the United Nations' System of National Accounts.

8. The energy intensity is calculated from the data in figure 1.1, which is a simple sum of total energy in standard coal equivalents (sce) of the various fuels. That simple index of total use ignores the big differences in the prices of a kilogram of coal equivalent of each of the fossil fuels; 1 kgce (kilogram of coal equivalent) of natural gas is much more expensive than 1 kgce of coal, reflecting its greater ease of use and economic value. An *economic* index of total energy input would use the price information and generate a very different series from the simple index.

9. Combustion and process emissions of particulates are termed "soot" and "dust," respectively, in the statistical tables.

10. See the CNEMC Web site at http://www.cnemc.cn/.

11. One prominent use of the urban monitoring system is calculation of the Air Pollution Index (API) in Chinese cities, which at the time of writing has been upgraded to a broader Air Quality Index (AQI) on a trial basis in many cities (see MEP 2012b, 2013). The API or AQI assigns a rating to the mean daily concentrations of each of 3–6 pollutants—at minimum, PM_{10}, SO_2, and NO_2, with carbon monoxide, ozone, and $PM_{2.5}$ also included in the AQI— based on 0–500 scales, with rising thresholds indicating increasing categories of health risk (e.g., 101–150 is deemed "slightly polluted" and 301–500 is "severely polluted"). The poorest-ranking pollutant determines the overall API or AQI rating for the day, which is then publicly reported. As is the case in other countries, this system is designed chiefly for public communication of current environmental health risk, not scientific use. In 2011–2012, disparities between readings at Chinese urban networks and individual monitors operated at U.S. consulates and its embassy in China—deployed initially to judge hardship pay to employees, but with the readings later routed to a Twitter feed—and between the descriptive labels used for given pollutant ranges (with stronger terminology used by the United States) sparked a social media–based citizen campaign for greater environmental transparency and inclusion of $PM_{2.5}$ in new air quality standards; they also resulted in a diplomatic dispute (Larson 2012; Bradsher 2012).

12. The rationale for observational sites in such locations, as opposed to within cities themselves, is discussed in section 1.5.2.

13. This method, however, does not control for the fact that measurements of the monitoring network in each city may become more or less representative of average pollutant levels in that city over time. Even assuming a static set of monitoring stations, the characteristics

represented by an average will change as a city grows and evolves over time, e.g., as stations originally sited to sample air near the margins of the city are surrounded by intensifying development. Stations can also be moved, closed, or added to the system—sometimes for defensible technical, scientific, or logistical reasons—affecting the citywide averages.

14. The rationale for switching to PM_{10} is clear, in that it is more closely associated with health impacts than TSP; the rationale for switching to NO_2 is less obvious.

15. The WHO interim and guideline levels for 24-hour average PM_{10} concentrations are 150, 100, 75, and 50 $\mu g/m^3$ (WHO 2006). The WHO interim and guideline concentrations for annual and 24-hour average $PM_{2.5}$ are, in each case, exactly half of the corresponding level for PM_{10}.

16. "Top-down" has another meaning to atmospheric scientists, in which emissions are inferred from concentrations measured in the atmosphere.

17. The authors of chapters 5 and 6 led several of these studies: Lei et al. (2011); Zhao, Nielsen, et al. (2011*); and Zhao, Zhang, and Nielsen (2013*).

18. Other entities such as the International Energy Agency and the World Bank also estimate Chinese emissions of CO_2; these estimates also rely on top-down methods and yield results generally comparable to those cited here.

19. These methods include compiling probability distributions from field tests reported in Chinese (or, if unavailable, foreign) field studies for all input parameters, including emission factors and activity levels. This step is then followed with statistical reproductions of the bottom-up emissions in so-called Monte Carlo simulations (Zhao, Zhang, and Nielsen 2013*; Zhao, Nielsen, and McElroy 2012*; Zhao, Nielsen, McElroy, et al 2012*; Zhao, Nielsen, et al. 2011*).

20. This method of uncertainty estimation is only a rough indicator and has weaknesses: the emission estimates of a country that does not revise erroneous data would be deemed more certain than those of a country that, like China, does try to correct equivalent errors.

21. These include an observatory built and operated 100 km north of Beijing by the Harvard-Tsinghua partnership conducting the current study (Y. X. Wang, McElroy, et al. 2008, 2010*).

22. The most renowned such station is Waliguan Baseline Observatory in Qinghai, operated by the Chinese Academy of Meteorological Sciences.

23. This vertical column density is in fact derived from slant column density, which is related to the angle of reflection of sunlight from the earth's surface to the satellite.

24. This is because NO_2 and NO are normally in chemical equilibrium determined by the state of the atmosphere, and because concentrations of NO_2 generally far exceed concentrations of NO.

25. Lin and McElroy (2011*) and Lin et al. (2010*) refer to this as "northeastern China" or "Northern East China," but it is almost identical to "East Central China" defined by Richter et al. (2005).

26. Recall that the record is a 12-month moving average, meaning the first data points on the left-hand side reflect the prior year.

27. Readers interested in satellite-based analyses of $PM_{2.5}$ from AOD are also directed to a global assessment published around the same time as Lin et al. (2010*), van Donkelaar

et al. (2010), which developed a new estimation technique validated by high-quality surface observations in North America and a chemical transport model. While it noted very high $PM_{2.5}$ concentrations over east China, this study assessed only a six-year average concentration for 2001–2006, and did not consider the temporal trends or effects of recent economic and policy developments evaluated for China by Lin et al. (2010*). A China-specific, peer-reviewed adaptation and application of the methods of van Donkelaar et al. (2010) to more recent data would be valuable.

28. Note that the comparison in growth rates must be made with respect to the two different *y*-axes' ranges, although this happens to be consistent with how the series appear at first glance.

29. We must note that the mean VCDs for 2010 indicated a reversal in the trend. However, the spatial pattern over China for 2010 is dramatically different from the consistent pattern of the prior four years, and at the time of writing, the institute that processes the data (Support to Aviation Control Service, http://sacs.aeronomie.be/archive/month/index_VCD.month.php) had recognized errors in the retrievals for 2010 in a personal communication and was reprocessing the SO_2 VCDs from the SCIAMACHY instrument for that year. These data are discussed in section 4.5 of Zhao, Zhang, and Nielsen (2013*), and the highly anomalous map for 2010 can be seen in the linked supporting material for that paper.

30. This is "good ozone," the gas that protects us from overexposure to ultraviolet radiation; the decline of this stratospheric layer over Antarctica causes the well-known ozone hole. This is to be contrasted with tropospheric ozone at the surface, the largely anthropogenic concentration that is harmful to human health and can damage plants.

31. T. Wang et al. (2006) measured 1-hour average ozone levels that exceeded 120 ppb on 13 out of 39 measurement days at a site 50 miles north of Beijing. It notably recorded 286 ppb on one day, China's highest recorded value in the open literature.

32. The Japanese space agency operates the currently active GOSAT satellite. The planned satellite is the Orbiting Carbon Observatory 2 (OCO2), to be launched by NASA as a duplicate of the original OCO, which crashed on launch in February 2009. Methods have also been developed to extract limited information on CO_2 from satellite-based instruments deployed for other purposes, such as TES aboard NASA's Aura Earth satellite.

33. For reviews of research on water pollution in China, afield from our primary interests, see World Bank–SEPA (2007) and Zhang et al. (2010).

References

Note that an * signifies a study produced by the Harvard-based program that produced this book.

Akimoto, H., T. Ohara, J. Kurokawa, and N. Horii. 2006. Verification of energy consumption in China during 1996–2003 by using satellite observational data. *Atmospheric Environment* 40(40):7664–7667.

Aunan, K., T. Berntsen, D. O'Connor, T. Hindman Persson, H. Vennemo, and F. Zhai. 2007. Benefits and costs to China of a climate policy. *Environment and Development Economics* 12:471–497.

Boden, T. A., G. Marland, and R. J. Andres. 2010. People's Republic of China fossil-fuel CO_2 emissions. Carbon Dioxide Information Analysis Center (CDIAC), Oak Ridge National Laboratory, U.S. Department of Energy, Oak Ridge, TN, U.S.A., doi:10.3334/CDIAC/00001_V2010. Available at http://cdiac.ornl.gov/trends/emis/tre_prc.html, last accessed June 19, 2011.

Bond, T. C., D. G. Streets, K. F. Yarber, S. M. Nelson, J.-H. Woo, and Z. Klimont. 2004. A technology-based global inventory of black and organic carbon emissions from combustion. *Journal of Geophysical Research* 109: D14203, doi:10.1029/2003jd003697.

Bradsher, K. 2012. China asks other nations not to release its air data. *New York Times*. 5 June. Available at http://www.nytimes.com/2012/06/06/world/asia/china-asks-embassies-to-stop-measuring-air-pollution.html?_r=0, last accessed January 9, 2013.

Cao, J., R. Garbaccio, and M. S. Ho. 2009*. China's 11th Five-Year Plan and the environment: Reducing SO_2 emissions. *Review of Environmental Economics and Policy* 3(2): 189–208.

Cao, J., and M. S. Ho. 2009*. Changes in China's energy intensity: Origins and implications for long-run carbon emissions and climate policies. Paper prepared for the Canadian International Development Research Centre, Economy and Environment Program for South-East Asia (EEPSEA), November.

Ding, A. J., T. Wang, V. Thouret, J.-P. Cammas, and P. Nédélec. 2008. Tropospheric ozone climatology over Beijing: Analysis of aircraft data from the MOZAIC program. *Atmospheric Chemistry and Physics* 8:1–13.

Eckholm, Erik. 2001. China said to sharply reduce emissions of carbon dioxide. *New York Times*, June 15, late edition-final, A1.

Fishman, J., A. E. Wozniak, and J. K. Creilson. 2003. Global distribution of tropospheric ozone from satellite measurements using the empirically corrected tropospheric ozone residual technique: Identification of the regional aspects of air pollution. *Atmospheric Chemistry and Physics* 3:893–907.

Fridley, D., and N. Aden, eds. 2008. *China Energy Databook Version 7.0*. Lawrence Berkeley National Laboratory, China Energy Group. Berkeley, CA: LBNL.

Garbaccio, Richard, Mun S. Ho, and Dale W. Jorgenson. 1999. Why has the energy output ratio fallen in China? *Energy Journal* 20(3):63–91.

Gregg, J. S., R. J. Andres, and G. Marland. 2008. China: Emissions pattern of the world leader in CO_2 emissions from fossil fuel consumption and cement production. *Geophysical Research Letters* (35):L08806, doi:10.1029/2007GL032887.

He, K., Y. Lei, X. Pan, Y. Zhang, Q. Zhang, and D. Chen. 2010. Co-benefits from energy policies in China. *Energy* 35(11):4265–4272, doi:10.1016/j.energy.2008.07.021.

Ho, M. S., and C. P. Nielsen, eds. 2007*. *Clearing the Air: The Health and Economic Damages of Air Pollution in China*. Cambridge, MA: MIT Press.

Huo, H., Q. Zhang, M. Q. Wang, D. G. Streets, and K. B. He. 2010. Environmental implication of electric vehicles in China. *Environmental Science and Technology* 44(13):4856–4861, doi: 10.1021/es100520c.

International Energy Agency (IEA). 2007. *World Energy Outlook 2007: China and India Insights*. Paris: OECD/IEA.

Joint Economic Study (JES). 2007. *U.S.-China Joint Economic Study: Economic Analyses of Energy Saving and Pollution Abatement Policies for the Electric Power Sectors of China and the United States—Summary for Policymakers* (project report). December. Beijing: State Environmental Protection Administration and U.S. Environmental Protection Agency.

Kahn, Joseph, and Jim Yardley. 2007. Choking on growth part I: As China roars, pollution reaches deadly extremes. *New York Times*, August 26.

Klimont, Z., J. Cofala, W. Schopp, M. Amann, D. G. Streets, Y. Ichikawa, and S. Fujita. 2001. Projections of SO_2, NO_X, NH_3 and VOC emissions in East Asia up to 2030. *Water Air and Soil Pollution* 130(1–4):193–198.

Klimont, Z., J. Cofala, J. Xing, W. Wei, C. Zhang, S. Wang, K. Jiang, P. Bhandari, R. Mathur, P. Purohit, P. Rafaj, A. Chambers, and M. Amann. 2009. Projections of SO_2, NO_X and carbonaceous aerosols emissions in Asia. *Tellus Series B—Chemical and Physical Meteorology* 61(4):602–617.

Larson, C. 2012. China's Ma Jun on the fight to clean up Beijing's dirty air. *Yale Environment 360*. 10 April. Available at http://e360.yale.edu/feature/chinas_ma_jun_on_the_fight_to_clean_up_beijings_dirty_air/2515/, last accessed April 13, 2012.

Lei, Y., Q. Zhang, K. B. He, and D. G. Streets. 2011. Primary anthropogenic aerosol emission trends for China, 1990–2005. *Atmospheric Chemistry and Physics* 11:931–954, doi:10.5194/acp-11-931-2011.

Li, C., Q. Zhang, N. A. Krotkov, D. G. Streets, K. B. He, S.-C. Tsay, and J. F. Gleason. 2010. Recent large reduction in sulfur dioxide emissions from Chinese power plants observed by the Ozone Monitoring Instrument. *Geophysical Research Letters* 37:L08807, doi:10.1029/2010GL042594.

Lin, J. T., and M. B. McElroy. 2011*. Detection from space of a reduction in anthropogenic emissions of nitrogen oxides during the Chinese economic downturn. *Atmospheric Chemistry and Physics* 11:8171–8188, doi:10.5194/acp-11-8171-2011.

Lin, J. T., C. P. Nielsen, Y. Zhao, Y. Lei, Y. Liu, and M. B. McElroy. 2010*. Recent changes in particulate air pollution over China observed from space and ground: Effectiveness of emission control. *Environmental Science and Technology* 44(20):7771–7776, doi:10.1021/es101094t.

Lu, Z., D. G. Streets, Q. Zhang, S. Wang, G. R. Carmichael, Y. F. Cheng, C. Wei, M. Chin, T. Diehl, and Q. Tan. 2010. Sulfur dioxide emissions in China and sulfur trends in East Asia since 2000. *Atmospheric Chemistry and Physics* 10:6311–6331, doi:10.5194/acp-10-6311-2010.

Lu, Z., Q. Zhang, and D. G. Streets. 2011. Sulfur dioxide and primary carbonaceous aerosol emissions in China and India, 1996–2010. *Atmospheric Chemistry and Physics* 11:9839–9864, doi:10.5194/acp-11-9839-2011.

Ma, C., and D. I. Stern. 2007. China's changing energy intensity trend: A decomposition analysis. *Energy Economics* 30(3):1037–1053, doi:10.1016/j.eneco.2007.05.005.

Matus, K., K.-M. Nam, N. E. Selin, L. N. Lamsal, J. M. Reilly, and S. Paltsev. 2012. Health damages from air pollution in China. *Global Environmental Change* (22):55–66, doi: 10.1016/j.gloenvcha.2011.08.006.

Ministry of Environmental Protection (MEP). 2009–2012a. *China Environment Yearbook* (2008–2011 editions). Beijing: China Environment Yearbook Press.

————. 2012b. Technical regulation on ambient air quality index (on trial). Available at http://kjs.mep.gov.cn/hjbhbz/bzwb/dqhjbh/jcgfffbz/201203/t20120302_224166.htm, last accessed January 9, 2013.

————. 2013. Beijing-Tianjin-Hebei Province and 74 cities to start monitoring by new air quality standard. Available at http://english.mep.gov.cn/News_service/infocus/201301/t20130106_244699.htm, last accessed January 9, 2013.

Mintzer, I., J. A. Leonard, and I. D. Valencia. 2010. *Counting the Gigatonnes: Building Trust in Greenhouse Gas Inventories from the United States and China* (World Wildlife Fund report). Washington, DC: World Wildlife Fund.

National Bureau of Statistics (NBS). 2005–2012. *China Energy Statistical Yearbook* (2004–2011 editions). Beijing: China Statistics Press.

National Bureau of Statistics (NBS) and Ministry of Environmental Protection or State Environmental Protection Administration (MEP/SEPA). 2004–2012. *China Statistical Yearbook on Environment* (2003–2011 editions). Beijing: China Statistical Press.

National Development and Reform Commission (NDRC). 2007. *Medium- and Long-Term Development Plan for Renewable Energy in China* (government document). September. Beijing: NDRC.

————. 2012. The second national communication on climate changes of the People's Republic of China. 8 November. Non–Annex I national communications, UNFCCC. Available at http://unfccc.int/essential_background/library/items/3599.php?rec=j&priref=7666#beg, last accessed January 9, 2013.

Ohara, T., H. Akimoto, J. Kurokawa, N. Horii, K. Yamaji, X. Yan, and T. Hayasaka. 2007. An Asian emission inventory of anthropogenic emission sources for the period 1980–2020. *Atmospheric Chemistry and Physics* 7(16):4419–4444.

Oliver, H. H., K. S. Gallagher, D. G. Tian, and J. H. Zhang. 2009. China's fuel economy standards for passenger vehicles: Rationale, policy process, and impacts. *Energy Policy* 37(11): 4720–4729.

Olivier, J. G. J., G. Janssens-Maenhout, and J. A. H. W. Peters. 2012. Trends in global CO_2 emissions; 2012 report. PBL Netherlands Environmental Assessment Agency report no. 500114022. The Hague/Bilthoven. Available at http://www.pbl.nl/en/publications/2012/trends-in-global-co2-emissions-2012-report, last accessed January 20, 2013.

PBL (Netherlands Environmental Assessment Agency). 2010. No growth in total global CO_2 emissions in 2009. Available at http://www.pbl.nl/en/publications/2010/No-growth-in-total-global-CO2-emissions-in-2009, last accessed June 19, 2011.

Qu, W. J., R. Arimoto, X. Y. Zhang, C. H. Zhao, Y. Q. Wang, L. F. Sheng, and G. Fu. 2010. Spatial distribution and interannual variation of surface PM_{10} concentrations over eighty-six Chinese cities. *Atmospheric Chemistry and Physics* 10:5641–5662, doi:10.5194/acp-10-5641-2010.

Richter, A., J. P. Burrows, H. Nuss, C. Granier, and U. Niemeier. 2005. Increase in tropospheric nitrogen dioxide over China observed from space. *Nature* 437:129–132.

Saikawa, E., V. Naik, L. W. Horowitz, J. F. Liu, and D. Mauzerall. 2009. Present and potential future contributions of sulfate, black and organic carbon aerosols from China to global air quality, premature mortality and radiative forcing. *Atmospheric Environment* 43: 2814–2822.

State Development Planning Commission (SDPC). 2004. The People's Republic of China initial national communication on climate change. 10 December. Non–Annex I national communications, UNFCCC. Available at http://unfccc.int/essential_background/library/items/3599.php?such=j&symbol=CHN/COM/1%20B#beg, last accessed January 9, 2013.

Shi, Y. Y. 2012. Air quality gets better test. *China Daily*. 16 November. Available at http://usa.chinadaily.com.cn/epaper/2012-11/16/content_15936064.htm, last accessed November 18, 2012.

Sinton, J. E. 2001. Accuracy and reliability of China's energy statistics. *China Economic Review* 12(4):347–354.

Sinton, J. E., and D. G. Fridley. 2000. What goes up: Recent trends in China's energy consumption. *Energy Policy* 28:671–687.

Sinton, J. E., and D. G. Fridley. 2003. Comments on recent energy statistics from China. *Sinosphere* 6(2):6–12.

Solomon, S., D. Qin, M. Manning, Z. Chen, M. Marquis, K. B. Averyt, M. Tignor, and H. L. Miller, eds. 2007. *Contribution of Working Group I to the Fourth Assessment Report of the Intergovernmental Panel on Climate Change*. Cambridge, UK: Cambridge University Press.

Streets, D. G., T. C. Bond, G. R. Carmichael, S. D. Fernandes, Q. Fu, D. He, Z. Klimont, S. M. Nelson, N. Y. Tsai, M. Q. Wang, J. H. Woo, and K. F. Yarber. 2003. An inventory of gaseous and primary aerosol emission in Asia in the year 2000. *Journal of Geophysical Research* 108 (D21):8809, doi:10.1029/2002JD003093.

Streets, D. G., K. J. Jiang, X. L. Hu, J. E. Sinton, X. Q. Zhang, D. Y. Xu, M. Z. Jacobson, and J. E. Hansen. 2001. Recent reductions in China's greenhouse gas emissions. *Science* 294 (November 30):1835–1837.

Tang, G., X. Li, Y. Wang, J. Xin, and X. Ren. 2009. Surface ozone trend details and interpretations in Beijing, 2001–2006. *Atmospheric Chemistry and Physics* 9:8813–8823.

United Nations Framework Convention on Climate Change (UNFCCC). 2011. National reports. Available at http://unfccc.int/national_reports/items/1408.php, last accessed February 15, 2011.

U.S. Department of Energy (U.S. DOE). 2009. Science Challenges and Future Directions: Climate Change Integrated Assessment Research. Office of Science, Office of Biological and Environmental Research. Report #PNNL-18417.

U.S. Energy Information Administration (U.S. EIA). 2011. Total carbon dioxide emissions from the consumption of energy (million metric tons). Available at http://www.eia.gov/cfapps/ipdbproject/IEDIndex3.cfm?tid=90&pid=44&aid=8, last accessed June 19, 2011.

van Donkelaar, A., R. V. Martin, M. Brauer, R. Kahn, R. Levy, C. Verduzco, and P. J. Villeneuve. 2010. Global estimates of ambient fine particulate matter concentrations from satellite-based aerosol optical depth: Development and application. *Environmental Health Perspectives* 118(6):847–855.

Vennemo, H., K. Aunan, J. W. He, T. Hu, and S. T. Li. 2009. Benefits and costs to China of three different climate treaties. *Resource and Energy Economics* 31:139–160.

Wang, L. T., C. Jang, Y. Zhang, K. Wang, Q. Zhang, D. Streets, J. Fu, Y. Lei, J. Schreifels, K. B. He, J. M. Hao, Y.-F. Lam, J. Lin, N. Meskhidze, S. Vorhees, D. Evarts, and S. Phillips. 2010a. Assessment of air quality benefits from national air pollution control policies in China.

Part I: Background, emission scenarios and evaluation of meteorological predictions. *Atmospheric Environment* 44(28):3442–3448.

———. 2010b. Assessment of air quality benefits from national air pollution control policies in China. Part II: Evaluation of air quality predictions and air quality benefits assessment. *Atmospheric Environment* 44(28):3449–3457.

Wang, S. W., D. G. Streets, Q. Zhang, K. B. He, D. Chen, S. C. Kang, Z. F. Lu, and Y. X. Wang. 2010. Satellite detection and model verification of NO_X emissions from power plants in Northern China. *Environmental Research Letters* 5(4):044007, doi:10.1088/1748-9326/5/4/044007.

Wang, T., A. J. Ding, J. Gao, and W. S. Wu. 2006. Strong ozone production in urban plumes from Beijing, China. *Geophysical Research Letters* 33:L21806, doi:10.1029/2006GL027689.

Wang, T., X. L. Wei, A. J. Ding, C. N. Poon, K. S. Lam, Y. S. Li, L. Y. Chan, and M. Anson. 2009. Increasing surface ozone concentrations in the background atmosphere of Southern China, 1994–2007. *Atmospheric Chemistry and Physics* 9:6217–6227.

Wang, X. 2011. On China's energy intensity statistics: Toward a comprehensive and transparent indicator. *Energy Policy* 39(11):7284–7289, doi:10.1016/j.enpol.2011.08.050.

Wang, Y. J., and W. Chandler. 2011. Understanding energy intensity data in China. Report for Carnegie Endowment for International Peace. Available at http://carnegieendowment.org/files/chinese_energy_intensity.pdf, last accessed September 24, 2011.

Wang, Y. X., J. M. Hao, M. B. McElroy, J. W. Munger, H. Ma, D. Chen, and C. P. Nielsen. 2009*. Ozone air quality during the 2008 Beijing Olympics: Effectiveness of emission restrictions. *Atmospheric Chemistry and Physics* 9(14):5237–5251.

Wang, Y. X., M. B. McElroy, J. W. Munger, J. M. Hao, H. Ma, and C. P. Nielsen. 2010*. Year-round measurements of O_3 and CO at a rural site near Beijing: Variations in their correlations. *Tellus B: Chemical and Physical Meteorology* 62(4):228–241, doi:10.1111/j.1600-0889.2010.00464.x.

Wang, Y. X., M. B. McElroy, J. W. Munger, J. M. Hao, H. Ma, C. P. Nielsen, and Y. S. Chen. 2008*. Variations of O_3 and CO in summertime at a rural site near Beijing. *Atmospheric Chemistry and Physics* 8(21):6355–6363.

Wang, Y. X., J. W. Munger, S. C. Xu, M. B. McElroy, J. M. Hao, C. P. Nielsen, and H. Ma. 2010*. CO_2 and its correlation with CO at a rural site near Beijing: Implications for combustion efficiency in China. *Atmospheric Chemistry and Physics* 10:8881–8897, doi:10.5194/acp-10-8881-2010.

Wang, Y. X., Q. Q. Zhang, K. B. He, Q. Zhang, and L. Chai. 2013. Sulfate-nitrate-ammonium aerosols over China: Response to 2000–2015 emission changes of sulfur dioxide, nitrogen oxides, and ammonia. *Atmospheric Chemistry and Physics* 13:2635–2652.

Wee, S.-L., and H. Li. 2013. Politics of pollution: China's oil giants take a choke-hold on power. *Reuters*. 2 February. Available at http://www.reuters.com/article/2013/02/02/us-china-pollution-oilcompanies-idUSBRE9110F620130202, last accessed April 16, 2013.

Wong, E. 2013. As pollution worsens in China, solutions succumb to infighting. *New York Times*. 21 March.

World Bank. 1992. China Statistical System in Transition. Report No. 9557, September. Available at http://www.worldbank.org/research/1992/09/736924/china-statistical-system-transition, last accessed October 28, 2011.

————. 2012. World development indicators. Available at http://data.worldbank.org/ indicator, last accessed December 19, 2012.

World Bank and State Environmental Protection Administration (SEPA). 2007. *Cost of Pollution in China: Economic Estimates of Physical Damages* (Conference report). Washington, DC: World Bank.

World Health Organization (WHO). 2006. WHO Air quality guidelines for particulate matter, ozone, nitrogen dioxide and sulfur dioxide—Global update 2005—Summary of risk assessment. Geneva. Available at http://whqlibdoc.who.int/hq/2006/WHO_SDE_PHE_ OEH_06.02_eng.pdf.

Yang, F., J. Tan, Q. Zhao, Z. Du, K. He, Y. Ma, F. Duan, G. Chen, and Q. Zhao. 2011. Characteristics of $PM_{2.5}$ speciation in representative megacities and across China. *Atmospheric Chemistry and Physics* 11:5207–5219.

Xue, L. K., A. J. Ding, J. Gao, T. Wang, W. X. Wang, X. Z. Wang, H. C. Lei, D. Z. Jin, and Y. B. Qi. 2010. Aircraft measurements of the vertical distribution of sulfur dioxide and aerosol scattering coefficient in China. *Atmospheric Environment* 44:278–282.

Zhang, J. F., D. L. Mauzerall, T. Zhu, S. Liang, M. Ezzati, and J. V. Remais. 2010. Environmental health in China: Progress towards clean air and safe water. *The Lancet* 375: 1110–1119.

Zhang, J. F., and K. R. Smith. 2007. Household air pollution from coal and biomass fuels in China: Measurements, health impacts, and interventions. *Environmental Health Perspectives* 115:848–855.

Zhang, Q., D. G. Streets, G. R. Carmichael, K. B. He, H. Huo, A. Kannari, Z. Klimont, I. S. Park, S. Reddy, J. S. Fu, D. Chen, L. Duan, Y. Lei, L. T. Wang, and Z. L. Yao. 2009. Asian emissions in 2006 for the NASA INTEX-B mission. *Atmospheric Chemistry and Physics* 9:5131–5153.

Zhang, Q., D. G. Streets, and K. B. He. 2009. Satellite observations of recent power plant construction in Inner Mongolia, China. *Geophysical Research Letters* 36, L15809, doi:10.1029/2009GL038984.

Zhang, Q., D. G. Streets, K. B. He, Y. X. Wang, A. Richter, J. P. Burrows, I. Uno, C. J. Jang, D. Chen, Z. L. Yao, and Y. Lei. 2007. NO_X emission trends for China, 1995–2004: The view from the ground and the view from space. *Journal of Geophysical Research* 112, D22306.

Zhao, Y., C. P. Nielsen, Y. Lei, M. B. McElroy, and J. M. Hao. 2011*. Quantifying the uncertainties of a bottom-up emission inventory of anthropogenic atmospheric pollutants in China. *Atmospheric Chemistry and Physics* 11:2295–2308, doi:10.5194/acp-11-2295-2011.

Zhao, Y., C. P. Nielsen, and M. B. McElroy. 2012*. China's CO_2 emissions estimated from the bottom up: Recent trends, spatial distributions, and quantification of uncertainties. *Atmospheric Environment* 59:214–223, doi:10.1016/j.atmosenv.2012.05.027.

Zhao, Y., C. P. Nielsen, M. B. McElroy, L. Zhang, and J. Zhang. 2012*. CO emissions in China: Uncertainties and implications of improved energy efficiency and emission control. *Atmospheric Environment* 49:103–113, doi:10.1016/j.atmosenv.2011.12.015.

Zhao, Y., J. Zhang, and C. P. Nielsen. 2013*. The effects of recent control policies on trends in emissions of anthropogenic atmospheric pollutants and CO_2 in China. *Atmospheric Chemistry and Physics* 13:487–508, doi:10.5194/acp-13-487-2013.

2

Summary: Sulfur Mandates and Carbon Taxes for 2006–2010

Chris P. Nielsen, Mun S. Ho, Yu Zhao, Yuxuan Wang, Yu Lei, and Jing Cao

2.1   Introduction

China's decision makers and ordinary citizens face a difficult confluence of problems in economic development, energy use, environmental degradation, and greenhouse gas emissions, as laid out in chapter 1. A key feature of the challenge is the intimate links between fossil fuel use, the impact of air pollution on public health and agriculture, and carbon dioxide emissions. The remarkable complexity of this confluence not only raises a multidecade policy-making challenge to Chinese leaders, but also poses a related challenge to researchers seeking new insights to help inform those pressing policy decisions. It requires building research capacities both within specialized fields of knowledge, and between them, to generate an appropriately integrated understanding of the tradeoffs.

The work of this book may be the most extensive effort to date to take on this interdisciplinary research challenge. This chapter presents a summary of both the methods used and the results generated in our analyses of the costs and benefits of two national emission control policy options set in the recent past, the years 2006–2010:

- *Scenario P1*, the technology mandates actually implemented in China's 11th Five-Year Plan (11th FYP) to control power sector emissions of sulfur dioxide (SO_2); and

- *Scenario P2*, a hypothetical, economy-wide tax of 100 yuan per ton of carbon (or 27 yuan per ton of carbon dioxide, CO_2) emitted by Chinese fossil fuel use.

The "P" denotes "past," to distinguish these scenarios from ones in the next chapter that are set in the future, labeled "F."

The deeper research in each of the areas that lie behind the integrated results summarized here are reported in detail in the six chapters of part II of this book, and the sections of this summary are intended to guide the reader to all of the original scholarship of chapters 4–9. Chapter 3 gives a summary of our evaluations of a separate set of carbon tax options applying the same framework, but instead looking to the future years of 2013–2020.

This chapter thus serves two readerships. First, it is a summary for nonspecialists seeking a general understanding of the aims, methodologies, and conclusions of the 2006–2010 policy evaluations. Second, it is an introduction to the integrated framework and results for those who plan to read more of the fuller research descriptions presented component by component in chapters 4–9.

Of course we encourage all readers who find this summary compelling to explore the more detailed discussions in chapters 4–9, as their interests dictate. Each is written to be accessible to lay readers, by defining terms, limiting jargon, and assuming no prior disciplinary expertise. By taking this approach we hope to expose readers to research topics that they currently may know little about. This will help nonspecialists gain awareness of the state of current knowledge in the relevant fields, including what is and is not currently known. We stress this last point especially to nonacademic readers. The wisest decision making on the policy issues of this book should recognize uncertainties and the limits of current understanding. Indeed, advisers to policy makers should consider whether support for new research might be justified to narrow uncertainties and to strengthen the basis for future policy choices.

Because this chapter is a summary and introduction, to ease reading we omit citations. All relevant citations and references can be found in the in-depth discussions of appropriate chapters.

Following this introduction, section 2.2 defines the two past policy scenarios and the base case evaluated in chapters 4–9. Section 2.3 provides an integrated overview of the interdisciplinary research framework developed to analyze the costs and benefits of the two policy options. The next four sections (2.4–2.7) then describe the research components of the integrated analysis in sequence: economy and energy use, atmospheric emissions, air quality, and avoided health and crop damages of air pollution. The results for each policy case will be reported within each of these study components, allowing readers to be clear about how outputs of one research element become inputs to the next one. Section 2.8 contrasts and compares the costs and benefits of the two policies, and section 2.9 provides a summary of key implications and conclusions of the 2006–2010 policy cases.

2.2   Definition of the Two Policies and Base Case

In this section we introduce the first two policy cases evaluated in this book, and the base case against which their effects are measured.

2.2.1   11th Five-Year Plan SO₂ Controls

China's Eleventh Five-Year Plan (11th FYP) included two targets directly affecting the atmospheric environment. One required that national emissions of SO_2 in 2010 be 10% lower than 2005 levels. The second mandated that energy consumption per unit GDP in 2010 be 20% lower than the value in 2005. The first policy that we analyze, P1, consists of the measures to meet the SO_2 target, which focused on the electric power sector and included both direct emission-control mandates and efficiency measures that also serve the objectives of the second, energy intensity, target.

First, all new coal-fired power plants, projected for 2006–2010 to be 354 giga-watts (GW), were required by the 11th FYP to install flue gas desulfurization (FGD) systems. (These projections were made at the time of research, before data were available covering the entire 11th FYP, but the actual construction of new plants was very close to these projections, as discussed in chapter 4.) An additional 125 GW of existing plants were to be retrofitted so that by the end of 2010 more than 80% of coal-fired capacity would include desulfurization, consistent with a subsequent official estimate of 82.6%.[1] By comparison, in 2005 only 52 GW of coal-fired capacity were equipped with FGD equipment.

Second, both to reduce SO_2 and to improve energy efficiency, the FYP also required many small, inefficient power plants to close; the original target was to close 50 GW of capacity by 2010, but was estimated by our team at the time of research to have actually totaled 59 GW (close to 62 GW, as reported in 2011 by the China Electricity Council).[2] The demand for power originally supplied by these small plants was to be met by the installed power generating capacity that remained in place or had been added during the 11th FYP.

Note that of the three 2010 scenarios evaluated (one base case and two policy cases), this SO_2 policy estimates what actually happened in China between 2005 and 2010. The SO_2 controls were actual policy measures that available evidence indicates were largely implemented successfully.

2.2.2   Carbon Tax

Our second case evaluates a hypothetical tax on carbon emissions implemented during the same years as the first policy, 2006–2010.

Readers may wonder why we consider a carbon tax set hypothetically in the past when in chapter 3 we will use the same framework to evaluate the options for an implementable policy of the same kind in the future. One reason is to develop our carbon tax assessment framework using the actual data that are available when researching the past, while examining future policies depends more on projections. Testing our methods first with real data can strengthen our analyses of the future.

But more importantly, we also want to compare a carbon tax to the 11th FYP SO_2 controls. First, relating a carbon tax to an existing policy will make it easier to place its effects in a familiar context. Second, we want to explore one of the central questions in emission control policy making: how a technology mandate policy (P1) might compare to a market-based mechanism (P2). We want to do this even though the two policies target different emissions, a domestic air pollutant under the first policy (SO_2) and a global GHG in the second (CO_2). We choose this course because both domestic air pollution policies and global carbon policies mainly affect fossil fuel use and thus will have impacts beyond the emissions that they specifically target. Indeed we want to see to what extent the effects of such varied policies coincide, to explore how domestic and global environmental priorities might be pursued simultaneously with integrated strategies.

The P2 policy consists of imposing a tax of 100 yuan per ton of carbon on the use of coal, oil, and gas, including imported fuels. This tax is implemented every year from 2006 to 2010. To facilitate comparisons to other carbon prices that may be more familiar, 100 yuan per ton of carbon is approximately 27 yuan per ton of CO_2 and, at exchange rates in 2010, about US$4 per ton of CO_2. It amounts to roughly 14% of the mine-mouth price of coal in China. The level is around 20% of carbon prices observed during the subject period in global markets like the European Union Emission Trading System; the average EU Allowance (EUA) price in 2010 of around €15 per ton of CO_2 was roughly equivalent to 130 yuan per ton. We consider the relatively low tax of our carbon tax case a reasonable starting point for analysis because China is not obligated to act as aggressively as richer countries under the principle of common but differentiated responsibilities of the United Nations Framework Convention on Climate Change (UNFCCC), and because if China were to implement a carbon tax, it would almost surely begin at a small tax level.

How the tax revenue is used affects the impact of the policy, especially on the economy. We simulated two alternatives: one recycles all revenue in lump sums back to households to maintain the base case level of government spending, while the other uses the revenue to cut existing tax rates. We subjected only the first alternative to the full, integrated analysis, and emphasize it here. The second was analyzed only by economic criteria, and we discuss it briefly.

Note that we impose the tax only on emissions of CO_2 from fossil fuel combustion, and not on emissions from chemical processes in sectors such as cement, iron and steel, and agriculture. Doing so would require a more complex policy implementation on the part of the government, and we want to consider a policy that would be as simple as possible to put in place. Compared to a more comprehensive carbon tax that includes process emissions, our results will understate somewhat the impact on these industries, especially cement. The effect on the overall results and conclusions, however, will be small.

2.2.3 Base Case

In order to evaluate and compare the benefits and costs of the two policies, we need a common base case consisting of what would have occurred in China's economy, energy use, and emissions without either policy. In other words, the base case assumes that the specific SO_2 controls of the 11th FYP were not implemented, while assuming that all other actual policies (and more limited control of SO_2 that would have been expected) were implemented. The base case is therefore identical to the SO_2 policy case except for two major elements: (1) we assume FGD systems would not have been retrofitted to any preexisting power plants, although they would have been required in all plants newly built during the 11th FYP; and (2) we assume that no early retirements of small power plants would have occurred during 2006–2010.

We first simulate the economy under these base case assumptions, not only from 2005 through 2010 but also beyond to 2030, to facilitate comparison to economic projections of other researchers. This projection has GDP growing at 7.6% annually over 2005–2030. During these 25 years, total primary energy use would rise only 3.7% per year, with coal use growing at 3.4% per year but natural gas use at a rapid 7.2%. Electricity use is projected to rise at a rate of 7.4% in the first 10 years. These projections are similar to the forecasts in the *World Energy Outlook* of the International Energy Agency (IEA) for 2008. By 2010, in this no-policy scenario, SO_2 emissions would have been 10.2% higher compared to the 2005 level, with coal-fired power plants contributing 46% of all SO_2 emissions.

Because of changes in the energy mix, the CO_2 emissions from fossil fuels are projected to grow slightly more slowly than total energy use. Over 1990–2006, the carbon intensity in China fell from 179 tons of carbon per million yuan of GDP (in constant 2000 yuan) to 95 tons, declining at 4.0% per year. This projection gives a similar rate of decline in intensity from 2005 to 2030. Despite this rapid decline, the carbon intensities per yuan of output for individual sectors in China are still very high compared to the United States, which illustrate the potential for improved carbon efficiency in China's future.

2.3   Overview of Integrated Framework to Assess Policies

For all policies, we evaluate both costs and benefits. The costs naturally include the direct costs of a given policy—for instance, the installation and operation of FGD systems on plant stacks. We also include the more important indirect costs, which are more difficult to evaluate. For example, we consider the indirect effects of higher electricity prices resulting from the SO_2 controls on the production and consumption of iron and steel, its energy use, and investment.

The benefits include reduced emissions of CO_2 and avoided health and crop damages from reduced concentrations of key local pollutants, in particular fine particles ($PM_{2.5}$) and ozone (O_3). Estimating the benefits involves a number of steps. It begins with estimating emissions and then simulating atmospheric chemistry and meteorology to evaluate the ambient concentrations of pollutants. These include "primary" pollutants, which are emitted directly, and also "secondary" ones, which form chemically in the air from precursors. The $PM_{2.5}$ found in the air can be either primary or secondary particles (largely sulfates formed from gaseous SO_2 or nitrates from nitrogen oxides, NO_X). Ozone is a secondary pollutant formed from NO_X, carbon monoxide (CO), and volatile organic compounds (VOCs), in the presence of sunlight. The next step is to combine the estimated pollutant concentrations with population distributions to assess human exposures, to which we then apply epidemiological relationships to evaluate the human health impacts of those exposures. The impacts of ozone levels on the productivity of three dominant food crops (rice, wheat, and maize) are quantified similarly, using ozone exposure indices and agricultural concentration–response functions.

To assess these costs and benefits, we have advanced a number of separate research capacities and developed a framework for integrating them. It improves on the previous study by members of the current research team, *Clearing the Air*, as introduced in section 1.6. Five components of the research framework are represented in figure 2.1 for the two policies. Highlighting enhancements of the framework of the earlier book, the components consist of the following:

- A 34-sector model of the economy, for incorporating the direct costs of a given policy and estimating the indirect ("general equilibrium") effects on the structure and growth of the economy, generating sector-level energy consumption paths

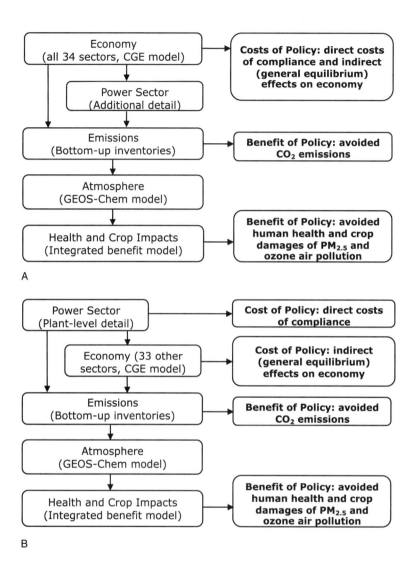

Figure 2.1
Framework for assessing costs and benefits of policies. (A) Economy-wide carbon tax. (B) SO₂ controls of the 11th Five-Year Plan.

- Newly added plant-by-plant databases of major emitting industries, with especially detailed information on the critical coal-fired electricity and cement sectors

- Newly incorporated inventories of all emissions across the economy that are now estimated "bottom-up," based in part on the databases for key industries just noted; these translate energy use into emissions of GHGs, primary pollutants, and secondary pollutant precursors, all mapped over a gridded domain

- A newly added, advanced global atmospheric transport and chemistry model (GEOS-Chem), enhanced to include a "nested" high-resolution window over China; this allows us to simulate concentrations of a wide range of primary and secondary pollutant species throughout the country that result from the emission scenarios

- A benefit assessment model used to estimate the changes in total damages to health resulting from changes in $PM_{2.5}$ and ozone concentrations, and now a new component evaluating impacts of ozone on productivity of key grain crops

These five components are used together to quantify costs and benefits of any given emission control policy option relative to the base case, and then in turn to compare costs and benefits across policy options.

Before turning to a discussion of the components, a few points about the integrated framework should be noted. First, the overarching methodological breakthrough of the current approach is bringing the power of a 3-D atmospheric chemical transport model, one used in fundamental atmospheric studies published in leading scientific journals, together with the advanced economic modeling framework that structured our previous integrated assessment, *Clearing the Air*. Members of our team developed the atmospheric and economic models independently over more than a decade of parallel research, based at Harvard University with collaborations in China. Their long-anticipated linkage became possible only after development at Tsinghua University of a number of species- or sector-specific emission inventories with improved technology and spatial resolutions. This allowed our team to integrate these into a complete emission inventory matched to the sector structure of the economic model and the spatial and chemical requirements of our atmospheric model.

Second, the effects of policies on all sectors (33 production sectors and one household consumption sector) are estimated. The plant-level databases in the second component support an extra focus on key industries, most importantly power generation. This provides greater detail on the technologies, fuels, emissions, and costs in the industry that plays the largest role in China's emissions trajectory. It not only enhances analysis of this crucial sector in the current study, but it establishes a research basis for future assessments of other types of policy strategies that target emissions from electricity generation. These could include, for instance, even higher capacity targets in renewable, nuclear, or hydroelectric power generation than China already has, or swift substitution of natural gas for coal in the power sector, as has been transforming the U.S. industry in recent years.

Third, the sequence in which the components are linked differs slightly depending on the nature of the policy. For a market-based policy that is implemented economy-wide, like a carbon tax, the assessment begins with analysis by the economic model, as shown in figure 2.1A. Its estimates of output and energy consumption by sector are then joined to the plant databases and inventories to generate gridded emissions, followed by the rest of the benefit analysis including atmospheric simulation. The SO_2 controls of the 11th FYP represent a command-and-control policy targeting the power sector, and so, as illustrated in figure 2.1B, it is first used to determine emissions at every power plant in China. The costs of the policy are translated into increases in the price of electricity, which the economic model then simulates to determine the indirect effects of the policy on energy use and emissions throughout the rest of the economy. (This difference in sequencing poses a small complication in the ordering of our presentation here. The sequence of sections in this chapter follows that of the carbon tax evaluation, as in figure 2.1A, starting with the economic component and then followed by the plant-level analysis and the wider emission assessment. The 11th FYP SO_2 policy evaluation reverses the first two steps, as we will note again later.)

Finally, we must emphasize that our current methodology captures many critical effects of emission controls, but cannot address all questions of interest. For example, the economic model is designed to evaluate the national economy and does not distinguish the economic impacts of policies at subnational scales (although the spatial resolution of our emission inventories allows us to evaluate air quality locally). Nor does our framework attempt to estimate the benefits to China of reducing GHG emissions—that is, the local benefits of limiting climate change, which even the best climate models are unable to estimate with much certainty. Instead we simply report the simulated change in carbon emissions. Nevertheless, the estimates we make are more comprehensive than those of similar initiatives and should provide valuable information for evaluating and comparing national emission control options in China.

2.4   The Economy and Energy Use

To assess the economic implications of the 11th FYP SO_2 measures and the carbon tax, we employ a multisector growth model of China's economy and energy use. The model has been developed, refined, updated, and applied to policy analyses over more than a decade of prior research. It is introduced in detail in chapter 9, and we summarize its central features in box 2.1.

It is important to stress that the goal of this model is not to predict the course of the economy or total energy use, but rather to study the effects of policies. As such, our focus is on the *percentage changes* in variables of interest between a given policy simulation and the base case. Economic variables include GDP, investment, and consumption, while other variables include emissions, air quality, and

Box 2.1
Economic growth and environment model

Economic growth in the model is mainly driven by labor-force growth, capital accumulation, and productivity growth, with additional drivers being improvements in the quality of labor and capital. The main agents are households, producers, government, and the rest of the world. The model recognizes that the central plan still plays some role in setting some prices and quantities, as discussed in chapter 9.

On the production side, 33 industries are identified, including six for energy. Each of the producers uses capital, labor, and intermediate goods to produce output. The production technology changes over time. There are thus 33 markets for commodities, that is, 33 endogenously determined prices that equate supply with demand for the domestic commodities. The household maximizes a utility function considering all 33 commodities. The parameters determining the demand for consumption goods are allowed to change over time to represent the "income effect," that is, the observed pattern of allocating bigger shares to services and smaller shares to food as households become richer. Labor is supplied inelastically by households and is mobile across sectors.

The private savings rate is set exogenously. Total national savings are made up of household savings and retained earnings of enterprises. These savings, plus allocations from the central plan, finance the following: national investment, the government deficit, and the current account surplus. Investment in a given period increases the stock of capital that is used for production in future periods.

The government imposes taxes on value added, sales, and imports, and also derives revenue from a number of miscellaneous fees. On the expenditure side, it buys commodities, makes transfers to households, pays for plan investment, makes interest payments on the public debt, and provides various subsidies.

A more detailed description of the model is presented in chapter 9.

health and crop damages. The base case projection is determined in part by exogenous elements from outside the model such as the initial stocks of debt and capital, the growth of the labor force, and improvements in technology. The base case itself is not the primary interest, and the absolute levels of the variables in the base case only slightly affect the percentage changes due to the types of policies that we analyze.[3]

2.4.1 Base Year Industry Output and Energy Use

In order to assess the costs and benefits of policies during 2006–2010, our analyses are rooted in a base year, 2005, for which data representing economic activity and energy consumption have been compiled and published by the government.

The economic model is based on the official input-output table for 2005 that gives the interindustry flow of commodities. In particular, it gives the yuan value of each fossil fuel and electricity input to each sector. We estimate the quantity of fuel consumed from this value, and the results for major emitting sectors are given in table 2.1.

Of the 33 production sectors represented in the model, construction had the highest level of gross output in 2005 at 4256 billion yuan, followed by agriculture, 3936 billion yuan, and metals smelting (from here on called "iron and steel") with 3143 billion yuan. This reflects the unusual nature of the 2005 economy, characterized by a huge investment boom. Of these big three industries, only iron and steel is a substantial direct consumer of energy. The biggest user of coal is electricity, steam, and hot water (from here on "electricity"), followed by iron and steel and nonmetal mineral products ("cement"). The biggest users of oil for combustion are transportation and the chemical industry. The biggest users of natural gas by far are the electricity and chemical sectors.

2.4.2 Effects of the 11th FYP SO_2 Controls on the Economy and Energy Use

Recall from section 2.2.1 that the 11th FYP case included two measures to reduce SO_2 emissions, both targeting the electric power sector: sharply expanded deployment of FGD and early retirement of small, inefficient power plants.

These are command-and-control technology mandates, and so the first step in analyzing these policies is to impose them in the electricity sector using our power plant database (see figure 2.1B). As noted earlier, in this case the sequence of research differs from that of a market-based policy, and the assessment in fact begins with analysis that will be described in section 2.5.3. That research generates estimates of the total technology changes required to meet the 11th FYP mandates.

Table 2.1
Emissions, fuel use, and output, 2005

Sector		SO$_2$ (kiloton)	Gross Output (billion yuan)	Coal Use (million tons)	Oil Use (million tons)	Gas Use (million cubic meters)
1	Agriculture	73	3,935.7	38.49	7.16	0.0
2	Coal mining and processing	296	792.4	66.71	2.85	0.0
3	Crude petroleum mining	44	567.4	11.05	5.51	2.0
4	Natural gas mining	1	36.3	0.06	0.28	247.0
5	Nonenergy mining	238	550.9	8.19	7.96	16.3
6	Food products, tobacco	519	2,587.8	23.36	1.37	87.6
7	Textile goods	416	1,586.0	23.30	1.83	211.6
8	Apparel, leather	51	1,222.2	3.49	0.70	0.0
9	Sawmills and furniture	72	602.4	14.48	1.24	0.0
10	Paper products, printing	613	1,085.2	20.60	1.75	30.3
11	Petroleum refining and coking	996	1,262.0	26.89	17.35	0.0
12	Chemical	1,982	2,872.0	116.92	52.31	12,900.6
13	Nonmetal mineral products	1,948	2,667.1	178.56	12.63	2,891.1
14	Metals smelting and pressing	1,694	3,143.4	219.93	21.14	1,803.7
15	Metal products	36	1,063.2	9.86	2.56	361.9
16	Machinery and equipment	124	2,509.6	42.58	4.94	1,189.1
17	Transport equipment	58	1,757.4	18.70	1.97	1,160.6
18	Electrical machinery	38	1,657.1	7.11	2.30	214.2
19	Electronic and telecom equipment	24	2,804.9	4.23	1.78	212.2
20	Instruments	18	359.6	0.67	0.21	1.2
21	Other manufacturing	378	496.5	12.90	0.87	0.1
22	Electricity, steam, hot water	16,241	1,845.3	1,051.98	19.45	14,337.8
23	Gas production and supply	26	74.4	10.51	3.23	97.9
24	Construction	308	4,256.4	11.07	13.15	0.0
25	Transportation	545	2,445.8	23.03	74.69	58.7
26	Communications	70	1,060.3	5.68	0.38	0.0
27	Trade	154	2,908.5	9.51	7.63	0.0
28	Accommodation and food	146	1,028.3	11.76	1.03	1,333.9
29	Finance and insurance	21	1,026.2	1.38	0.87	0.0
30	Real estate	116	1,025.0	9.54	0.48	6.8
31	Business services	296	1,820.0	22.37	5.16	417.2
32	Other services	918	2,873.3	74.48	3.73	185.6
33	Public administration	143	1,281.4	10.83	2.63	28.6
	Households	835		47.59	4.16	3,284.1
	Total	29,439		2138	285	41,080

Note: Fuel use is combustion, excluding the transformation to secondary fuels and products.

Sources: 2005 input-output table (NBS 2008, referenced in chapter 9), authors' calculations.

These estimates are then used to determine how the investment costs affect electricity prices and, in turn, the rest of the economy.

The capital cost of an FGD system for a new 600 MW plant is estimated at 3.8% of the cost of construction, while operating costs would increase an average of 2.4%, as described in chapter 9. However, the average cost per kilowatt hour (kWh) generated by small plants that are shut down in the other part of the 11th FYP SO_2 policy is almost three times that of big plants, so this element reduces costs. Together, these measures are estimated to have raised the cost of delivered electricity by about 0.4% only, as shown in figure 2.2A (see chapter 9 for details).

With the fuel requirements of electric utilities reduced as a result of the plant shutdowns, the implicit government subsidies to this sector (i.e., to compensate for the higher fuel costs of inefficient and uncompetitive small plants) are likewise reduced. This outcome is equivalent to a small positive productivity shock and allows the government to cut taxes equivalent to the eliminated subsidies, permitting higher investment.

The policies reduce coal consumption in two ways: power from inefficient small plants is replaced with that from efficient large plants, and the rise in electricity prices reduces the use of electric power. The reduction in demand for fuels by the power sector lowers their prices and encourages their use by other sectors. The net effect of more expensive electricity and cheaper coal is to raise the output of most sectors other than energy-producing ones, as illustrated in figure 2.2B.

As a result of the small positive productivity shock, initial GDP rises with a 1.1% boost to investment. The higher investment accumulates into a bigger capital stock, and thus GDP by 2010 is significantly larger, 0.74%. There are costs to help workers laid off by small plant closures and to build new transmission lines that we have not explicitly counted, but these are likely much smaller than the economic benefits of these efficiency gains.

2.4.3   Effects of the Carbon Tax on the Economy and Energy Use

Recall that this policy simulation consists of imposing a tax of 100 yuan per ton of carbon on the use of coal, oil, and gas, including imported fuels.

The tax is imposed every year from 2006 to 2010. The carbon tax raises the price of coal by 16% and the price of oil by 2% in the first year, as shown in figure 2.3A. These increases reduce the demand for these fuels proportionately. They raise the costs of producing carbon-intensive products such as primary metals and cement, and thus reduce their output, as the sectors with the biggest reductions are shown in figure 2.3B. These products are also the biggest emitters of PM, SO_2, and NO_x,

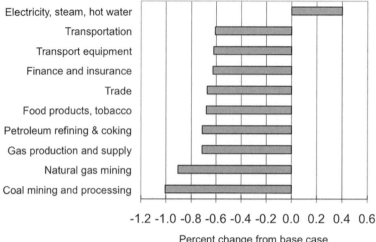

-1.2 -1.0 -0.8 -0.6 -0.4 -0.2 0.0 0.2 0.4 0.6

Percent change from base case

A

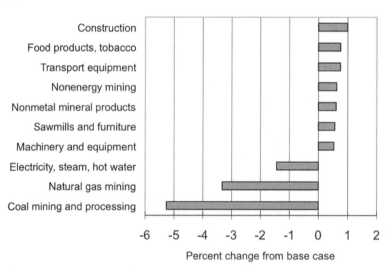

-6   -5   -4   -3   -2   -1   0   1   2

Percent change from base case

B

Figure 2.2
11th FYP SO₂ control policy: ten biggest changes in 2010 of industry prices and industry output. (A) Industry prices. (B) Industry output.

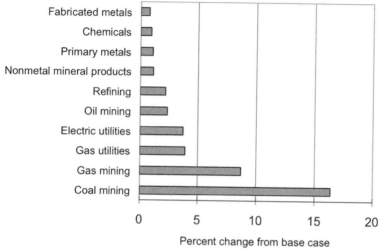

A

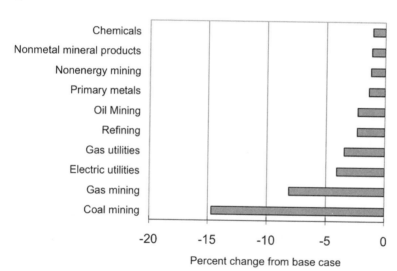

B

Figure 2.3
Carbon tax policy: ten biggest changes in 2010 of industry prices and industry output.
(A) Industry prices. (B) Industry output.

and along with CO_2, these emissions are thus sharply reduced, to be detailed in the next section. There is also a second effect that is due to the changes over time: investment goods (e.g., buildings and machinery) are more carbon-intensive, so their prices rise relative to the prices of consumption goods. This price rise reduces investment in each period, leading to a 0.1% smaller stock of capital by 2010.

The revenue from this new tax comes to 3.1% of total government revenue. How the tax revenue is used affects the impact of the policy. In the scenario of lump sum rebates to households, consumption is raised at the expense of investment while maintaining the base-case level of government spending.

The smaller stock of capital and the distortions due to the carbon tax lead to a small 0.19% fall in 2010 GDP. Coal use in 2010 is 14.6% lower as a result of the lower GDP and the price-induced lower demand. Electricity use falls by 4.1% because of the higher price of electricity and the reduced demand from lower output of carbon-intensive products. This large reduction of coal use, accompanied by a smaller decline in electricity use, is due to the substitutability of energy inputs in the electric utility sector. The output of the refining sector falls by 2.3%.

We also briefly considered a second scenario in which the carbon tax revenues are not directly returned to consumers, but used to cut existing distortionary taxes. We evaluated only the economic impacts. The comparison with the lump sum rebate case is given in box 2.2; the tax cut produces a slightly smaller loss of 2010 GDP and correspondingly higher energy use and emissions. Though these differences are small over this short five-year span, they would grow if extended.

2.5   Emissions

Representing all the emissions that affect the atmospheric environment in China—both air quality and CO_2—in a way that not only informs atmospheric simulation but also reflects the swiftly evolving use of energy in China's dynamic economy is a challenging and data-intensive research undertaking. Three chapters of this book reflect the scale of this effort, and are developed from prior and ongoing emission inventory work. Chapters 4 and 5 (adapted from previously published journal articles) describe in detail the research required to estimate emissions of two key industries, electric power and cement production, and include discussion of uncertainties. Chapter 6 incorporates those results and analogous work on all other sectors of the economy to provide the comprehensive inventory of emissions that is required by the atmospheric model for simulating the complex atmospheric processes that cause major air pollution hazards. Space does not allow us to detail the

Box 2.2
Second revenue case

The main carbon tax case returns the revenues to consumers as a lump sum transfer. If the revenues are used to cut existing taxes instead, reducing the distortions in capital allocation, then the cost to GDP is lower. In this second carbon tax simulation, the reduced taxes on enterprise income allow a greater rate of retained earnings and investment. The cumulative effect of the higher investment, along with lower consumption, leads to a smaller cost to GDP by 2015 and a slightly smaller reduction in emissions, as shown in the following table.

Variable	Effect of Carbon Tax Versus Base Case in 2010	
	Lump Sum Transfer	Reduction of Other Taxes
	Percent Change	
GDP	−0.19	−0.03
Consumption	0.13	−0.14
Investment	−0.25	0.28
Energy use	−11.5	−11.3
Coal use	−14.6	−14.4
CO_2 emissions	−12.2	−12.0
Pollution tax revenue/total tax revenue (percent)	3.07	3.09

This simulation is not to suggest that enterprises should benefit at the expense of consumption by households but to spotlight that choices on use of revenues matter. Although this recycling regime would be economically beneficial judged by effect on GDP, such strategies are generally considered more difficult politically.

This revenue case is discussed in more detail in chapter 9.

data and methods used to estimate emissions from sectors other than electric power and cement, but much of the underlying research is cited in chapter 6.

2.5.1 Emission Estimation

In this research summary we focus on these key atmospheric emissions: SO_2, NO_x, CO_2, and primary PM_{10} and $PM_{2.5}$ (particle matter smaller than 10 and 2.5 microns in aerodynamic diameter, respectively). Chapters 4–6 describe other species that also play important roles in both atmospheric chemistry and radiative forcing.[4] In general, the emissions of each species are calculated based on the following:

- Activity levels—fuel consumption or industrial production—by sector and region
- Unabated emission factors, expressed as the mass of pollutant per unit fuel consumption or per unit industrial production, prior to emission control
- Removal efficiencies of applicable emission control technologies

The sources, assumptions, and calculations generating these inputs for the many sectors and pollutants are described in detail in chapters 4–6.

The sources of anthropogenic emissions of SO_2, NO_X, PM_{10}, $PM_{2.5}$, and CO_2, are classified as either combustion or noncombustion sources. Our focus is on five major sectors, as categorized by official statistics:

- Energy industries, including thermal power plants, heating supply, coking plants, oil refineries, and gas works
- Other manufacturing industries, which includes industrial boilers and kilns, and processes like the production of cement, lime, and chemicals, and smelting of iron, steel, and nonferrous metals
- Transport, including on-road vehicles, railway transportation, construction machinery, agricultural vehicles, and inland shipping
- Residential and commercial use (a single sector in official statistics often translated as "domestic" use)
- Open burning of biomass, chiefly the combustion of agricultural wastes in the fields

Note that the sectors differentiated by the economic models of section 2.4 and chapter 9 are not the same, notably including many subdivisions of what official statistics lump together as "industry," meaning manufacturing, mining, and utilities.

In the detailed source data, coal-fired power plants, cement plants, and iron and steel plants are treated as large point sources. The fuel consumption, output, production technologies, and plant location in these industries are investigated to estimate emissions. Figure 2.4 gives a diagrammatic example of the bottom-up, plant-level estimation of emissions in chapter 4 for the key coal-fired power sector. Other sectors, including both mobile and stationary source types, are modeled as area sources.

We should note that air quality is affected not only by emissions from energy use and industrial activity, but also by those from naturally occurring processes. Volatile organic compounds emitted by pine forests, for instance, can play a significant role

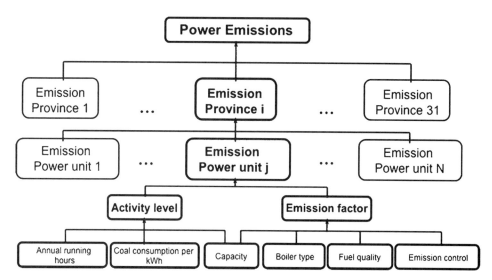

Figure 2.4
Flow chart of bottom-up emission estimation for the coal-fired power sector.

in the photochemistry that generates ground-level ozone and urban smog. These must also be included in the emission inventory for accurate simulation of ambient air pollution levels. Our inventories include such sources, but we will not recount them, as they are irrelevant to the policy analyses. Estimates of naturally occurring emissions are drawn from atmospheric science literature, and are cited in the series of scientific publications on which chapter 7 is based.

2.5.2 Emissions in the Base Year of 2005 and Base Case for 2010

On the basis of the calculations just described, total SO_2 emissions in our base year of 2005 are estimated to have been 29.4 million metric tons (Mt, equivalent to teragrams, Tg), 5.0 Mt higher than the official estimate. Emissions of PM_{10} are estimated to have been 18.1 Mt, $PM_{2.5}$ 12.5 Mt, NO_X 18.8 Mt, and CO_2 from fossil fuel combustion 5516 Mt. How these estimates compare to those of other studies and the official data is discussed in chapters 1 and 6.

The sector distributions of the emissions in 2005 for SO_2 and PM are shown in figure 2.5. The big coal users, especially electricity, iron and steel, cement, and commercial and residential use were the biggest contributors to SO_2 emissions in 2005. By our estimates, the electricity sector alone was responsible for 16.1 Mt of the total 29.4 Mt of SO_2 emissions in 2005.

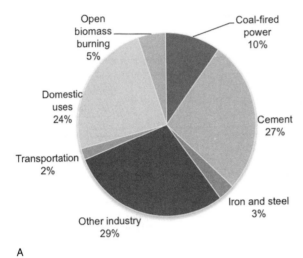

A

B

Figure 2.5
Bottom-up emission inventories for 2005 base year. (A) PM emissions. (B) SO_2 emissions.

The emissions of PM_{10}, SO_2, and NO_X for 2005 are allocated to the 0.5° by 0.67° grid that is used in the atmospheric model described in the next section. These maps of emissions are shown in figure 2.6. As described in chapter 6, the spatial allocation is done using administrative boundaries matched to the categorization in the underlying data: we use the exact location for large point sources, and for other sources we use provincial-level data that we disaggregate spatially using social and economic characteristics. (See chapter 6 for discussion of the regional distribution of emissions in the base case.)

For the base case scenario in 2010, national SO_2 emissions would have increased 10% compared to 2005, driven by the rapid growth of consumption of energy, including electric power. Emissions of primary PM_{10} and $PM_{2.5}$, in contrast, would have declined by 6.3% and 6.7%, chiefly because of the transition in the cement industry from shaft kilns to precalciner kilns (section 5.4.3). One of the most significant changes projected is the large 29% rise in NO_X emissions in the base case due to the growth in electric power and manufacturing with minimal NO_X controls. This huge increase helps to justify a new policy in the 12th FYP (2011–2015) to begin addressing this crucial pollutant.

2.5.3 Emissions under the 11th Five-Year Plan SO_2 Controls

By analyzing how the 11th FYP measures affected the thermal power sector plant by plant (in chapter 4) we determine the technology requirements of compliance and the emissions of each plant. These are then used in the cost assessment described in section 2.4.2. We estimate that the SO_2 measures were impressively successful at reducing emissions to 20% below the base case projection in 2010, or 11.6% lower than the 2005 national SO_2 emissions. This result indicates that the official target of 10% below 2005 was exceeded. (For comparison to an official estimate, in 2011 the Ministry of Environment reported that 2010 SO_2 emissions declined 14.3% from 2005 levels.)[5] Unsurprisingly, these reductions were driven foremost by reduction of emissions by the power sector, at 46% compared to 2005, aided by reductions in cement production, at 17%. These more than compensate for substantial increases in SO_2 emissions in iron and steel, other industries, and residential and commercial use.

National emissions of primary PM_{10} are estimated to have declined 3.7% compared to the 2010 base case and 10% compared to 2005, thanks mainly to the side effects of FGD systems in PM control. No similar side benefit accrues to emissions of NO_X, however; they are estimated to have declined by a meager 1.4% compared to the base case, mostly because of the slightly higher electricity prices dampening

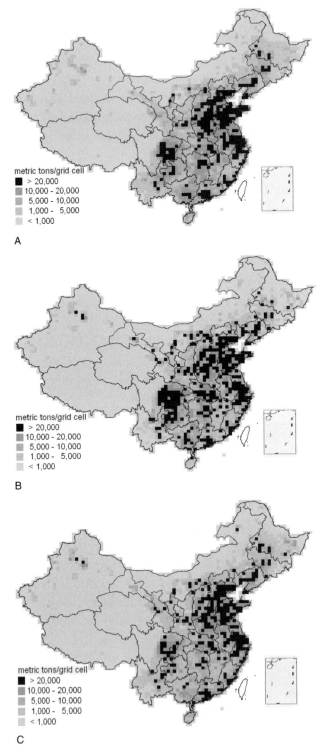

Figure 2.6
Emissions in 2005 base year.
(A) PM_{10}. (B) SO_2. (C) NO_X.

power consumption and output of energy-intensive manufacturing. This small reduction resulting from the policy is inconsequential compared to the huge 29% increase in projected overall NO_X emissions from 2005 to 2010 in the base case noted earlier.

A nontrivial 4.6% reduction in CO_2 emissions, compared to the base case, is estimated to have resulted from the 11th FYP. This result may at first surprise many, considering the well-known energy penalty that the deployment of FGD systems engenders. To understand this outcome, recall first that our base case does assume considerable FGD deployment, but none retrofitted at preexisting plants as under the policy case. Then note that the increase in power use resulting from the FGD systems that were added (to newly built plants during 2006–2010) is more than compensated for by the gains of the other component of the 11th FYP policy—replacement of 59 GW of highly inefficient small plants—and by modest suppression of power demand of other sectors resulting indirectly from a rise in electricity prices shown in figure 2.2A and described in chapter 9.

The spatial allocations of the SO_2, PM_{10}, and NO_X emission reductions are illustrated in figure 2.7, where the reductions closely map the locations of individual coal-fired power plants described in chapter 4. The atmospheric model of section 2.6 uses these spatial emission estimates to determine the resulting air pollution concentrations.

2.5.4   Emissions under the Carbon Tax

As with the 11th FYP SO_2 control policy, the carbon tax would have similarly reduced its target emissions, of CO_2, by 12% compared to the base case. This reduction is achieved through substantially lower use of fossil fuels, especially coal, across most major sectors of the economy including electricity, cement, iron and steel, other industries, and households, as described in chapter 9. One exception is transportation, where the estimated effects are modest because of limited transportation use of coal and a relatively small tax on oil.

Such broad, economy-wide, reduction of fossil fuel use would bring large side benefits in reduced emissions of conventional pollutants. Emissions of SO_2 and primary PM_{10} in 2010 would have declined 14% and 11% versus the base case, respectively, and 5% and 17% compared to 2005. A notable contrast to the 11th FYP SO_2 controls is the effect a carbon tax would have had on emissions of NO_X: an 11% reduction in 2010 compared to the base case.

Figure 2.8 shows the spatial allocation of the SO_2, PM_{10}, and NO_X emission reductions resulting from the carbon tax. It shows broader geographical reductions

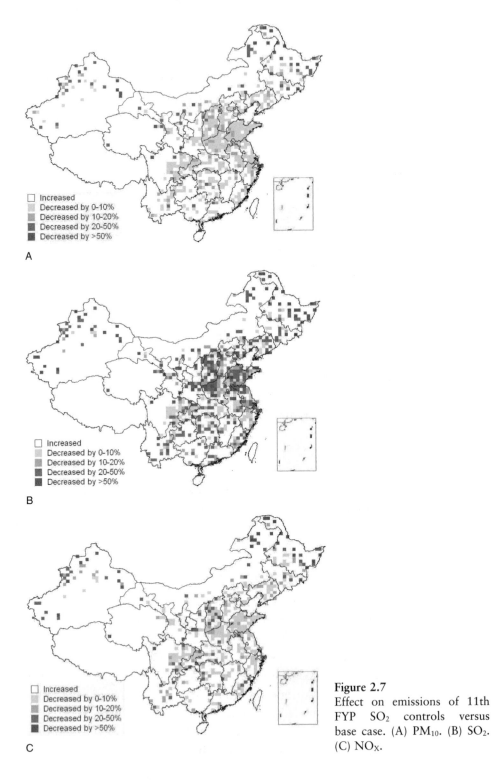

Figure 2.7
Effect on emissions of 11th FYP SO₂ controls versus base case. (A) PM₁₀. (B) SO₂. (C) NOₓ.

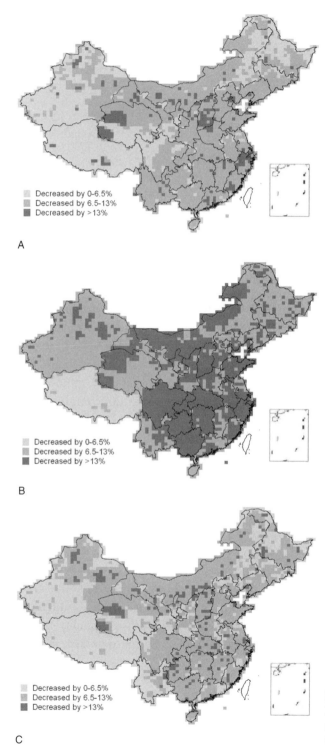

A

Decreased by 0-6.5%
Decreased by 6.5-13%
Decreased by >13%

B

Decreased by 0-6.5%
Decreased by 6.5-13%
Decreased by >13%

C

Decreased by 0-6.5%
Decreased by 6.5-13%
Decreased by >13%

Figure 2.8
Effect on emissions of carbon tax versus base case. (A) PM_{10}. (B) SO_2. (C) NO_X.

than the largely point-source-based reductions of the SO_2 controls mapped in figure 2.7.

2.6   Pollution Concentrations

2.6.1   GEOS-Chem Atmospheric Model of China

Once we have estimated emissions under the base and policy cases and mapped them geographically, the next step in the benefit assessment represented in figure 2.1 is to evaluate how emitted pollutants are transported and chemically transformed in the atmosphere to result in concentrations that can affect public health and crop productivity. To do so we employ an advanced model of regional air quality over China developed and tested against scientific observations over years of prior research.

We introduce here the key features of the atmospheric model, an extension of the global model GEOS-Chem developed specifically for research on China and reported in chapter 7, where the details are given. GEOS-Chem has a global resolution of 4° by 5°. It takes background concentrations and emissions in each grid cell for a given time period and uses meteorological fields to simulate atmospheric transport and chemistry to yield estimated concentrations of a full complement of trace gases and suspended particulate matter (a type of "aerosol," in the terminology of atmospheric scientists) throughout a domain. It includes interactions of all significant pollutant gases and aerosols, including 100 chemical species and more than 200 reactions, in 15-minute time increments throughout the year. It is driven by global meteorological fields constructed in a continuing time series by NASA, assimilating a state-of-the-art weather and climate model with measurements taken worldwide each day by surface stations, aircraft, balloons, ships, buoys, and satellites.

Our version of GEOS-Chem nests a window over China and East Asia of higher resolution than the global model, 0.5° by 0.67° (roughly 50 km by 67 km), shown in figure 2.9. This resolution is determined by the available meteorological data.

The nested grid framework allows us to simulate China's air quality in full regional, continental, and global context. The advantages of this approach, in addition to the embedded detail of the chemistry and meteorology that we have described, include the following:

- Built-in inclusion of background and boundary conditions (concentrations), essential for both accurate chemistry and source attribution

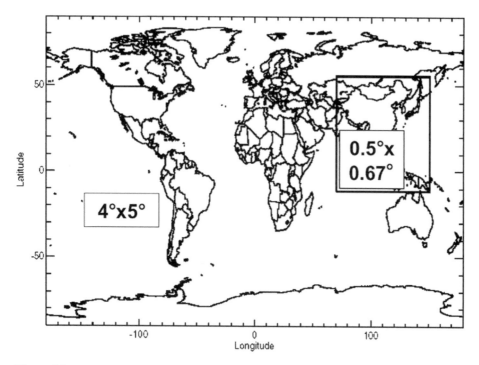

Figure 2.9
Domain and resolution of GEOS-Chem China model.

- A record of rigorous testing of simulations against observations of scientific ground stations, satellites, and aircraft

The use of these tools by atmospheric scientists with thorough knowledge of the chemical and physical processes behind both observations and the model allows us to go beyond mechanical calculation of concentrations and analyze the chemistry and dynamics behind the results, including unexpected outcomes, as discussed in section 2.6.3.

While the model simulates a suite of atmospheric species of scientific interest, for purposes of policy analyses we are focused on the two that dominate health and crop damages. The first is primary and secondary $PM_{2.5}$, which can be classified from GEOS-Chem outputs on the basis of chemical composition.

The second is surface ozone, which is a secondary species formed by complex nonlinear photochemistry involving NO_X, VOCs, CO, and solar radiation. The main ingredient of Los Angeles–style photochemical smog, ozone is a more "modern" form of air pollution that is still poorly measured, researched, and understood in

China. It is nevertheless growing into a critical air pollution hazard because of China's exploding motor vehicle population and weak controls on emissions of NO_X and VOCs.

The GEOS-Chem model simulates concentrations of pollutants throughout the year, recognizing the crucial role of meteorology and regional climate on pollution levels, such as monsoonal circulation in the summer and cold fronts in winter. For the health damage assessment we are most interested in annual average pollution concentrations because the epidemiological relationships that we use to estimate health risk are calculated on the basis of annual average exposures (as detailed in section 2.7). For ozone impacts on crops, however, we must look at growing seasons of different crops and concentration changes with the much higher temporal resolution that GEOS-Chem provides.

2.6.2   Pollution Concentrations under the 11th Five-Year Plan SO_2 Controls

The SO_2 controls of the 11th FYP are estimated to have had very large beneficial impacts on $PM_{2.5}$ concentrations, mainly through reductions of SO_2, the precursor of secondary sulfate particles, but also by reducing primary PM, as described previously. Figure 2.10A maps the effect of the SO_2 controls on sulfate $PM_{2.5}$ concentrations, compared to the base case, as simulated by the atmospheric model. The results for four seasonally representative months are shown, with blue representing reductions in concentrations and tan and red representing increases. Integrating the results over the year, the annual average reduction is large, reaching around 5 micrograms per cubic meter ($\mu g/m^3$) in some highly populated areas of the east (compared to the base case levels of 20–25 $\mu g/m^3$), and yielding major health benefits described in the next section.

The effect of the SO_2 controls on ozone, however, is mixed, reflecting the chemical and meteorological complexity of ozone formation. Figure 2.10B shows ozone increases in a few places and seasons compared to the base case, but reductions over broader regions in summer and fall; in general the effects are much more spatially varied than the $PM_{2.5}$ effects. More noteworthy, however, is the very modest scale of the changes, reflecting the minor impact of the policy on emissions of ozone precursors, including NO_X. Annual average variations in many populated areas are only ±0.5 parts per billion by volume (ppbv); the national annual effect is a very minor net ozone reduction.

2.6.3   Pollution Concentrations Under the Carbon Tax

The impact of the carbon tax policy on $PM_{2.5}$ concentrations is even larger than that of the SO_2 policy case, because of its broader effect not only on SO_2 but also

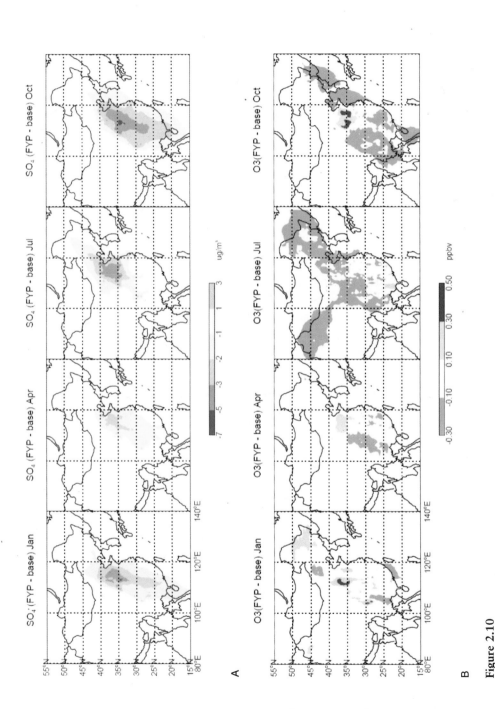

Figure 2.10
Simulated changes in surface concentrations of PM$_{2.5}$ and ozone due to the 11th FYP SO$_2$ controls, by season. Monthly mean concentrations (January, April, July, and October) in 2010 resulting from SO$_2$ control scenario P1 versus the base case. Negative values are reductions. (A) PM$_{2.5}$ (in the form of SO$_4$) in μg/m³. (B) O$_3$ in ppbv.

on NO_X (thus reducing both secondary sulfate and nitrate aerosols) and primary $PM_{2.5}$. The reductions versus the base case by season are shown in figure 2.11A, and the annual average effect, reaching as high as $7.7\,\mu g/m^3$ in Henan, is shown in figure 2.12A.

The impact of the carbon tax on ozone is larger than that of the SO_2 controls and is again mixed, as shown in figure 2.11B. It is more smoothly distributed spatially, however, reflecting the broader reduction of emissions compared to the SO_2 controls targeting power plants. An interesting result is that despite these reductions, ozone levels would rise throughout a large swath of densely populated China in the month of January, and rise even as an annual average in a region centered on Henan, Hebei, and Shandong. The phenomenon of rising ozone levels when precursors are reduced is very familiar to atmospheric chemists—another example is the "ozone weekend effect" observed in many cities around the world, where ozone concentrations sometimes rise on the weekends when there are fewer vehicles generating NO_X[6]—again reflecting the complex, nonlinear photochemistry of surface ozone formation and removal in the atmosphere.

Overall, however, the tax policy would lead to an annual net national reduction as shown in figure 2.12B, especially in the south and southwest, with the peak reduction around 1.1 ppb. This mixed result in ozone concentrations—that the broad reductions in emissions brought about by a carbon tax could nevertheless increase ozone levels in one region, even as it decreases them nationally—is explored in some detail in chapter 7, and illustrates the need to analyze the complex atmospheric effects of such policies with the most advanced research capacities available.

2.7   Health and Agricultural Benefits

2.7.1   Health Benefit Framework

The tools of population risk assessment are used to evaluate the effects of air pollution on human health across China. A convenient system for assessing population risk is the Environmental Benefits Mapping and Analysis Program (BenMAP) developed by the U.S. Environmental Protection Agency, which also sponsored its initial adaptation to China (see section 8.1). As shown in figure 2.13, the key inputs to BenMAP are as follows:

1. Air pollutant concentrations estimated by an air quality model

2. Population distribution corresponding to the gridded domain of the atmospheric model output

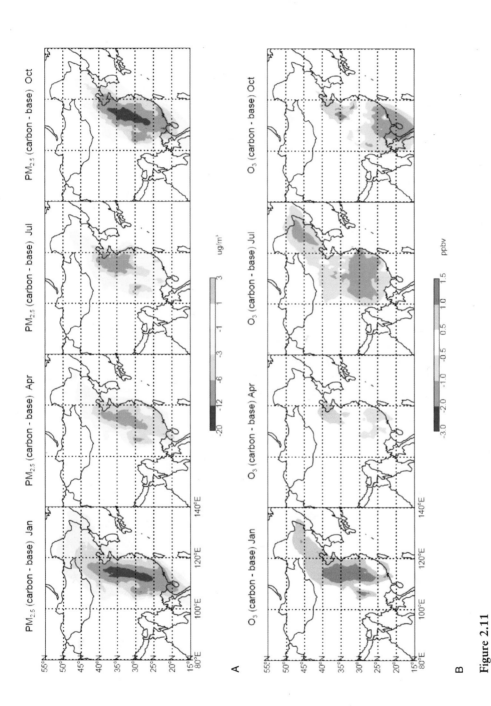

Figure 2.11
Simulated changes in monthly surface concentrations of PM$_{2.5}$ and ozone as a result of the 2006–2010 carbon tax, by season. Monthly mean concentrations (January, April, July, and October) in 2010 resulting from carbon tax scenario P2 versus the base case. Negative values are reductions. (A) PM$_{2.5}$ in µg/m^3. (B) O$_3$ in ppbv.

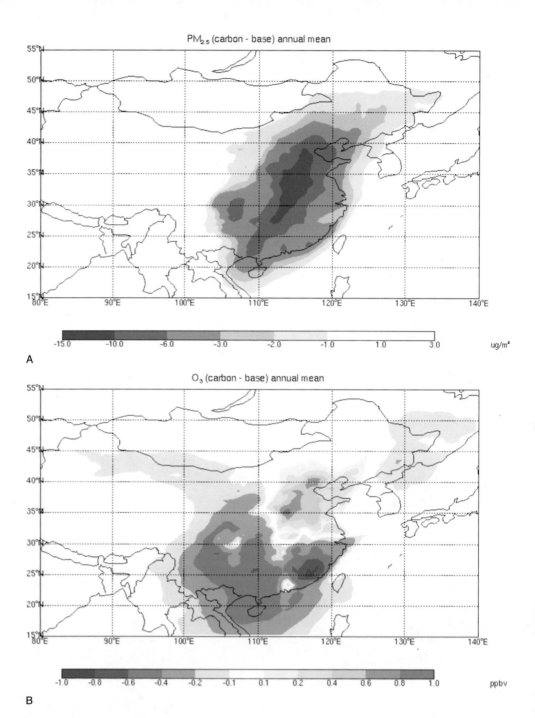

Figure 2.12
Simulated changes in annual surface concentrations of $PM_{2.5}$ and ozone as a result of the 2006–2010 carbon tax. Annual mean concentrations in 2010 resulting from carbon tax scenario P2 versus the base case. Negative values are reductions. (A) $PM_{2.5}$ in μg/m³. (B) O_3 in ppbv.

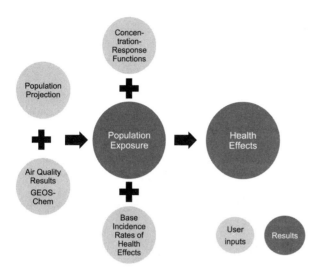

Figure 2.13
Framework for estimating health impacts of $PM_{2.5}$ and ozone exposures.

3. "Concentration-response" functions for various endpoints and pollutants, which give, for example, the change in the incidence of acute mortality resulting from an increase in the concentration of PM_{10} by $1\,\mu g/m^3$

4. Base incidence rates of health effects in China

Our study considers both mortality and morbidity endpoints, as described in chapter 8, but for the purposes of explaining methods and key results we limit the discussion here only to the largest health effect: premature mortality. Chapter 9 also considers options for monetizing health risks and incorporating them into the analyses, but we do not report those results here because they are of secondary importance to the main results emphasized in this summary.

We refine the existing BenMAP model of China for our purposes. For input 1, we use the $PM_{2.5}$ and ozone concentrations simulated by GEOS-Chem as described in the prior section. For input 2, we scale up and project to 2010 a 1 km by 1 km population distribution developed by the Chinese Academy of Sciences. These data are aggregated up to the 0.5° by 0.67° grid scale of the atmospheric model, and shown in figure 2.14. For input 4, we rely on results from the *Third National Review* on causes of death in 2006, which reports the incidence of all-cause, nonaccident mortality.

Determining input 3 is one of the most critical but uncertainty-laden steps of the entire analysis and is discussed in chapter 8. Concentration-response (C-R) functions

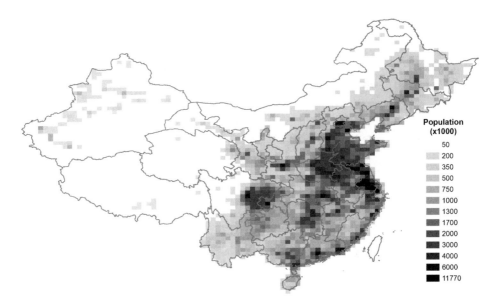

Figure 2.14
Population distribution of China for pollution exposure assessment, 2010.

Table 2.2
Summary of concentration-response functions for acute and chronic mortality resulting from exposure to $PM_{2.5}$

Mortality Endpoints	Relative Risk	95% Confidence Interval	Source
Acute mortality	0.65	(0.35, 0.94)	8 studies in Chinese cities
Chronic mortality	10.0	(1.0, 20.0)	U.S. studies only

Note: Relative risk is the increase in percentage of mortality rate per $10\,\mu g/m^3$ increase of $PM_{2.5}$ concentration.

for acute mortality due to $PM_{2.5}$ exposures are available from a growing body of Chinese epidemiological research, and it is relatively straightforward to apply them to our results. "Acute" refers to the short-run impact on health that is distinct from the "chronic" effect over many years of exposure to a particular risk factor. The chronic effect is estimated from "cohort" studies that track people over many years. No such studies have been conducted in China, however, and estimates are only available from research in the United States, where atmospheric and life conditions differ considerably from those in China.

Given the importance of the chronic mortality effect of $PM_{2.5}$ in U.S. studies, we report both acute and chronic mortality based on C-R functions shown in table 2.2, based on reviews of the literature described in chapter 8. We take explicit note of the inherent research uncertainties and the difficult methodological decisions discussed in that chapter. Keep in mind that these two mortality pathways are not additive, and in fact cohort studies should pick up acute effects in addition to the generally larger chronic effects. The two $PM_{2.5}$ mortality results for each policy case should instead be interpreted as based on epidemiological assumptions that are either conservative about the risks but more certain (acute effect, using results from China) or less conservative but more uncertain (chronic effect, applying results from the United States). As such, they help us define a range of possible health impacts.

2.7.2   Agricultural Benefit Framework

Of all air pollutants, ozone is generally thought to cause the greatest damage to agricultural productivity, as discussed in chapter 8. Different crop types respond differently to ozone exposures, and it is unrealistic to try to assess the effects of ozone on all types of agricultural production in China. Instead we focus on the most important food crops, the three major grains: rice, maize, and wheat.

The framework used to estimate effects of changes of ozone levels on grain yields due to policies versus the base case is largely analogous to that used to assess health benefits, as described in the prior section. It combines results of the atmospheric simulations with county-level geographical representations of grain production, and C-R (or, more precisely, exposure–relative yield) functions for each of the three grains that have been reported in the scientific literature (preferably in studies of Asian grain cultivars).

One difference from the health framework is that the temporal dimension of exposure assessment is necessarily more complex, in that it must take account of the very strong seasonal and diurnal (24-hour) variations in concentrations of ozone itself, varying crop seasons, and the likelihood that damage to crops may occur only above concentration thresholds and may be dependent on cumulative exposures. A number of ozone exposure indices have been developed to capture these complexities, as described in chapter 8, and each C-R relationship in the literature is specified for one of these exposure indices. Applying the C-R functions thus requires first calculating the appropriate exposure index for all locations where the grain is grown, which in turn requires simulated ozone concentrations resolved in hourly, not yearly, averages.

For each grain type, several pairs of exposure indices and available C-R functions are available, resulting in more than one estimate of the increased (or decreased)

Table 2.3
Central estimates of avoided premature mortality from $PM_{2.5}$
and ozone exposures due to the two policies

Policy Case	Mortality Endpoint	Central Estimate of Avoided Mortality
11th FYP SO_2 Control	Acute $PM_{2.5}$	12,300
	Chronic $PM_{2.5}$	73,900
	Acute O_3	120
Carbon tax	Acute $PM_{2.5}$	17,200
	Chronic $PM_{2.5}$	103,000
	Acute O_3	1,380

national yields that result from the changes in ozone from the policy. We take the arithmetic mean of each set of results as our central estimate of the effect on production of that grain.

2.7.3   Health and Agricultural Benefits of the 11th Five-Year Plan SO_2 Controls

As shown in table 2.3, the number of estimated cases of acute mortality avoided due to $PM_{2.5}$ reductions from the 11th FYP SO_2 controls amount to approximately 12,300 a year in 2010. If we use chronic mortality epidemiology from U.S. cohort studies, we estimate that the policy reduced $PM_{2.5}$ premature deaths by a very large number of 73,900 cases per year. Only another 120 acute mortalities per year were avoided by the very modest net ozone reductions in 2010.

The avoided mortalities from $PM_{2.5}$ and ozone reductions occur across most of the country, especially in Henan, Hebei, Shandong, and Jiangsu, as illustrated in figure 2.15A. (The figure shows acute mortality. Applying chronic mortality C-R functions for $PM_{2.5}$ instead would yield nearly the same geographical distribution but raise the scale substantially.)

Based on the minimal estimated health impact of the net national ozone reduction, the impact on grain productivity would be similarly negligible. Given uncertainties, this possibility does not justify the additional research effort needed to quantify the crop impacts, and we conclude that under this policy there was effectively no net impact of ozone on grain production.

2.7.4   Health and Agricultural Benefits of the Carbon Tax

Also shown in table 2.3, the health benefits due to reductions of $PM_{2.5}$ from the carbon tax are estimated to be somewhat larger than those due to the 11th FYP policy. We estimate a reduction of 17,200 cases of acute mortality a year due to avoided $PM_{2.5}$. If we use chronic effect estimates imported from Western literature, the carbon tax would have reduced $PM_{2.5}$ premature mortality by 103,000 deaths per year by 2010. The reduction in ozone concentrations would have reduced acute mortality an additional 1380 cases.

As shown in figure 2.15B, these health benefits would have been somewhat more broadly distributed geographically than those of the 11th FYP SO_2 controls, concentrated in central east China (Henan-Hebei-Shandong) but also further south in Hubei, Hunan, and Guangdong, and west in Sichuan. Chapter 8 explains why the spatial distributions of avoided health damages of the two policies differ.

The ozone reductions induced by this policy lead to notable improvements in grain productivity. Total annual maize and rice yields rise by more than 1.4 million and 990,000 tons, respectively, along with a modest increase in wheat production. Based on market prices for the grains, the total monetary benefit is estimated at more than 5 billion yuan.

2.8   Comparison of the Two Policies

The total national emissions of the key species—SO_2, PM_{10}, NO_X, and CO_2—in the 2005 base year and base and policy cases for 2010 are shown in figure 2.16. This comparison illustrates several useful points about the effectiveness of the two policies.

First, compared to the base case, the 11th FYP measures appear highly successful at reducing SO_2 in particular, as per the policy objective. A 100 yuan per ton carbon tax would substantially reduce its target emissions, of CO_2, but would also have broader effects across other emission types, reducing PM_{10} and NO_X more than the 11th FYP measures and also reducing SO_2 considerably. While the 11th FYP SO_2 controls appear effective at their primary goal, a carbon tax of this scale would function as a stronger multipollutant policy.

Second, while both policies reduce total emissions of SO_2 and PM_{10} over time, they would only reduce *the growth rate* of CO_2 and NO_X (barely, in the case of NO_X under the 11th FYP). These points emphasize that actual reductions of SO_2 and PM_{10} appear to be relatively tractable under current policy and technology options, while reductions of NO_X and especially CO_2 will be far more challenging.

Figure 2.17 puts this contrast in emission control efficacy in the context of the economic impacts of the policies, spotlighting that in neither case do the emission reductions come with major economic losses in the aggregate, even if particular

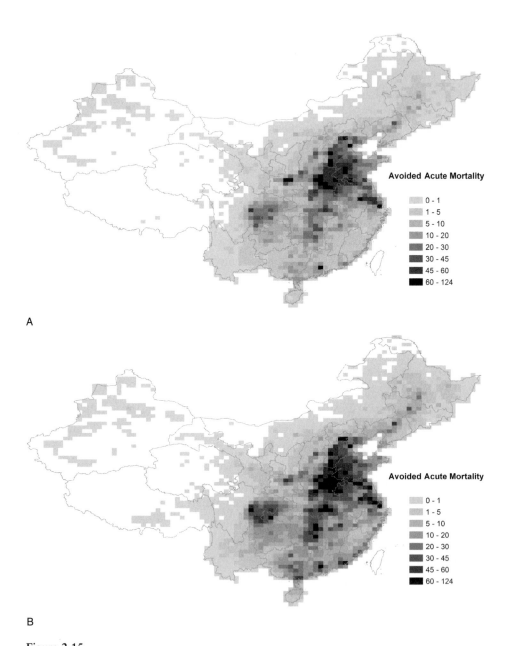

Figure 2.15
Geographic distribution of avoided acute mortality under the two policies. (A) 11th FYP SO$_2$ control policy. (B) Carbon tax policy.

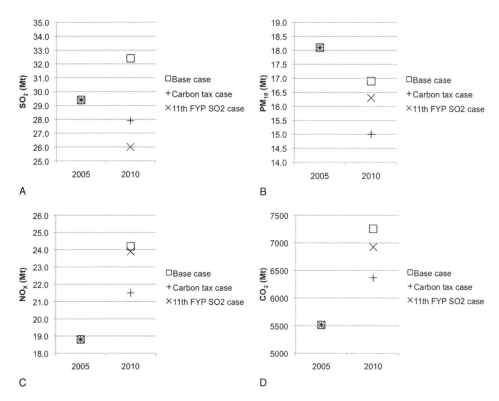

Figure 2.16
Comparison of emissions of the base case (2005 and 2010) and the two policy cases. (A) SO_2. (B) PM_{10}. (C) NO_X. (D) CO_2.

industries shrink significantly. In both cases consumption increases somewhat, while in the 11th FYP the effects on investment and GDP are somewhat positive (1.1% and 0.74%, respectively) and in the carbon tax those effects are modestly negative (−0.25% and −0.19%).

We should stress here, however, that these comparisons are snapshots of the effects at the fifth year of implementation only. The economic benefits of the SO_2 controls derive chiefly from the one-time opportunity to replace so many small, inefficient power plants with more efficient ones. While FGD systems will be required in new power plants and can be deployed in other industries in the future, these will not yield the net economic benefits of the two-part strategy of the 11th FYP, nor can this strategy be scaled up further. The costs of the modeled carbon tax, by contrast, are small over a longer run than the five-year period of the current analysis, and, more importantly, such a policy could be expanded in the future. The

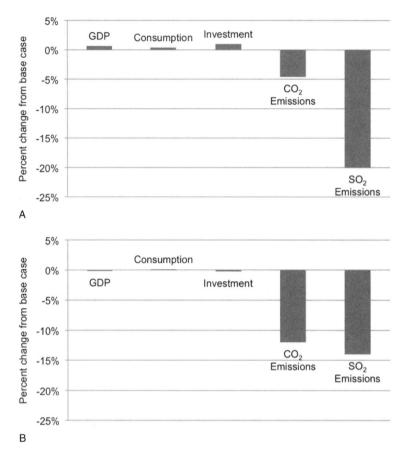

Figure 2.17
Effects on major economic and emission indicators in 2010 of the two policy cases. (A) 11th
FYP SO$_2$ control policy. (B) Carbon tax policy.

simulations of future carbon taxes in chapter 3, which cover a longer horizon,
illustrate this point.

We summarize the most important effects of the policies, including those just
discussed, and their implications for public health, in table 2.4. In terms of total
health benefits—a function chiefly of the transformation of primary emissions into
PM$_{2.5}$ and ozone through atmospheric processes—the number of avoided premature
mortalities of the two policies is very large in both cases although somewhat higher
under the tax. This difference is in part due to the broader effects of the carbon
tax, leading not only to reductions of sulfate PM$_{2.5}$ as in the 11th FYP policy, but

Table 2.4

Comparative summary of major effects of the two policies versus the base case

Variable	Effect of SO$_2$ Policy versus Base Case in 2010	Effect of Carbon Tax versus Base Case in 2010
	Percent Change	
GDP	0.74	−0.19
Consumption	0.47	0.13
Investment	1.07	−0.25
Coal use	−5.6	−14.6
CO$_2$ emissions	−4.6	−12.2
PM$_{10}$ emissions	−3.6	−10.6
SO$_2$ emissions	−19.8	−14.0
NO$_X$ emissions	−1.3	−11.1
	Number of Cases	
Avoided premature deaths		
Acute effect PM$_{2.5}$ (PRC epidemiology)	12,300	17,200
Acute effect ozone (PRC epidemiology)	120	1,380
Chronic effect PM$_{2.5}$ (U.S. epidemiology)	73,900	103,000
	Million Tons or Billion Yuan	
Increased grain production		
Wheat (Mt)	~0	0.174
Rice (Mt)	~0	0.993
Maize (Mt)	~0	1.454
Total value (billion yuan)	~0	5.1

also greater reductions of primary PM and of secondary nitrate PM$_{2.5}$ through the abatement of NO$_X$.

These differences aside, the most notable feature about the avoided health damages attributable to both policies is their very large scale: each has been, or would have been, a powerfully effective public health strategy for the government to undertake. The reader should also keep in mind, especially in the carbon tax case, that this health benefit is in addition to a sizable global benefit from reduced CO$_2$.

It is quite common for analysts to compare policies employing technology mandates (like P1) and market-based tools (like P2) to make the case for one type of policy instrument over the other. While we expected only rough comparative insights from examining a single example of each in this chapter, it is interesting, and no less valuable, to find that our results cannot be said to support one approach over the other. The reason should be encouraging to those seeking effective policy solutions. Judged by their original criteria—the extent to which each policy achieved or would have achieved its intended environmental objective at modest economic cost, in the conditions that prevailed during 2006–2010—both policies should be understood to have performed quite well in our assessment.

2.9   Conclusions

To summarize our key conclusions, starting with the SO_2 control policy of the 11th FYP:

- The policy is estimated to have reduced emissions of SO_2 compared to 2005 by 11.6%, exceeding the national target of 10% under the 11th FYP, and to have reduced atmospheric concentrations of $PM_{2.5}$ substantially compared to a business-as-usual base case.

- Under the policy, primary PM_{10} and NO_X emissions may have decreased slightly, and changes in ozone concentrations were likely very modest.

- As a result, the policy may have reduced premature mortality from $PM_{2.5}$ exposures in 2010 by 12,300 cases (applying a conservative but more certain epidemiological assumption based on Chinese studies) to as many as 73,900 cases (applying a more uncertain epidemiological assumption on chronic effects derived from U.S. studies). An additional 120 premature mortalities may have been avoided by reduction of ozone.

- The FGD component of the policy helped produce this enormous health benefit, at a small positive cost to the economy. The small-plant shutdowns, however, not only added to the health benefit but also led to net economic benefits (negative costs) for the policy as a whole, by boosting GDP, investment, and consumption.

- The plant shutdowns also reduced coal use, more than compensating for increased electricity (and coal) use of FGD systems, leading to a small net reduction in CO_2 emissions overall versus the base case.

While adjustment costs should also be considered, such as the welfare of workers laid off by plant shutdowns, the 11th FYP SO_2 strategy appears to have been a major policy success judged by environmental, public health, and economic criteria.

Regarding a hypothetical 100 yuan per ton carbon tax implemented in lieu of the SO_2 policy during the same years (2006–2010):

- It would have reduced emissions of all species analyzed—CO_2, PM_{10}, SO_2, and NO_X—in 2010 by large amounts compared to business as usual.
- It would have reduced net concentrations of $PM_{2.5}$ very substantially, and of ozone somewhat.
- Along with the large intended benefit in carbon control, estimated premature deaths avoided from reduced $PM_{2.5}$ in 2010 would have totaled 17,200 cases (applying Chinese epidemiology on acute mortality effects) to as many as 103,000 cases (applying U.S. epidemiology on chronic mortality effects). An additional 1380 premature mortalities would have been avoided as a result of reductions of ozone.
- The reduction of ozone would have yielded an additional benefit in improved grain productivity, especially rice and maize, estimated to value around 5 billion yuan in 2010.
- The net effects on the economy (GDP, investment, and consumption) would have been modest in all cases, and could have been either positive or negative depending on the timing and use of tax revenues. Under all assumptions, the cost to GDP would have been small compared to the large benefits to public health of reduced air pollution and considerable reduction in the growth of CO_2.

In short, tax-based policies can be a highly effective approach to controlling a variety of pollutants along with CO_2.

The 11th FYP SO_2 policy demonstrates the effectiveness of well-targeted technology mandates for pollution control in major industries. This finding is encouraging for the prospect of effective NO_X-control technology mandates in the 12th FYP, although matching the scale of the health and economic benefits of the 11th FYP SO_2 control policy will be very difficult.

The use of tax mechanisms may not only be a cost-effective way to lower the carbon intensity of the Chinese economy over time, but also a potent multipollutant strategy. It could integrate China's primary objectives in domestic air quality (regarding $PM_{2.5}$ and ozone) with the newer goal of reducing China's carbon trajectory to limit global climate change.

Notes

1. This estimate of the Ministry of Environmental Protection (MEP) is reported by the *Xinhua News Agency* at http://news.xinhuanet.com/english2010/china/2011-08/29/c_131081860.htm, last accessed October 10, 2012. See chapter 4 on the role of the operation rate of FGD systems, represented as the removal efficiency, in the results.

2. See endnote 2 in chapter 4.

3. Note that the absolute values of the variables would have a bigger influence on an assessment of certain types of policy aims, including China's 2020 carbon intensity target, but the current study is neither intended nor designed to evaluate that objective. China's official intensity target is discussed further in sections 3.1 and 3.4.1.

4. These include black carbon, organic carbon, nonmethane volatile organic compounds, and ammonia. We plan to add evaluation of radiative forcing to future applications of this research framework.

5. See the source in endnote 1. As noted earlier in this chapter, our projections of the capacities of FGD deployments and small power plant retirements are close to official estimates at the end of the 11th FYP. The basis of the MEP estimate, e.g., assumptions about removal efficiency of FGD systems, is unclear.

6. This weekend effect in California is discussed at http://www.arb.ca.gov/research/weekendeffect/weekendeffect.htm.

3

Summary: Carbon Taxes for 2013–2020

Chris P. Nielsen, Mun S. Ho, Jing Cao, Yu Lei, Yuxuan Wang, and Yu Zhao

3.1 Introduction

The Chinese government asserts that its contribution to greenhouse gas (GHG) control should be based on the principle of "common but differentiated responsibilities" agreed to under the 1992 United Nations Framework on Climate Change (UNFCCC), by which wealthier nations are to take the lead. It recognizes that some level of explicit GHG control is inevitable, however, and that China must move toward controls that can be credited in any international regime that succeeds the Kyoto Protocol.

Indeed, at the 2011 UNFCCC negotiations in South Africa, China signed on to the Durban Platform, in which all nations agreed to establish a common legal regime in GHG control for both developing and industrialized countries. (Under the 1997 Kyoto Protocol, only the latter face binding emission constraints.)[1] China's position at Durban built on its unilateral commitment in 2009 to reduce the carbon intensity of its economy—carbon dioxide emissions divided by GDP—by 40–45% in 2020, compared to 2005 levels.

In this chapter, we examine a number of carbon policy options for China, applying the integrated research framework introduced in the last chapter and described in the rest of the book. We first developed the research tools by evaluating emission control policies set in the recent past; here we use them to analyze policy alternatives set in the future, to 2020. We narrow our focus to policies designed to limit China's carbon emissions, but making full use of the breadth of our integrated framework. This allows us to assess the effects of future policies not only on carbon emissions, but also on China's economy, energy use, air pollution emissions, domestic air quality, and the damages to public health and agricultural productivity that result.

Many observers would agree, at least in theory, that the most straightforward approach to limit emissions of carbon is to place a price on it. The rationale is simple, that introducing economic disincentives to activities that emit carbon, notably combustion of fossil fuels, would encourage lower emissions. The policy imposes a higher financial penalty if the emission intensity of the activity is higher; burning coal, for instance, generates roughly twice as much CO_2 per unit of delivered energy as burning natural gas. By putting a price on each kilogram of CO_2 emitted, the use of coal would incur an additional cost that is twice that imposed on use of gas. This will shift energy consumption at the margin toward gas, and even more so toward noncarbon energy sources, and incentivize improved energy efficiency. Over the longer run, it will raise investment in lower-carbon infrastructure and gradually restructure the energy system as a whole.

Harnessing the tax system for CO_2 control may be the simplest and most direct approach to pricing carbon. New revenues are also generated that the government can use in a variety of ways, such as compensating those disproportionately impacted by the tax or reducing existing taxes that may distort the economy. In political terms, a carbon tax would be instantly credible to the international community, even envied in some nations where carbon taxes are rejected chiefly on the grounds of political infeasibility. A tax-based approach could have an additional advantage over command-and-control mandates like those of the 11th and 12th Five-Year Plans (FYPs): moving from unpredictable, sequential, and piecemeal emission controls to a consistent mechanism that can be adjusted incrementally over time, thus reducing the costs of policy uncertainty. Such desirable properties of carbon taxes have been discussed in many studies, including Nordhaus (2010) and Aldy et al. (2010).

We should note, as many readers will know, that an alternative way to place a price on carbon is a cap-and-trade policy. In this case a maximum level of annual emissions is set for an economy (or segment of an economy), and a system of emission allowances is created where the total number of allowances is equal to the cap. Emitters are required to hold one unit of allowance for each unit of emissions. These allowances are allocated to economic actors—for free and/or through auctions—and the actors, usually enterprises, can trade them. The relative scarcity of the allowances created by the cap, compared to demand for the activities that produce carbon emissions, results in a positive market price on those emissions. The cap can be tightened over time, yielding the same effect as an increase in a carbon tax rate.

There is a relatively mature carbon cap-and-trade system now in place in the European Union and others being implemented in a number of other jurisdictions around the world, including a large and comprehensive one in the U.S. state of California.

Local carbon trading systems are also being developed in seven pilot cities or provinces of China. These carbon markets, the first ones launched in 2013, are policy experiments mandated by the environmental title of the 12th FYP but designed with quite different architectures and sector coverages by local authorities. National officials have expressed hope that they can be developed into a national system during the 2016–2020 13th FYP (Xinhua News Agency 2012). Skeptics, including many within China, question whether it has the institutional and legal capacities to create functioning markets in CO_2, which turns this invisible and intangible gas into a priced commodity. Doing so has proven challenging even in the most mature market economies, with much more established legal traditions and property rights.[2] At the time of writing, moreover, most of the pilot systems seek to transform the cap component of cap-and-trade from one that is set simply on the quantity of emissions, as such systems were originally conceived, to a more convoluted cap on emission intensity. This effort may further complicate implementation.[3]

While we believe that a national carbon tax may be more feasible to implement and enforce in China than a national carbon trading system, our objective here is neither to make this case nor to debate other relative merits of the two carbon-pricing approaches. Placing the implementation challenges aside, the two mechanisms have nearly the same theoretical basis and the same primary aim—placing a price on carbon—and their first-order effects on the economy and emissions will be very similar. Indeed most of the lessons derived from our research of carbon taxes would also pertain in principle to cap-and-trade versions of the same policies. Advocates of both approaches should find much of value in the research that follows.

Carbon tax options have been much discussed by Chinese officials, and the fiscal title of the 12th FYP includes a mandate to initiate environmental taxes on pollutants that many experts indicate was designed for possible extension to carbon. High-level Chinese officials have since indicated that some form of tax on carbon is possible by 2015 (Wei 2012). (The coexistence of pilot carbon markets and carbon taxes in China would not be unusual; China has a strong tradition of concurrent, overlapping policy experiments.) There have been many analyses of carbon taxes by Chinese researchers, in addition to the more interdisciplinary collaborative studies discussed in chapter 1. Important among these is Su et al. (2009), a study by the Research Institute for Fiscal Science, a center affiliated with the Ministry of Finance, which pushes for carbon taxes within government deliberations on carbon policy. Academic studies include Liang, Fan, and Wei (2007), Wang, Li, and Zhang

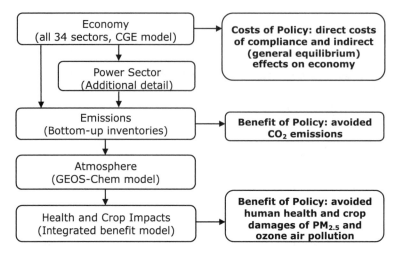

Figure 3.1
Framework for assessing costs and benefits of carbon tax policies.

(2011), Yi et al. (2011), and Lu, Tong, and Liu (2010). While these analyses have their individual strengths, they are not integrated with a comprehensive accounting of emissions and simulation of local air quality that our research program provides. They are therefore unable to quantify potentially large ancillary benefits of the tax policies to China's domestic environment, public health, and agriculture. We consider in some detail various carbon price options using the research framework of this book to examine the costs and benefits of such policies comprehensively. A complete overview of this research framework is given in chapter 2 and is represented in figure 3.1 for analysis of carbon tax policies. In this chapter, we discuss methods only in brief reminders of descriptions from chapter 2 (which will then be presented in depth in the chapters that follow), focusing here on the new results.

Many other analysts have noted that there is no best option to implement a carbon tax; different policy packages have different impacts on the various segments of society (e.g., Fischer and Fox 2011; Liang, Fan, and Wei 2007). The size of the tax obviously matters, as well as the timing. Given the broad-based nature of most carbon tax proposals, substantial new revenues can be generated, and their use is a key area of emphasis and contention. A well-designed policy to compensate affected taxpayers may substantially improve the political feasibility of the tax while reducing the economic costs. Another area of concern in countries that have considered carbon price policies is the protection of energy-intensive industries that are exposed to international competition.

To illustrate the main issues China faces in placing a price on carbon emissions, in this chapter we examine six different carbon tax options. As noted, instead of placing these policies hypothetically in the past, as we did for our first carbon tax case in the preceding and following chapters, we analyze them as if implemented in the future, from 2013 to 2020.

The choice of 2020 as the final year of the modeled policy term is not arbitrary. Our reasons relate both to the policy context and our research needs. First, as noted at the outset, 2020 is the target year of China's domestic commitment to improve the overall carbon intensity of its economy by 40–45%, measured against the 2005 level. While our methods are designed to emphasize the effects of different policy options on economic growth, household consumption, and air quality impacts, and not against such an intensity metric,[4] the relative effects of policies on carbon emissions and the GDP for the year 2020 in particular are obviously pertinent. Second, 2020 is also the target year of the Durban Platform when, in principle, all countries will become subject to a common legal regime in GHG control. China's anticipated CO_2 emissions in the period leading up to 2020 could influence its negotiating positions, and China's carbon path under different policy scenarios is of interest. Third, 2020 is a primary target year used in China's recent energy planning, including, for instance, capacity targets in hydroelectric, nuclear, wind, and solar power generation. These plans serves our analytical needs because they help to inform our understanding of what would happen in China in a "business-as-usual" base case—that is, a scenario assuming no additional policy—against which we estimate the effects of an added carbon tax.[5]

The choice of 2013 as the first year of the tax term is less critical. We might have alternatively chosen 2016, the first year of the 13th FYP, and produced results that would likely be similar in effect, if smaller in scale. In any case our goal is not to recommend a specific policy for actual implementation during specific years, but rather to investigate the general effects of the policy options, whenever they might be implemented. Since a carbon tax will exert lasting influences on an economy, it is useful to evaluate them over longer periods if possible, and analyzing eight years of tax imposition from 2013–2020 is somewhat more informative than a shorter period such as the 13th FYP term.

To evaluate the comprehensive effects of the carbon tax options, we choose three of the six options, and the base case assuming no additional policy, for analysis with the complete framework presented in chapters 2 and 4–9, and add a fourth for more limited analysis. These are the focus of this chapter. The final two cases, of narrower economic interest and judged mainly by economic criteria, will be discussed in box 3.1 (see page 123).

In chapter 2 we considered two policies during the 11th FYP (2006–2010) period. Here we are looking to the future instead of the past, and it requires more assumptions because we can no longer draw on empirical data from the target period. Indeed we chose first to analyze policy cases in 2006–2010 to develop and demonstrate our integrated research capabilities in a more amenable data context, and then to apply them to the future.

We should be clear that while the policies of interest are set in the future, our modeling analysis still begins in empirical data of the past. Specifically, we adopt a base modeling year of 2007—for which relatively complete data on the economy, energy use, emissions, and the atmosphere are available—and project forward to 2013, when a tax is first imposed, and then beyond, to 2020.

We could report the 2013–2020 tax cases in detail comparable to that of chapters 4–9, but to do so would be highly repetitive. Instead, we summarize the results of the analyses in this chapter in a way that is accessible to both nonexpert and expert readers, as we did in chapter 2 for the 2006–2010 cases. This summary will draw from descriptions and figures in chapter 2 that serve the same informative purposes here.

Fellow researchers need to know details of new methods and assumptions required for the analyses of future policies. Changes in methods and assumptions from the rest of the book that relate generally to the new analyses are presented together, component by component, in appendix C. Methods narrowly related to individual carbon tax cases are placed instead in endnotes to this chapter, so that they can be easily related to the discussion of those specific tax cases. In this way, the results of research reported in this chapter should be understood as supported by chapters 4–9, but modified by information in the endnotes and appendix C. This form of presentation is chosen to present the relevant information in the most efficient manner, without burdening the reader with too much detail.

In the next section we introduce the base case and first four carbon tax options. In section 3.3 we report the simulated effects of these carbon tax policies on the economy, energy use, and CO_2 emissions, and in box 3.1 we consider two additional tax policies judged by these same criteria. The subsequent sections present full benefit analyses for our main carbon tax scenarios: section 3.4 describes the resulting changes in all relevant atmospheric emissions; section 3.5 analyzes the changes in critical air pollution levels; and section 3.6 calculates the environmental benefits in terms of public health and crop productivity for those cases. A summary comparison of all the scenarios is given in section 3.7. We conclude the chapter in section 3.8 with observations about the integrated research framework in general and the specific carbon tax policies in particular.

3.2   Carbon Tax Options and the Base Case

3.2.1   Base Case

Recall from chapter 2 that we measure the effects of policies against a base case that omits those policies but includes our best estimates about the implementation of all other anticipated policies. For the future carbon tax scenarios of this chapter, we must set a new base case that differs from that of chapter 2 in that it looks beyond 2010. It therefore includes the effects of the now-implemented SO_2 controls of the 11th FYP and, as noted earlier, other existing energy and environmental policy commitments up to 2020 (as described sector by sector in appendix C).[6]

The base case is also the foundation of all of our economic modeling, which, in this chapter, is based on the 2007 official input-output tables of China (NBS 2009). The economic model has the same structure as the version for our 2006–2010 cases that will be presented in detail in chapter 9 and in appendix A, but is updated to the new 2007 base year. We briefly summarize it in table 3.1 and figure 3.2.

We are implementing each policy over the period 2013–2020, but we start our projections in 2007. We report the projections in the table and figure both to 2020 and beyond, to 2032, to give the reader a full sense of the underlying economic trajectory. The GDP growth between 2007 and 2020 is projected in the base case at a rapid 8.4% per year. The population growth, while slowing, averages 0.54% for this period, but the working-age population is essentially flat. Coal consumption is projected to grow at 3.3% per year, while oil, and especially gas, use are expected to grow faster. This rise in oil use is expected even as we include the projection of rising world oil prices from EIA (2009, table 12), shown in figure 3.2 (gray line marked with squares). As a result, total fossil energy use is projected to grow at 3.9% per year. Given the rapid GDP growth, however, this figure corresponds to a substantial decline in the energy-to-GDP ratio.[7] In figure 3.2 we can see the much less steep profile of energy consumption compared to the GDP line. The emissions of CO_2 are also rising at less than half the rate of GDP growth, at 3.8% per year.

We also project a rapid growth of electricity consumption, 6.9% per annum, and gas consumption at 8.1% per year over the period 2007–2020. These rates are close

Table 3.1
Base case projection

Variable	2007	2013	2020	2032	2007– 2020 Growth Rate (percent)	25-Year Growth Rate (percent)
Population (million)	1,316	1,360	1,411	1,437	0.54%	0.35%
Effective labor supply (billion 2007 yuan)	10,019	10,387	10,043	9,423	0.02%	−0.24%
GDP (billion 2007 yuan)	27,260	45,985	80,777	150,894	8.4%	7.1%
Consumption/ GDP	0.37	0.41	0.47	0.55		
Fossil energy use (million tons of standard coal equivalent)	2,523	3,386	4,195	5,096	3.9%	2.9%
Coal use (million tons)	1,013	1,295	1,554	1,784	3.3%	2.3%
Oil use (million tons)	845	1,139	1,348	1,669	3.6%	2.8%
Gas use (billion cubic meters)	89	154	255	445	8.1%	6.6%
Electricity use (TWh)	2,246	3,399	5,500	9,749	6.9%	6.0%
CO_2 emissions (fossil fuel combustion, million tons)	6,703	8,915	10,966	13,136	3.8%	2.7%
PM_{10} (million tons)	19.05		13.96		−2.4%	
SO_2 emissions (million tons)	30.53		24.66		−1.6%	
NO_X emissions (million tons)	23.63		26.21		0.80%	

Note: There is 0.714 ton standard coal equivalent (sce) in 1 ton of coal.

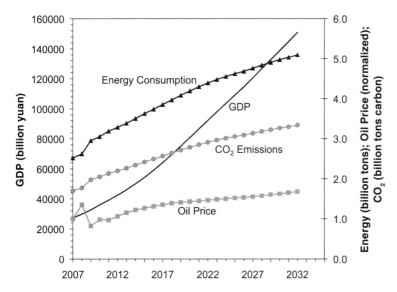

Figure 3.2
Base case projection of GDP, energy consumption, oil price, and CO_2 emissions. GDP is in billion 2007 yuan (left scale); oil price is the world price projected by EIA, normalized to 1 in 2007 (right scale); energy consumption is in billion tons of oil equivalent (right scale); and CO_2 emissions are in billion tons of carbon (right scale). The carbon tax cases run only through 2020, but the base case is given for more years to give the reader a sense of the underlying projections.

to forecasts of the International Energy Agency well known to researchers (IEA 2010, Annex A). This rapid growth of electricity output and the much slower growth of coal consumption mean that we (and IEA) are projecting a substitution toward other fuels and renewable energy sources, and an improvement in productivity in electricity production. Thus in the base case we are already projecting a steady reduction, for instance, in the amount of coal used to generate a kilowatt-hour of electricity. A carbon price would accelerate these trends in productivity and fuel substitution.

During this period of rapid income growth and structural change, the composition of the economy is expected to change substantially. The share of consumption in GDP rises from 37% in 2007 to 47% in 2020.[8] The structure of output shifts toward services, which are less energy-intensive than manufacturing. We emphasize that this shift only means that the share of resources devoted to services is higher; the absolute quantity of steel and other manufactures produced continues to rise with capital accumulation and productivity growth.

Table 3.2

Carbon tax options examined

Scenario	Time Path of Carbon Tax	Use of Revenues
F1	Tax of 30 yuan/ton of CO_2 (in 2007 yuan) applied for all years during 2013–2020	Rebated lump sum to households
F2	Tax of 10 yuan/ton in 2013 rising to 50 yuan/ton in 2020; same cumulative CO_2 reduction as scenario F1	Rebated lump sum to households
F3	Tax of 10 yuan/ton in 2013 rising to 100 yuan/ton in 2020	Rebated lump sum to households
F4	Same as F2, 10 yuan/ton in 2013 rising to 50 yuan/ton in 2020	Cut tax rates on enterprise income
F5	Same as F2, 10 yuan/ton in 2013 rising to 50 yuan/ton in 2020	Subsidies to EITE industries; rest rebated lump sum to households
F6	Same as F2, 10 yuan/ton in 2013 rising to 50 yuan/ton in 2020, assuming action by the rest of the world	Rebated lump sum to households

Note: EITE industries are energy-intensive, trade-exposed industries.

3.2.2   Carbon Tax Options

As noted previously, a carbon tax policy is a package of related but distinct rules: the magnitude of the carbon price, the date of implementation, the use of revenues, and sometimes the protection of vulnerable sectors. In total we examine six policy scenarios, summarized in table 3.2, to explore these issues.

The size and timing of a carbon tax are important because economic actors need time to adjust to the policy. Firms need to change production processes and capital equipment to use less carbon-intensive fuels and other intermediate inputs, and households also need to acquire new equipment that uses less energy. A tax that is announced in advance and implemented gradually will ease these costs of adjustment. A delay in implementing the carbon price, of course, means a delay in achieving emission reductions. It is less valuable to compare a tax of 10 yuan per ton of CO_2 starting today with the same tax starting in 10 years' time than to compare two different time profiles of tax rates that will generate the same cumulative reduction in emissions at some date. Given the nature of the climate risks of GHG concentrations and the long atmospheric lifetime of CO_2, the cumulative reduction is the most relevant point of comparison; the emissions in any particular year do not have a large impact on the climate risks in, say, 2100. For further discussion of this issue, see Newell and Pizer (2003).

Before we introduce our tax scenarios, note that unless otherwise noted, all values in this chapter are expressed in constant 2007 yuan, meaning in terms of prices that prevailed in the 2007 base year. This practice allows us to ignore the effects of inflation and eases comparison of economic effects across years.

Scenario F1   (The "F" denotes future, in contrast to the scenarios of the past discussed in chapter 2.) In the first scenario, we begin with a simple tax of 30 yuan per ton of CO_2 for all years in the policy period 2013–2020. This is almost identical in structure and tax rate to the hypothetical carbon tax for 2006–2010 analyzed in chapters 2 and 4–9.[9] Because it would be quite repetitive as a result, we do not choose F1 as one of the cases for the substantial additional research of full atmospheric simulation and damage assessment. The main purpose of this scenario is the economic comparison with the next scenario, F2, for reasons that will soon become clear. Thirty yuan was roughly equivalent to US$3.90 in 2007. At this price, we project that the cumulative emissions of fossil fuel CO_2 from 2013 to 2020 would be 64.79 gigatons (Gt), a substantial reduction compared to the 74.23 Gt projected for the base case with no additional policy to those already announced.

Scenario F2   In our second scenario, we choose a different profile of carbon prices that achieves the same level of cumulative reductions: we start with a low tax of 10 yuan per ton of CO_2 in 2013 and increase it steadily to roughly 50 yuan per ton by 2020. The exact tax rate in 2020 and thus the profile of taxes are calculated such that the cumulative CO_2 emissions from fossil fuel combustion from 2013 to the end of 2020 are identical to those of F1, 64.79 Gt. Holding total CO_2 emissions constant over the two cases will allow us to explore the effect on the economy of implementing a tax gradually over time compared to setting a static rate from the outset.

Scenario F3   We have chosen the preceding carbon tax levels even if they seem low compared to prices in international carbon markets. These are imaginable rates that are not excessive relative to prevailing energy prices in China, but that still would generate significant emission reductions. Note, however, that carbon prices in the EU Emission Trading System (EU ETS) were about US$25–30 per ton of CO_2 in our base year of 2007, although they were under US$4 in the spring of 2013 (in 2013 currency) due to economic recession. When the U.S. House of Representatives passed a cap-and-trade bill in 2009 (which later died in the U.S. Senate), a U.S. Environmental Protection Agency analysis projected an initial carbon price of US$13–17 per ton of CO_2, equivalent to more than 100% of the price of coal at the mine mouth. To show the effect of higher carbon prices, in F3 we begin with the same 10 yuan per ton of CO_2 in 2013 as in F2, but increase it instead to 100 yuan per ton by 2020. This is much closer to historical prices in international carbon markets and will show the effects on China's economy, carbon emissions, and air quality of such a price convergence assuming the international price recovers to pre-recession levels.

Importantly, in these first three scenarios, the carbon tax revenues are given back to households in a lump sum fashion. By this, economists mean that households receive a transfer from the government, the size of which does not depend on any household decision[10] such as the expenditures on fuels. The revenue use assumption is changed in our next scenario.

Scenario F4   This scenario has the same path of tax rates as in F2—rising from 10 yuan per ton of CO_2 to roughly 50 yuan per ton—but the new revenues are now used to cut the enterprise income tax rate.[11] The result is that the proportion of profits that enterprises keep after taxes is higher, and this changes the incentive to invest. It is expected to lead to slightly higher growth in GDP compared to the lump sum transfer of revenues to households, as in F2.

3.3   Effects on the Economy and Energy Use

We implement all scenarios by imposing a tax on the three fossil fuels (coal, oil, and natural gas) in proportion to their carbon content, as detailed in section 9.5.1. The setting of tax rates ignores CO_2 emissions from other sources such as process emissions and combustion of biofuels and agricultural wastes, to keep implementation as simple as possible.[12] We also recycle the new revenues completely so that aggregate government consumption of goods and services remains unchanged at base case levels. At this stage we can estimate emissions of CO_2 under the scenarios because they are generated directly from consumption of fossil fuels projected in the economic model and, because CO_2 is a global pollutant, we are unconcerned with its spatial distribution; this characteristic is in contrast to all of the other species, which affect air quality and must be estimated bottom-up.[13]

The results for the first carbon tax year of 2013 are given in table 3.3 and for 2020 in table 3.4. We will focus especially on F2 but describe the outcomes in order,

Table 3.3
Effects of a carbon tax in the first year, 2013

Variable	Base Case 2013 (levels)	Scenario F1: 30Y/Ton All Years — Lump Sum Offset	Scenario F2: 10Y Rising to 50Y/Ton — Lump Sum Offset	Scenario F3: 10Y Rising to 100Y/Ton — Lump Sum Offset	Scenario F4: 10Y Rising to 50Y/Ton — Tax Cut Offset	Scenario F5: 10Y Rising to 50Y/Ton — Subsidies to EITE	Scenario F6: 10Y Rising to 50Y/Ton — Lump Sum; Multilateral
				Percent Change from Base Case			
GDP (billion yuan 2007)	45,985	-0.09	-0.02	-0.02	-0.02	-0.01	-0.01
Consumption (billion yuan 2007)	18,148	0.29	0.11	0.11	-0.03	0.05	0.11
Investment (billion yuan 2007)	21,455	-0.08	-0.02	-0.02	0.10	0.01	0.00
Government consumption (billion yuan 2007)	5,171	0.00	0.00	0.00	0.00	0.00	0.00
Exports (billion yuan 2007)	14,308	-0.69	-0.28	-0.28	-0.22	-0.19	-0.42
Fossil energy use (million tons of standard coal equivalent)	3,386	-12.1	-4.51	-4.51	-4.44	-3.96	-4.38
Coal use (million tons)	1,378	-15.6	-5.81	-5.81	-5.74	-5.13	-5.68
Oil use (million tons)	1,942	-1.6	-0.54	-0.54	-0.48	-0.34	-0.36
Gas use (billion cubic meters)	153	-4.5	-1.54	-1.54	-1.48	-1.20	-1.49
Electricity (billion kWh)	3,399	-2.8	-0.98	-0.98	-0.95	-0.82	-0.98
CO_2 emissions (fossil fuel; million tons)	8,915	-13.0	-4.82	-4.82	-4.74	-4.25	-4.69
Carbon tax as percent of total tax revenue		2.4	0.89	0.89	0.89	0.89	0.89	0.89

Table 3.4
Effects of a carbon tax in 2020

Variable	Base Case 2020 (levels)	Scenario F1: 30Y/Ton All Years — Lump Sum Offset	Scenario F2: 10Y Rising to 50Y/Ton — Lump Sum Offset	Scenario F3: 10Y Rising to 100Y/Ton — Lump Sum Offset	Scenario F4: 10Y Rising to 50Y/Ton — Tax Cut Offset	Scenario F5: 10Y Rising to 50Y/Ton — Subsidies to EITE	Scenario F6: 10Y Rising to 50Y/Ton — Lump Sum; multilateral
				Percent Change from Base Case			
GDP (billion yuan 2007)	80,777	-0.09	-0.14	-0.33	0.02	-0.07	-0.11
Consumption (billion yuan 2007)	34,906	0.17	0.27	0.39	0.00	0.15	0.27
Investment (billion yuan 2007)	33,343	0.01	0.01	-0.04	0.57	0.10	0.05
Government consumption (billion yuan 2007)	8,267	0.00	0.00	0.00	0.00	0.00	0.00
Exports (billion yuan 2007)	26,655	-0.69	-1.12	-2.11	-0.81	-0.75	-1.27
Fossil energy use (million tons of standard coal equivalent)	4,195	-11.6	-17.7	-28.8	-17.5	-16.2	-17.5
Coal use (million tons)	1,653	-15.2	-23.0	-37.2	-22.9	-21.3	-22.9
Oil use (million tons)	2,297	-1.3	-2.11	-4.01	-1.91	-1.37	-1.94
Gas use (billion cubic meters)	252	-4.5	-7.37	-13.65	-7.13	-6.11	-7.32
Electricity (billion kWh)	5,500	-2.4	-3.86	-6.82	-3.66	-3.40	-3.85
CO_2 emissions (fossil fuel; million tons)	10,966	-12.4	-18.9	-30.8	-18.6	-17.4	-18.8
Carbon tax as percent of total tax revenue	1.9	2.91	2.91	5.03	2.93	2.89	2.91

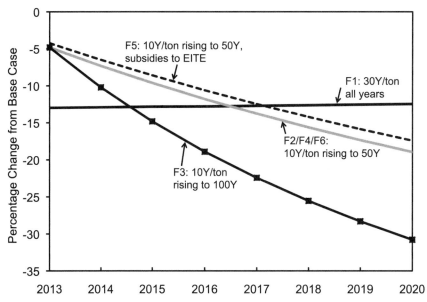

Figure 3.3
Change in CO_2 emissions resulting from different carbon tax options, compared to the base case. *Note:* F2 (lump sum rebate), F4 (tax cut offset), and F6 (multilateral tax) have the same tax rate profile and nearly identical emission paths. F5 (subsidies to EITE) has the same tax rate profile as F2/F4/F6, but a somewhat different emission path.

beginning with F1. We also introduce and discuss our more narrow analyses of two more scenarios, F5 and F6, in box 3.1 on pages 123–126.

3.3.1   Effects of Tax Rates and Time Paths: Scenarios F1, F2, and F3

In F1, with the same 30 yuan per ton tax in all years, the annual reduction in CO_2 emissions compared to the base case is about 12–13% in each year of the period, as plotted by the plain black line in figure 3.3. In the first year, coal use falls by 15.6% while oil use falls only 1.6%, leading to a 12.1% fall in total energy use. This causes a reduction in output of the energy intensive industries and a small 0.09% reduction in aggregate GDP. With the transfer of carbon tax revenues to households their consumption does not fall. The important change is in the terms of trade, which appreciates by 0.1% as a result of a fall in import demand, especially for oil imports.[14] This appreciation is accompanied by a fall in total exports of 0.69%. Thus the reduction in GDP falls mainly on net exports, cushioning the shock to domestic investment. Over time, the lower investment results in a smaller capital stock and lower consumption. The fall in capital would have normally lowered

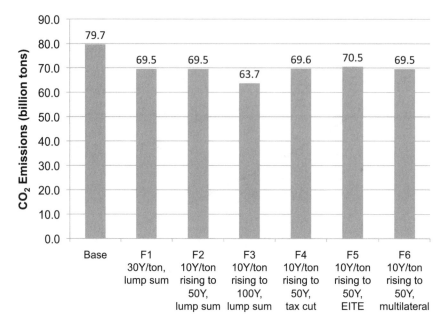

Figure 3.4
Cumulative 2013–2020 CO_2 emissions for the different carbon tax scenarios.

GDP; the continued (small) appreciation of the terms of trade, however, keeps the GDP loss to 0.09% even in 2020 (see table 3.4).

Scenario F2 allows a more gradual adjustment, with the price of carbon rising from 10 yuan per ton in 2013 to around 50 yuan per ton in 2020. As noted, the exact path of carbon tax rates is chosen such that the cumulative emissions over the period 2013–2020 are the same 64.79 Gt achieved in F1. The cumulative emissions are plotted in figure 3.4 for all scenarios. In F2, the small initial tax leads to a modest 4.5% reduction in fossil energy use in 2013, and a corresponding 4.8% reduction in CO_2 emissions (table 3.3). In comparison, the CO_2 reduction versus the base case is 13.0% under the static 30-yuan tax of F1; the reductions are not proportional given the nonlinearities in the system. The first-year fall in GDP is a tiny 0.02%, with a slight increase in consumption, a slight decrease in investment, and a bigger decrease in exports. The relative contributions of these factors to the change in GDP—consumption, investment and net exports—are the same as in F1.

Over time, as the carbon tax rate rises, the reduction in fossil fuel CO_2 emissions versus the base case grows, as plotted in figure 3.3 (the gray line, representing F2, F4, and F6); by 2020, fossil CO_2 emissions are 18.9% lower. The paths of carbon

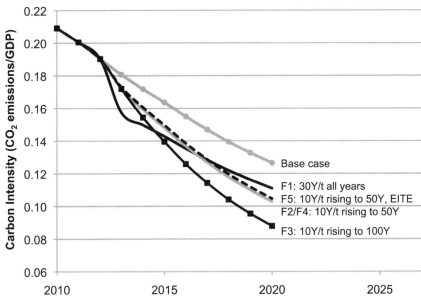

Figure 3.5
Fossil fuel carbon intensities of the base case and carbon tax scenarios. Carbon intensity here is defined as the emissions of CO_2 from fossil fuel combustion per unit of GDP. The carbon intensities of F2 (lump sum rebate) and F4 (tax cut offset) are indistinguishable, because the effects of the policies on both emissions and GDP are very similar.

intensity (fossil fuel CO_2 emissions per unit GDP) of the two cases are shown in figure 3.5. In F1 (the plain black line), there is a big initial drop, followed by a path that is almost parallel to the base case. In F2 (plain gray line), there is a smaller initial fall in carbon intensity, but by 2020 it is lower than that of F1.

The changes in the macroeconomic variables over time for F2 are plotted in figure 3.6. The lower investment in all years result in a smaller capital stock and the reduction in GDP, compared to the base case, grows year by year. As the tax rate rises, the increasing carbon tax revenues imply an increasing transfer to households and increasing consumption. This significant fall in GDP and rise in consumption is accompanied by a small reduction in investment; these patterns are again due to the appreciation in the terms of trade, leading to a large 1.1% reduction in exports by 2020 (table 3.4). With the high tax of 50 yuan per ton in 2020, coal use is 23% lower than in the base case in 2020 and oil use is 2.1% lower, resulting in a total fossil fuel reduction of 17.7%.

Behind these small changes in the aggregate economy are bigger changes at the industry level, as a result of the diverse range of fossil fuel and electricity use. In

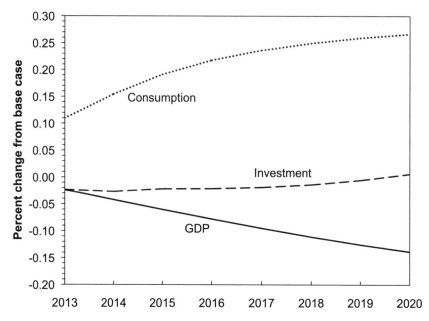

Figure 3.6
Change in GDP, consumption, and investment due to a rising carbon tax (scenario F2: 10 yuan per ton rising to 50 Y).

figure 3.7 we show the 10 largest percentage changes in commodity prices versus the base case, and figure 3.8 gives the 10 largest percentage changes in industry output. The light bars are for the first-year effects in 2013, and the dark bars are for 2020. With a first-year tax of 10 yuan per ton of CO_2, the after-tax price of coal (at the mine mouth) naturally rises more than any other commodity, a 5.7% increase. Electricity is fossil-fuel-intensive, and its price rises by 0.96%, while the energy-intensive manufacturing industries (nonmetal mineral products, chemicals, and primary metals) show price increases of about half a percent. These higher prices lead to a fall in sales, and the industry outputs fall by corresponding percentages: coal (−5.8%), electricity (−0.98%), and nonmetal mineral products (−0.43%).

By 2020, when the tax has risen to about 50 yuan per ton, the price shocks are much bigger. The price of coal rises by 28%, of electricity by 4.0%, of petroleum refining by 2.0%, and of the products of the three energy-intensive manufacturing industries by almost 2%. These price changes lead to a big reduction in output of coal (−23%), electricity (−3.9%) and petroleum refining (−2.4%). Our model takes a long-run view of market adjustment, which means that the labor and capital markets have time to clear and any workers initially laid off are reallocated to other

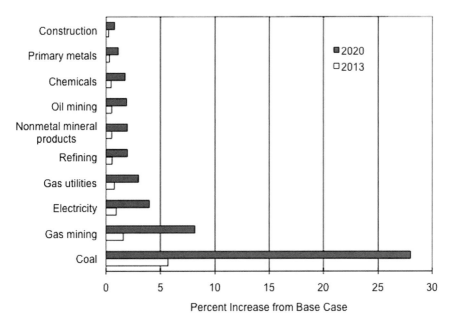

Figure 3.7
Effect of a carbon tax (scenario F2) on commodity prices, 2013 and 2020: top 10 percentage changes from base case.

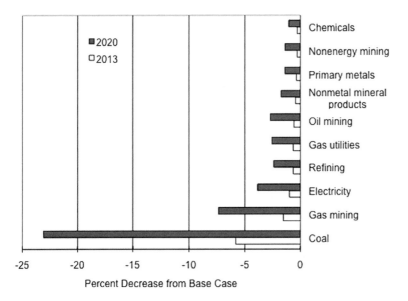

Figure 3.8
Effect of a carbon tax (scenario F2) on industry output, 2013 and 2020: top 10 percentage changes from base case.

industries. Capital resources also move from these shrunken industries to others. Which sectors expand depends on two factors: their energy intensity and the recycling of tax revenue. In F1 and F2, the lump sum transfer of the new revenues results in higher household consumption, so sectors producing consumption goods expand, particularly the less energy-intensive ones. By 2020, the output of finance grows by 0.22%, food manufacturing by 0.13%, and real estate and agriculture by 0.12%.

The last topic we examine in this section is the effect of more ambitious reduction targets. In F3, we adjust the profile of tax rates so that they rise from 10 yuan per ton of CO_2 in 2013, as in F2, to 100 yuan per ton by 2020, twice the level in F2 and closer to historical international carbon prices. As shown in figure 3.4, the cumulative fossil CO_2 emissions in this case are 59.35 Gt, or a 20.0% reduction from the base case, compared to the 12.7% reduction in F1 and F2. From table 3.4, we see that aggregate GDP in 2020 is 0.33% lower than the base case, compared with 0.14% in F2. The decline in 2020 coal use is 37% (versus 23% in F2), in oil use is 4.0% (2.1% in F2), and in electricity use is 6.8% (3.9% in F2). The reduction in annual fossil fuel CO_2 emissions is 31% in 2020, compared to 19% in F2.

These comparisons show that the economy does respond to higher carbon prices, but at a somewhat diminishing rate; doubling the tax rate does not double the emission reduction. There is a rising marginal cost in reducing CO_2 emissions, as shown in other analysis of carbon prices. (See, for example, the analyses of U.S. GHG control proposals in U.S. EPA 2010.) We should caution, however, that we have not yet considered attendant increases in health and crop benefits from ancillary reduction of air pollution, to be discussed in section 3.6. Clearly the choice of an emissions target, realized through a tax rate, must consider local environmental benefits as well as benefits of reducing climate change risks.

3.3.2   Effects of Alternative Revenue Uses: Scenario F4

Many analysts have emphasized the desirability of using new revenues to reduce the distortions caused by existing taxes. (See, for instance, Bovenberg 1999; Parry, Goulder, and Williams 1999.) Taxes introduce a wedge between the buyer and seller: a seller willing to sell for at least one yuan and a buyer willing to pay up to one yuan cannot complete a mutually beneficial transaction if there is also a sales tax. (Readers who have studied economics may recall that this is referred to as the "deadweight loss.") More specific to our analyses, economics recognizes distortions due to taxes in the largest markets, in labor and capital. The income tax on labor and capital introduces a sizable wedge between firms and workers, and between firms and capital owners.

To explore the prospect for a carbon tax to reduce preexisting distortions, in F4 we use the new tax revenues not for lump sum transfers, but to cut all existing tax rates in the same manner as in section 9.5.1. The path of carbon prices in this case is the same as that in F2, 10 yuan per ton of CO_2 rising to 50 yuan per ton. Looking at the 2020 effects in table 3.4, we see that this alternative use of revenues yields a big change in the composition of aggregate GDP. On one hand, households no longer receive a large transfer, and consumption is essentially unchanged from the base case, compared to the 0.27% increase in F2. On the other hand, enterprises receive a large cut in the tax on profits. With higher retained earnings, investment is 0.57% higher than in the base case, compared to no change in F2.

GDP falls slightly in the first year compared to the base case (−0.02%, table 3.3), but with the higher investment, GDP by 2020 is slightly higher (+0.02%, table 3.4). That is, the higher capital stock offsets the negative impact of the carbon tax. Given that the GDP in F4 is slightly higher than in F2, the energy use and emissions are

Box 3.1
Effects of two additional scenarios: F5 and F6

We consider two more carbon tax scenarios to evaluate their economic effects, without the fuller benefit assessment described in the rest of this chapter. These two cases explore the impacts of using revenues to protect vulnerable industries, and the effect of carbon pricing in the rest of the world on a carbon tax in China.

Scenario F5   In the European Union and the United States, representatives of energy-intensive industries—such as thermal power generation, iron and steel production, and cement manufacturing—have been among the most vocal in their opposition to carbon price policies because they can substantially raise their costs. For example, a modest tax of US$15 per ton of CO_2 in the United States has been projected to raise the costs for energy-intensive industries by 2–6% (Adkins et al. 2010). Such increases in costs are widely feared to reduce employment and profits significantly, particularly in industries competing in international markets. In response to such concerns, plans such as the Waxman-Markey cap-and-trade bill in the United States contained subsidies to the most vulnerable sectors, the energy-intensive, trade-exposed (EITE) industries. In that bill, EITE industries included any whose energy costs were projected to rise by more than 5% at a carbon price of US$20 per ton, and whose "trade intensity"—the value of exports and imports divided by the value of output of that industry—was more than 15%.[15]

Box 3.1 (continued)

In F5 we follow the same path of carbon prices as in F2, but use a part of the new revenues to subsidize the EITE manufacturing industries, calculating output subsidy rates similar to the Waxman-Markey mechanism. Determined as described in an endnote,[16] these are reported in the following table. The most EITE sector in China is production of nonmetal mineral products (including cement), which would receive a 0.25% subsidy rate, followed by the primary metal and chemical industries, as shown in the table. At these rates, the subsidies do not recycle all the carbon tax revenue, and the remaining revenue is transferred in lump sums back to households.

Subsidies given to energy-intensive industries in Scenario F5 in year 1

Number	Industry	CO_2 Embodied in Fuels and Electricity Used (million tons carbon)	Output in 2007 (billion yuan)	EITE Subsidy Rate (percent)
6	Food products and tobacco	34.47	4179	0.03
7	Textile goods	33.27	2520	0.05
8	Apparel and leather	7.44	1807	0.01
9	Sawmills and furniture	10.63	1099	0.04
10	Paper products and printing	31.09	1493	0.08
11	Petroleum refining and coking	66.87	2107	0.12
12	Chemicals	240.40	6200	0.14
13	Nonmetal mineral products	155.65	2280	0.25
14	Primary metals	276.52	6110	0.16
15	Metal products	30.48	1771	0.06
16	Machinery and equipment	40.62	3949	0.04
17	Transport equipment	18.26	3298	0.02
18	Electrical machinery	11.17	2716	0.01
19	Electronic and telecom equipment	14.95	4119	0.01
20	Instruments	1.49	488	0.01
21	Other manufacturing	7.56	1055	0.03

Notes: Based on the initial carbon price of 10 yuan per ton of CO_2; numbers in the left-hand column correspond to those shown in table 2.1 for all 33 production sectors in the economic model.

Box 3.1 (continued)

The subsidy rates look modest, but the impact compared to F2 is noticeable in table 3.5. The carbon-intensive industries do not need to raise prices to the extent shown in figure 3.6, and thus experience a much smaller output loss. Coal consumption falls by less (−21.3% in 2020 versus −23.0% in F2), and total energy consumption falls by only 16.2% compared to 17.7% without the subsidy. Annual CO_2 emissions are thus only 17.4% lower than the base case, compared to 18.9% in F2; cumulative fossil CO_2 emissions are 65.67 gigatons, versus 64.79 without the subsidy.

The smaller amount of transfers to households means that the change in consumption is smaller (0.15% versus 0.27%). The net effect of this smaller change in prices of energy-intensive goods and smaller consumption is a smaller loss in GDP; by 2020, GDP is only 0.07% smaller than the base case, compared to the 0.14% loss without the subsidies in F2. Other effects of the smaller change in prices are a weaker appreciation of the terms of trade and a fall in exports that is much less than that in F2.

Scenario F6   In the first five scenarios, we assume that the carbon price is imposed unilaterally by China. This action would make the prices of energy-intensive Chinese goods rise relative to their world prices. In the best case, however, carbon prices would be imposed within a multilateral framework, in which a number of countries would act in concert. Countries need not necessarily impose the same price on emissions, but to give a simple illustration of the impact of international carbon pricing, in F6 we raise the prices of goods traded between China and the rest of the world by the same amount that we estimate for the domestic prices.[17] F6 follows F2 in other respects, including the path of carbon prices and the recycling of revenues directly to households. While this topic would be better investigated using a global economic model, as in Adkins et al. (2010), this adjustment of the import and export prices allows a quick estimate of the impacts.

The results for the first year, 2013, are shown in the last column of table 3.3. When the prices of energy-intensive goods in the rest of the world are high, the impact of the Chinese carbon tax on their net exports is smaller; in F2 the higher prices of Chinese goods reduces net exports of those goods significantly, while this effect is dampened in the multilateral tax case. This case is complicated by the assumption that the trade balance is held fixed between the base case and policy case; this implies that any reduction in net exports of energy-intensive goods has to be offset by changes in the terms of trade and changes in the net exports of the non-energy-intensive goods. A simple scenario where only world prices are raised will generate an appreciation of the terms of trade and thus lower net exports.

The following figure shows the changes in industry output as a result of these interactions in the two scenarios; we show them only for the manufacturing and utility sectors to avoid a cluttered plot. The changes for F2 (white bars) for the energy-intensive industries—refining, chemicals, nonmetallic mineral products, primary metals—are bigger than those for the multilateral tax scenario, F6 (dark bars). The changes for the less energy-intensive industries (e.g., apparel, machinery, electrical

Box 3.1 (continued)

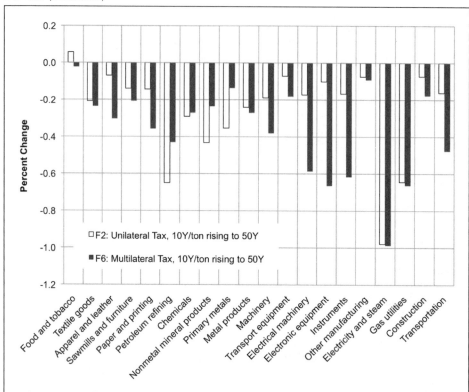

Changes in industry output in 2020, under a unilateral and multilateral carbon tax (scenario F2 versus F6).

machinery) are smaller in F2. The nontraded industries such as electric and gas utilities show only tiny differences between the two scenarios.

In the aggregate the effects are small, as shown in table 3.3: GDP, consumption and investment in 2013 in F6 are very similar to those of the unilateral carbon tax in F2. With the smaller reduction in the energy-intensive sectors, the fall in energy consumption is smaller in F6: 4.38% versus 4.51%. As a result carbon emissions are also a little smaller, 4.69% versus 4.82%. The key point of this simple case is that the industries affected most by a carbon price will suffer less in a multilateral tax system than under a unilaterally imposed tax.

slightly higher; the change in coal use in 2020 is –22.9% compared to –23.0% in F2, and the change in total fossil energy use is –17.5% versus –17.7%. While we might have expected larger differences in energy use, keep in mind that prices are not identical in the two scenarios. With a higher demand for oil than in F2, the price of oil is higher and the terms-of-trade effect is milder: exports decline 0.81% in F4 versus 1.12% in F2. These higher energy prices help to curb demand for energy, even as the composition of activity switches away from consumption toward investment, which has a higher proportion of energy-intensive goods such as cement and steel.

3.4   Effects on Emissions

The economic simulations of carbon tax policies in the prior section generate for each policy not only the costs, but also the consumption of energy, including fossil fuels, that are the starting point for evaluating the environmental benefits. The first step is estimating the emissions that result in each scenario (as illustrated in figure 3.1).

We focus on emissions of CO_2—any reduction of which is regarded as a benefit by itself—and of key air pollutant species: sulfur dioxide (SO_2), nitrogen oxides (NO_X), and particle matter less than 10 microns and 2.5 microns in diameter (PM_{10} and $PM_{2.5}$). We need to know emissions of the latter four species, along with others noted in chapters 6 and 7 not emphasized here, in order to analyze their effects on air quality, and ultimately on public health and crop productivity.

Note that in section 3.3 we already reported some estimates of fossil fuel emissions of CO_2, in terms of cumulative totals for the tax policies over 2013–2020, so that we could define several of the policies. Here we provide more detail about the CO_2 results.

We first recapitulate how we estimate emissions in general. (This procedure is summarized in section 2.5 and described in detail in chapters 4–6, amended by new assumptions noted in appendix C.) Emissions are calculated "bottom-up," combining information on (1) fuel consumption and industrial production by sector and region; (2) unabated emission factors, expressed in the mass of pollutant per unit fuel combusted or per unit output; and (3) removal efficiencies of relevant emission control technologies. This information must be compiled from the literature on all emitting processes, covering diverse industries, transportation modes, and household, commercial, and agricultural practices. All emission processes, additionally, must be geographically located so that they can serve as inputs in the atmospheric simulations described in section 3.5. This difficult and data-intensive research draws on much prior work by the authors of chapters 4–6 and others published in peer-reviewed literature, as cited in those chapters.

3.4.1   2007 Base Year and 2020 Base Case

The estimated emissions of CO_2 from combustion of fossil fuels are 6703 million tons (Mt) in the 2007 base year. For reference, total CO_2—including process emissions from cement and aluminum production, and combustion emissions from household biofuels and agricultural biomass—is estimated at 8476 Mt in 2007. In the tax simulations we only consider fossil fuel emissions.

These base year values are higher than often-cited top-down estimates of Chinese CO_2 emissions, as shown in figure 1.8 and discussed in section 1.5.1 of chapter 1. They result from application of the same databases and methods already introduced, but further developed by contributors to this book into what may be the first full bottom-up estimate of Chinese CO_2 in the literature (Zhao, Nielsen, and McElroy 2012); the study capitalizes on new field research of CO_2 emission factors and includes preliminary validation by instrumental measurements of CO_2 and CO. The CO_2 projections into the future, as noted before, are generated by fossil fuel consumption in the economic model, scaled to the bottom-up base year total.

The total SO_2 emissions in our base year of 2007 are estimated at 30.6 Mt, 5.9 Mt higher than the official estimate (NBS and MEP 2010). Estimated emissions of PM_{10} are 18.8 Mt, $PM_{2.5}$ 13.3 Mt, and NO_X 23.8 Mt. Specifically, SO_2 emissions from coal-fired power plants are estimated to have declined from 16.1 Mt in 2005 to 14.2 Mt in 2007, and PM_{10} emissions from 1.8 to 1.6 Mt. These declines are attributed to national policies in energy saving and emission controls during the 11th FYP that are analyzed elsewhere in this book. NO_X emissions from power plants, which were not targeted in the 11th FYP, are estimated to have increased from 7.0 to 8.2 Mt.

For the base case in 2020, the estimated emissions of all species are shown in table 3.5. National CO_2 emissions from fossil fuel combustion are projected to increase 63.6% to 10,966 Mt, from 6703 Mt in 2007. A notable but cautionary[18] result is that in terms of fossil fuel carbon intensity (CO_2 emissions divided by GDP), this projection translates to a 44.8% reduction compared to the base year of 2007, or a 49.7% reduction compared to 2005.[19] Recall that in 2009, China committed domestically to reduce its carbon intensity 40–45% by 2020, relative to 2005. While we only consider carbon emissions from fossil fuel combustion, this is the dominant source of CO_2 and it suggests, caveats duly noted, that existing policy as represented

Table 3.5
Emissions in the base case and scenarios F1–F4 for 2020 (million metric tons per year)

	Base	F1	F2	F3	F4
CO_2	10,966	9601	8888	7590	8908
SO_2	24.7	21.2	19.4	16.1	19.4
NO_X	26.2	23.3	21.9	19.2	21.9
$PM_{2.5}$	9.94	9.04	8.57	7.73	8.58
PM_{10}	14.0	12.5	11.8	10.4	11.8

in the base case—without any additional policy such as a carbon tax—may be sufficient to achieve China's target.

Sulfur dioxide emissions are projected to decline 19% compared to 2007, driven by extension of policies to control SO_2 to additional power plants and new applications of flue gas desulfurization in other industrial sectors such as iron and steel (see appendix C). The national share of emissions from coal-fired electric power will decline to 33%. Moreover, national emissions of primary PM_{10} and $PM_{2.5}$ will decline by 25% and 26%, respectively, chiefly because of the continued transition in the cement industry from shaft kilns to precalciner kilns (see Lei et al. 2011; and chapter 5 and appendix C).

Despite the proposed NO_X emission control policies that we incorporate (mainly in the electric power sector), national NO_X emissions are projected to increase by 10% in the base case as a result of the growth in energy consumption in industrial sectors. Specifically, NO_X emissions will decrease by 9% for coal-fired power plants but increase by 37% in other industries. The effects of the swift increase in fuel consumption in transportation will likely be counteracted by the falling emission factors that result from the implementation of stricter emission regulation for both on-road and nonroad sources (see appendix C). The shares of NO_X emissions from coal-fired electric power plants, industry, transportation, and residential and commercial combustion (including open burning of biomass) are projected at 28%, 42%, 21%, and 9%, respectively.

3.4.2 2020 Carbon Tax Cases

The imposition of carbon taxes would lead to major reductions compared to the base case in emissions not only of CO_2—the primary aim of any carbon tax—but also of all other key species of interest. To show the effects of the carbon tax policies on emissions of the key species, figure 3.9 gives the total emissions in the 2007

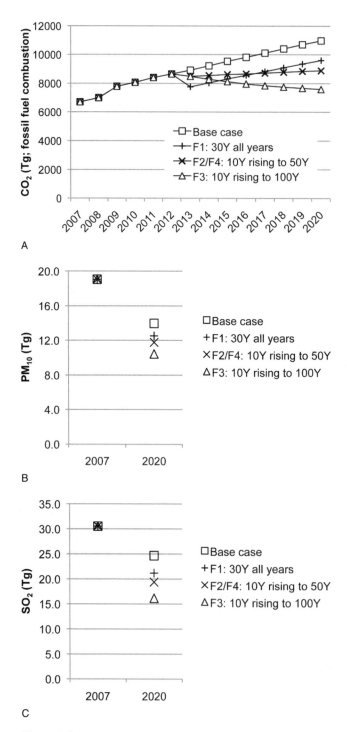

Figure 3.9
Total estimated emissions of CO_2 and key air pollutants under the base case and Scenarios F1–F4. (A) CO_2. (B) PM_{10}. (C) SO_2. (D) NO_X.

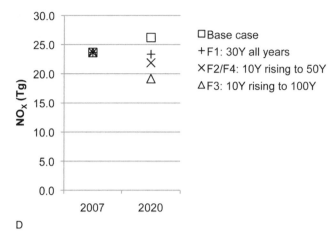

D

Figure 3.9 (continued)

base year and in 2020 for the base case and four main tax scenarios, and also for the intervening years for fossil fuel CO_2.[20]

In figure 3.9A we see the continued growth in fossil fuel CO_2 emissions already described for the base case, even as carbon intensity declines (as discussed in section 3.2.1). Beginning in 2013, the first year of tax imposition in all of our main policy scenarios (F1 to F4), the effect of the tax on CO_2 is obvious. While the sudden 30 yuan per ton tax in F1 would immediately drive annual emissions down in 2013, in subsequent years the growth rate would return almost back to the base case rate. By 2020, CO_2 emissions would be 11% higher than the 2012 level, but 12% lower compared to the base case 2020 level. The initially small, but rising, taxes of F2 and F4 would elicit sustained reductions of growth in annual fossil CO_2 emissions, leading to a 19% reduction compared to the base case in 2020, though still a 3% increase over the 2012 level. Only in F4, where the tax begins small and then approaches international carbon prices by 2020, would the policy drive total emissions down year on year, achieving a 12% reduction over 2012 emissions in 2020 and 31% reduction relative to the base case. Indeed in this case, 2012 would be China's peak year of fossil CO_2 emissions.

(Readers should keep in mind the suggestion in section 3.2 that, given the atmospheric lifetime of CO_2, a stronger analytical metric for comparing the CO_2 effects of the different carbon taxes is the cumulative emissions from 2013–2020, as discussed earlier and shown in figure 3.4. We report the effects on annual emission levels partly because of their relevance to policy metrics and debate.)

As for emissions of PM_{10} and SO_2 (figures 3.9B,and 3.9C), the effectiveness of existing policies on energy efficiency and emissions is reflected in the reduction of PM_{10} and SO_2 in the base case for 2020 versus 2007 discussed earlier, while figure 3.9D shows the difficulty reining in emissions of NO_X. Importantly, however, each of the carbon tax policies would sharply improve these pollutant trajectories by 2020. The expected declines in emissions of SO_2 and PM_{10} in the base case would be enhanced: F1 would reduce 2020 SO_2 emissions 14% and PM_{10} emissions 10% relative to the base case, F2 and F4 would reduce them 21% and 16%, respectively, and F4 would reduce them 34% and 25%. In the case of NO_X, moreover, the carbon taxes would arrest the anticipated growth in annual emissions: F1 would lead to a slight decline by 2020 compared to 2007 levels (−0.3 Mt) and an 11% reduction relative to the base case, F2 and F4 would generate a somewhat greater reduction from 2007 (−1.8 Mt) and 17% lower than the base case, and F3 a very significant reduction from 2007 (−4.5 Mt), 27% less than the base case.

We forgo representing our emission estimates geographically in this chapter because the results would appear quite similar to those presented in figures 2.6, 2.7, and 6.9.

3.5   Effects on Pollution Concentrations

To briefly recapitulate the next step in the benefit research, as represented in figure 3.1, we must translate gridded emission estimates to atmospheric concentrations, including both physical transport and chemical transformation in the air. Recall that we are interested not only in primary pollutants such as particles and SO_2, emitted directly from smokestacks and tailpipes, but also secondary pollutants formed chemically in the air from precursor species, such as sulfate particulates and ozone.

To conduct these analyses, we use a model of Chinese regional air quality that was developed by our program over more than a decade, nested within a global chemical transport model, GEOS-Chem. The China model has been tested against observations by ground stations, aircraft, and satellites, and employed in a number of studies of atmospheric chemistry in China, as reported in peer-reviewed scientific articles cited in chapters 1 and 7. Our focus is simulating the concentrations of SO_2, NO_X, $PM_{2.5}$, and ozone (O_3), the latter two causing the dominant damages to public health and crop productivity. The model is further introduced in section 2.1 and discussed in detail, along with important scientific considerations in our atmospheric research, in chapter 7.

Both the simulation of air quality and analysis of health and crop benefits of reduced pollution exposures are time-consuming research endeavors, and we do not aim to complete these steps for all our scenarios. Instead, we consider the value of different comparisons, concluding that the best set for full atmospheric assessment is the base case and F2, F3, and F4. As shown in table 3.5, however, our bottom-up emission assessment unexpectedly resulted in very similar emission profiles for F2 and F4 for all the concerned species, because these two scenarios yield similar paths in energy consumption. Indeed the emissions are far too close to merit separate atmospheric simulations for these two scenarios, and we can simply treat them as equivalent in this section and the next one.

3.5.1   Effects of Scenarios F2 and F4 on Pollution Concentrations

Figure 3.10 summarizes monthly mean changes in surface concentrations of (A) NO_X, (B) SO_2, (C) O_3, and (D) $PM_{2.5}$ for four months, representing the four seasons in 2020, between the 2020 policy cases F2 and F4 versus the base case. The changes are expressed as the level of the policy case minus the level of the base case.[21] Negative values indicate that the policy measures in the F2/F4 scenarios reduce pollutant levels at the surface compared to the base case.

The model results in figures 3.10A and 3.10B show that concentrations of both NO_X and SO_2 are reduced over all regions of China and in all seasons in F2/F4 compared to the base case, which is to be expected as total national emissions of NO_X and SO_2 are estimated to decrease by 17% and 21%, respectively. The changes in NO_X and SO_2 are significant in absolute terms over east China[22]; the annual mean decrease over this region is about 1.2 parts per billion by volume (ppbv, a standard concentration metric for pollutant gases) for NO_X and 1.5 ppbv for SO_2. To gain some perspective on these changes, compare them to the base concentrations in 2010 shown in figure 7.4. The reductions of NO_X and SO_2 are highest in the winter versus the base case, and lowest in the summer. As discussed in chapter 7, because the emission data used in the model are annual averages distributed uniformly across months, the seasonal variability of the induced differences in NO_X and SO_2 is driven entirely by the seasonal variability of the atmospheric processes—chemistry and transport (or dynamics)—that affect the lifetimes of these two gases.

Emissions of primary $PM_{2.5}$ are lower by 14% in the F2/F4 cases than in the base case. As a result of reductions of both primary and secondary $PM_{2.5}$ levels (the latter from lowered emissions of NO_X and SO_2, the chemical precursors of $PM_{2.5}$), $PM_{2.5}$ declines throughout China and in all seasons compared to the base case, shown in figure 3.10D. Relatively large reductions are simulated for all of east China

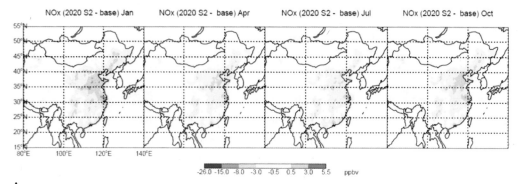

A

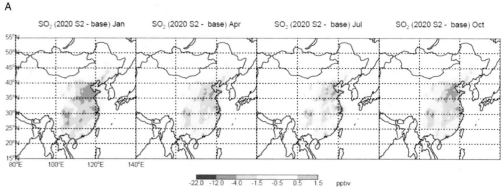

B

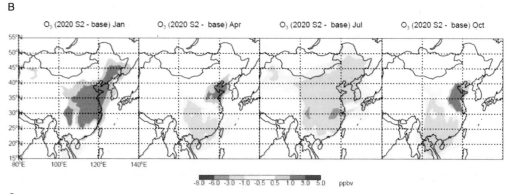

C

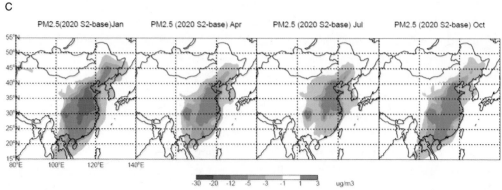

D

Figure 3.10
Simulated changes in surface concentrations of air pollutants due to carbon tax scenarios F2 and F4, by season. Concentrations in 2020 resulting from scenario F2 or F4 versus the base case, in January, April, July, and October. Negative values are reductions. (A) NO_x in ppbv. (B) SO_2 in ppbv. (C) O_3 in ppbv. (D) $PM_{2.5}$ in $\mu g/m^3$.

extending from the North China Plain to the Pearl River Delta, where the annual mean reduction of $PM_{2.5}$ is $4.2\,\mu g/m^3$ (micrograms per cubic meter) in F2 or F4 compared to the base case. The seasonal variations in these differences are again driven by atmospheric dynamics and chemistry affecting secondary $PM_{2.5}$ formation. The maximum reduction in absolute levels of $PM_{2.5}$ occurs in January, reaching a quite large $6.1\,\mu g/m^3$ average over east China.

The carbon tax policies F2 and F4 do not always lead to reductions in ozone concentrations, for reasons discussed in section 3.5.3 and in detail in chapter 7. Over east China, concentrations of O_3 rise in January and decline in July, as shown in figure 3.10C. The mean changes in ozone in F2/F4 over east China are $+0.8\,ppbv$ in January and $-1.7\,ppbv$ in July, about $+2\%$ and -3.5% of the base case level, respectively. The largest decrease in monthly mean ozone occurs around 30°N in east China in July, with a maximum of 3–4 ppbv reduction, or about 7.5% of the base case level. In both April and October, changes in ozone transition from increases over the North China Plain to decreases south of the Yangtze River.

3.5.2   Effects of Scenario F3 on Pollution Concentrations

The stronger atmospheric effects of the higher tax rates of F3 are revealed in figure 3.11, again summarizing monthly mean changes from the base case in surface concentrations of (A) NO_X, (B) SO_2, (C) O_3, and (D) $PM_{2.5}$ for the seasonally representative months of 2020.[23] Negative values indicate that the policy measures in the F3 scenario reduce pollutant levels at the surface compared to the base case.

Figures 3.11A and 3.11B show that both NO_X and SO_2 are again lower over all regions of China and in all seasons in F3, compared to the base case, decreasing nationally by 27% and 34.5%, respectively. The annual mean decrease over east China is 2 ppbv for NO_X and 2.4 ppbv for SO_2. The largest reductions of NO_X and SO_2 due to F3 occur in the winter, and the lowest in the summer.

Emissions of primary $PM_{2.5}$ are lower, by 22%, in F3 compared to the base case, and there is a decrease in concentrations of $PM_{2.5}$ throughout China and in all seasons (figure 3.11D). Relatively large reductions are simulated for all of east China extending from the North China Plain to the Pearl River Delta, where the annual mean reduction of $PM_{2.5}$ can reach a very significant $7.2\,\mu g/m^3$, 69% larger than the

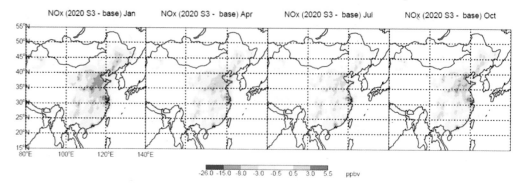

A

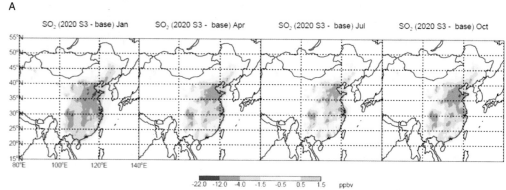

B

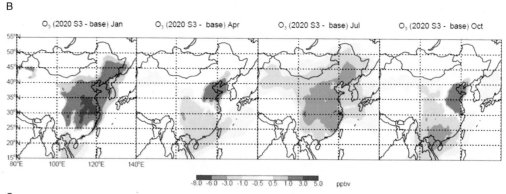

C

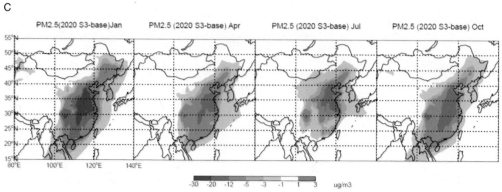

D

Figure 3.11
Simulated changes in surface concentrations of air pollutants due to carbon tax scenario F3, by season. Concentrations in 2020 resulting from scenario F3 versus the base case, in January, April, July, and October. Negative values are reductions. (A) NO_x in ppbv. (B) SO_2 in ppbv. (C) O_3 in ppbv. (D) $PM_{2.5}$ in $\mu g/m^3$.

reduction in the F2 case. The maximum reduction in absolute levels of $PM_{2.5}$ occurs in January, reaching an average of $10.3\,\mu g/m^3$ over east China.

As in other carbon tax cases, F3 does not always induce a decline in ozone concentrations. Over east China, ozone rises in January and declines in July compared to the base case, shown in figure 3.11C. The mean changes in F3 over east China are +1.3 ppbv in January and –3 ppbv in July, about +3.2% and –6% of the base case levels, respectively. The largest decrease in monthly mean ozone occurs around 30°N in east China in July, with a 6–7 ppbv reduction, or about 13% of the base case level. F3 shows the same geographical characteristics as F2/F4 in both April and October, where it transitions from increases over the North China Plain to decreases south of the Yangtze River.

3.5.3   Comparison of Scenarios F2 and F4 versus Scenario F3

Comparing the changes in annual mean SO_2 concentrations in F2/F4 versus F3, we find that the reduction corresponds almost linearly to emission reductions. The same linearity holds also for NO_X. The reason is that primary emissions are dominant contributors to the two species. The reductions in annual mean $PM_{2.5}$ over east China, however, are difficult to analyze similarly with respect to emissions because they result from reduced emissions not only of primary $PM_{2.5}$ but, to a larger extent, of SO_2 and NO_X (precursors to sulfate and nitrate forms of $PM_{2.5}$).

Concentrations of ozone similarly do not vary linearly with reductions of O_3-precursors, because of the chemical mechanisms presented in chapter 7. However, as only the emissions of NO_x are reduced in the 2020 policy cases from the base case levels, and not other O_3 precursors including volatile organic compounds (VOCs),[24] the change in ozone in F2/F4 or F3 can be used to indicate the chemical regimes of ozone production (specifically, whether they are NO_X-limited or VOC-limited, as presented in chapter 7). The simulations indicate that under the 2020 emission scenarios (including the emission levels of VOCs in the 2007 base year), the North China Plain region lies in a VOC-limited regime except in summer, and the Pearl River Delta region in a NO_X-limited regime except in winter. This is consistent with the results found for the 2010 emission scenarios in chapter 7. All of east China is in a NO_X-limited regime in summer.[25]

3.6   Effects on Health and Agricultural Damages of Air Pollution

In this section we estimate the net effect of the changes in the $PM_{2.5}$ and ozone concentrations due to the carbon tax policies on public health and productivity of major grain crops. This last step (see figure 3.1) follows the introduction to these topics in section 2.7 and detailed discussion in chapter 8. Some adjustments in methods and revisions to assumptions for both the health and crop benefit research are described in appendix C.

The framework for evaluating the health impacts, using the U.S. EPA assessment tool BenMAP, is illustrated in figure 2.13. In a nutshell, health damages—in terms of both premature mortalities and morbidity (disease) endpoints—are estimated taking account of (1) the simulated pollution concentrations; (2) the population distribution in China, updated to 2020; (3) the base incidence rates of nonaccident mortality and other health endpoints; and (4) concentration-response (C-R) relationships from the epidemiological literature. The same conclusions about applicable C-R functions drawn in chapter 8 are directly applied here. This is one of the most important steps in determining our estimates of the ancillary benefits of the policies.

The crop impact assessment focuses on ozone effects on three primary grains—wheat, rice, and maize—and uses a framework that is analytically similar to that of the health assessment. One important distinction is the need to consider greater temporal resolution in both pollutant concentrations (because ozone exhibits such strong daily and seasonal variations) and crop growth (which is, naturally, highly seasonal). Important judgments on crop exposure indices and exposure-relative yield relationships are made in chapter 8 based on published literature, and are again applied in this chapter.

3.6.1   Health Benefits

The estimated health benefits for each endpoint are summarized in table 3.6, with lower and upper bounds reflecting uncertainties in the underlying epidemiological studies. Compared with the base case, improvement of air quality in the carbon tax policy scenarios (F2/F4 and F3) would lead to 13,800 and 23,200 avoided premature deaths from acute effects of $PM_{2.5}$ exposures in 2020, respectively, and 5,500 and 9,900 premature deaths from acute effects of ozone exposures in 2020. The estimates of avoided deaths from chronic effects of $PM_{2.5}$ exposures are 83,200 and 139,000 for the emission reductions of F2/F4 and F3, respectively. (As discussed in chapter 8, health scientists consider the chronic impact pathway, with its larger

Table 3.6

Health benefits of carbon taxes in scenarios F2/F4 and F3

Units: Number of cases avoided compared to the base case in 2020. In addition to the mean estimates, lower and upper bounds representing the 95% confidence intervals (CIs) are also calculated based on the published CIs for relative risk in the corresponding literature, as discussed in chapter 8.

Pollutant	Health Endpoints	F2/F4			F3		
		Mean	Lower	Upper	Mean	Lower	Upper
$PM_{2.5}$	Acute mortality	13,800	8,500	19,100	23,200	14,300	32,100
	Chronic mortality	83,200	23,700	142,300	139,000	39,800	237,200
	Hospital admissions (cardiovascular)	24,100	5,900	42,200	40,500	10,000	70,800
	Hospital admissions (respiratory)	43,200	29,800	56,600	72,400	49,900	94,800
	Outpatient visits, all causes	5,320,000	2,120,000	8,520,000	8,930,000	3,560,000	14,310,000
Ozone	Acute mortality	5,500	1,500	9,500	9,900	2,700	17,100
	Hospital admissions (cardiovascular)	25,900	18,400	33,500	46,800	33,200	60,300
	Hospital admissions (respiratory)	29,600	24,000	35,200	53,200	43,200	63,300

effects, more indicative of total mortality risks. Estimating this effect in China, however, requires application of C-R results from U.S. studies. This introduces additional uncertainty compared to estimation of acute mortality, which is based on epidemiological studies of Chinese populations and conditions. We think it is best to acknowledge these uncertainties by reporting both sets of damage estimates for $PM_{2.5}$ exposures, one based on the acute effect and the other on the chronic effect.) The avoided premature deaths from acute effects of $PM_{2.5}$ and O_3 reductions due to the F2/F4 and F3 carbon taxes would account for 0.25% and 0.42%, respectively, of all-cause deaths; the percentages due to estimated chronic effects of $PM_{2.5}$ exposures are 1.06% and 1.78%.

The benefits of the policies in terms of population risks of morbidity are also significant, although not as striking. The reduced pollution levels due to the carbon taxes of F2 or F4 would lead to an estimated average of 123,000 avoided cases of hospital admissions and 5.3 million cases of outpatient visits, and those due to F3 would avoid 213,000 cases of hospital admissions and 8.9 million cases of outpatient visits.

We also represent the geographical distributions of avoided acute mortality due to changes in $PM_{2.5}$ exposures for F2/F4 and F3 in figure 3.12, and those due to changes in ozone exposures in figure 3.13. Compared to the base case, acute mortality due to $PM_{2.5}$ and ozone exposures is reduced in most inhabited areas of China. This statement is true even for ozone exposures, except for a few megacities including Beijing, Tianjin, and Shanghai, and the smaller city of Dalian, despite increases in ozone that result from the policies in some regions and seasons, as shown in figures 3.10 and 3.11, to be explained in the following paragraphs.

As shown in table 3.6, the health benefit of avoided acute mortality is mainly due to the decline of $PM_{2.5}$ exposures. The areas with the largest such benefits are unsurprisingly located in densely populated metropolitan areas such as Beijing, Tianjin, Shanghai, Guangzhou, and Chengdu. Regionally, the benefits of reduced $PM_{2.5}$ are very significant in areas centered on Henan-Hebei-Shandong, southern Jiangsu, and the Sichuan basin.

Compared with our previous estimates of health benefits in 2010 from a carbon tax policy (chapter 8), both the absolute and relative levels of avoided deaths due to reduced ozone are much larger. While the geographical patterns of ozone reductions or increases are similar in figure 7.6C compared to figures 3.10C and 3.11C, note that the scales differ and the reductions in 2020 are in fact quite a bit larger. We might attribute this difference to more significant reductions of NO_X emissions in the 2013–2020 cases, with 17% reductions in F2 and F4 and 27% in F3 versus

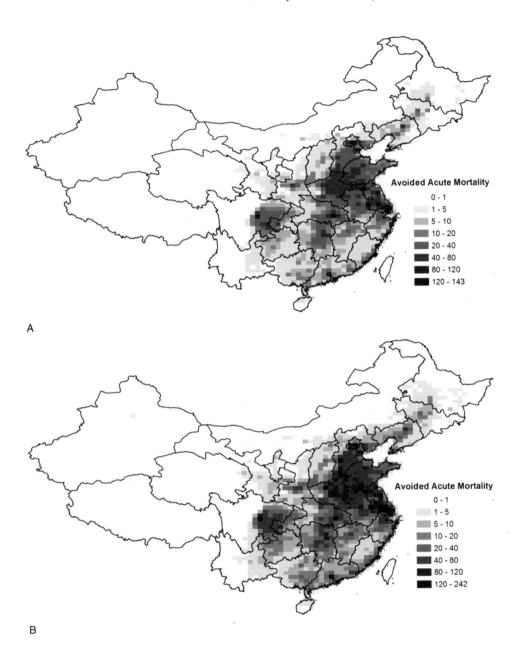

Figure 3.12
Cases of acute mortality avoided in 2020 from changes in $PM_{2.5}$ exposure under a carbon tax compared to the base case. (A) Scenarios F2/F4. (B) Scenario F3.

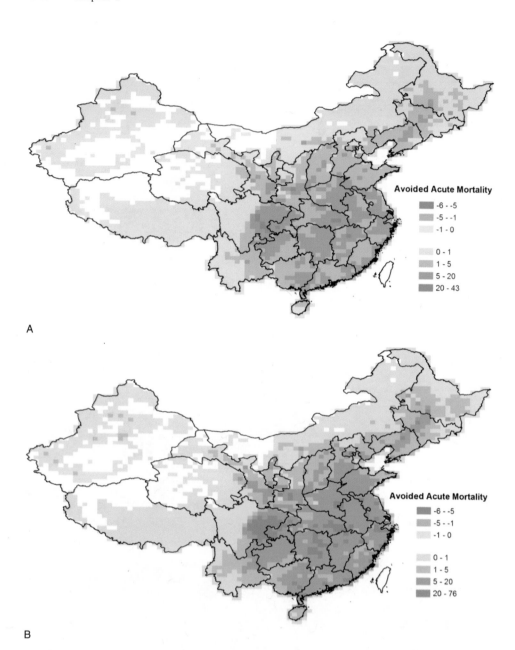

A

B

Figure 3.13
Cases of acute mortality avoided in 2020 from changes in ozone exposure under a carbon tax compared to the base case. (A) Scenarios F2/F4. (B) Scenario F3.

the base case (see section 3.4.2). These lead to annually averaged ozone decreases of 0.48 ppbv and 0.84 ppbv, respectively, in heavily inhabited areas of China.

3.6.2   Crop Benefits

Estimates of the benefits to rice, wheat, and maize production due to the carbon tax policies are summarized in table 3.7. The estimates are based on averaging the effects calculated using several ozone exposure indexes (M12, SUM06, W126, and AOT40) and corresponding exposure-relative yield relationships, as discussed in chapter 8. We also present the geographical distributions of the benefits to rice, wheat, and maize production in figures 3.14–3.16.

The estimates of benefits to wheat production of the carbon tax, depending on the exposure index, vary from 0.13 to 2.17 Mt in F2/F4 and 0.33 to 4.44 Mt in F3. The estimates for rice vary from 1.00 to 2.81 Mt in F2/F4 and 1.68 to 4.80 Mt in F3, and for maize from 2.50 to 5.36 Mt F2/F4 and 4.28 to 9.36 Mt in F3. Increased grain production would occur throughout most of China in both F2/F4 and F3 compared with the base case, noting that there would be reduced production of wheat in the Hebei-Henan-Shandong area. This interesting effect results from a coincidence of elevated ozone levels in this area and the growing season of winter wheat, namely, March through May.

We compare our estimates to China's total grain production in 2007. We find that the carbon tax would increase wheat, rice, and maize production in China by around 0.8%, 1.0%, and 2.7%, respectively, in F2/F4, and around 2.0%, 1.7%, and 4.7% in F3. These increases appear moderate but notable compared to total national production; in some regions the effect could exceed 10%, which would be very significant locally.

Finally, we also calculate the monetary value of the agricultural benefits from the carbon tax by multiplying the increased yields by projected prices for the grains. To estimate these prices, we use the projection of the average price for national agricultural production from the economic model described in section 3.3 and appendix C. The total estimated value of the grain yield benefits is 15.7 billion yuan in F2/F4 and 28.3 billion yuan in F3.

3.7   Comparison of the Effects of the Carbon Tax Scenarios

To summarize our results, we compare key effects in 2020 of the carbon tax policy options in table 3.8. Keep in mind that these results are snapshots of the impacts in the eighth year of the tax, and therefore do not capture all the effects that occur

Table 3.7
Grain production benefits of carbon taxes in Scenarios F2/F4 and F3
Units: Increase of crop yields (thousand tons) compared to the base case in 2020, calculated for four different ozone exposure indices (M12, SUM06, W126, and AOT40). See section 8.2 for discussion of the indices.

	F2/F4					F3				
	M12	SUM06	W126	AOT40	Average	M12	SUM06	W126	AOT40	Average
Wheat	133.0	708.8	2165.4	573.3	895.1	325.5	2529.5	4437.4	1303.7	2149.0
Rice	995.8			2808.1	1902.0	1682.4			4796.4	3239.4
Maize	2495.0	5362.3	4516.9		4124.7	4277.5	9363.9	7697.6		7113.0

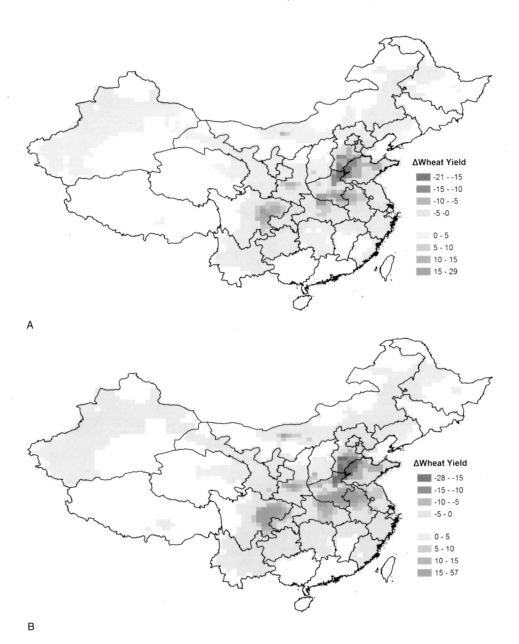

A

B

Figure 3.14
Changes of wheat yields in 2020 due to a carbon tax compared to the base case (in metric tons). (A) Scenarios F2/F4. (B) Scenario F3.

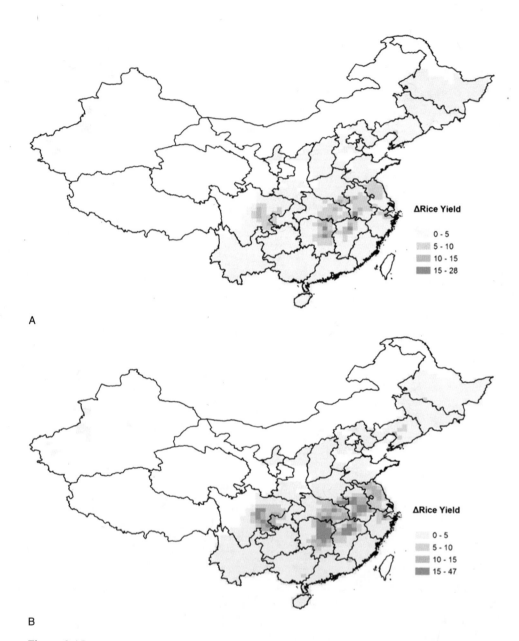

A

B

Figure 3.15
Changes of rice yields in 2020 due to a carbon tax compared to the base case (in metric tons). (A) Scenarios F2/F4. (B) Scenario F3.

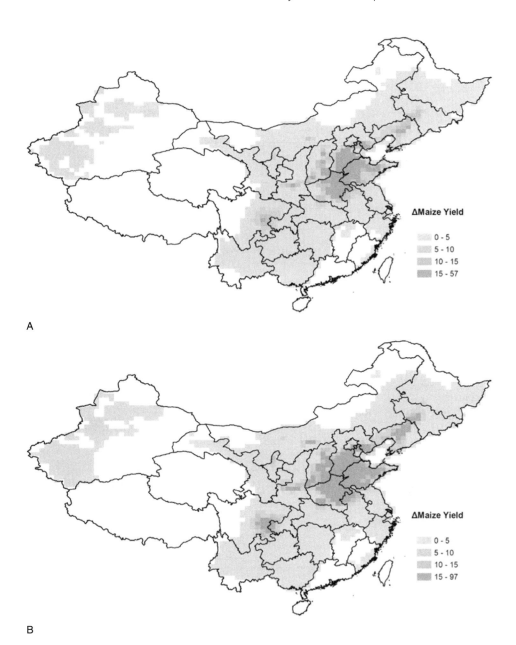

Figure 3.16
Changes of maize yields in 2020 due to a carbon tax compared to the base case (in metric tons). (A) Scenarios F2/F4. (B) Scenario F3.

Table 3.8

Comparative summary of major effects in 2020 of four carbon tax policies versus the base case

Variable	Scenario F1: 30Y/ton all years Lump Sum Offset	Scenario F2: 10Y rising to 50Y/ton Lump Sum Offset	Scenario F3: 10Y rising to 100Y/ton Lump Sum Offset	Scenario F4: 10Y rising to 50Y/ton Tax Cut Offset
	Percent Change			
GDP	−0.09	−0.14	−0.33	0.02
Consumption	0.17	0.27	0.39	0.00
Investment	0.01	0.01	−0.04	0.57
Coal use	−15.2	−23.0	−37.2	−22.9
CO_2 emissions	−12.4	−18.9	−30.8	−18.8
PM_{10} emissions	−10.3	−15.7	−25.3	−15.6
SO_2 emissions	−14.1	−21.4	−34.5	−21.3
NO_X emissions	−10.9	−16.6	−26.9	−16.4
Avoided premature deaths	Number of Cases			
Acute effect $PM_{2.5}$ (PRC epidemiology)	13,800		23,200	13,800
Acute effect ozone (PRC epidemiology)	5,500		9,900	5,500
Chronic effect $PM_{2.5}$ (US epidemiology)	83,200		139,000	83,200
Increased grain production	Million Tons or Billion Yuan			
Wheat (Mt)	0.895		2.149	0.895
Rice (Mt)	1.902		3.239	1.902
Maize (Mt)	4.125		7.113	4.125
Total value (billion yuan)	15.7		28.3	15.7

over the course of the analyzed policy term. Nor do they capture effects beyond 2020, should the carbon tax policy remain in place.[26]

Scenario F1, the static 30 yuan per ton tax with revenues rebated to consumers in lump sums, shows moderate economic effects compared to the base case in 2020, with a 0.09% reduction in GDP, a 0.17% increase in consumption, and a negligible effect on investment. Its impacts in 2020 on coal use and emissions would be modest compared to the other cases, with reductions from base-case levels of 10% to 15%. However, the unique challenges posed by this carbon tax structure are somewhat lost viewing it from a 2020 perspective, because it is the most abrupt policy analyzed. In this case readers should also revisit table 3.3 and figure 3.5, showing the effects in the first tax year of 2013. The table indicates impacts of F1 on the economy, energy use, and emissions that would be two- to fourfold larger than those of all the other cases. Because Chinese decision makers would seek to avoid a sudden disruption to GDP and notably export volumes, given an option of a smoother policy transition as shown in our other cases, we consider it unlikely that a carbon tax would be imposed with such a static time profile.

Scenario F2, a 10 yuan per ton tax that climbs to around 50 yuan in 2020, with lump sum rebates to consumers, is a likelier carbon tax policy. It would achieve the exact same cumulative CO_2 reduction as F1 over the eight-year policy term, but would do so by easing a policy into place incrementally. Its negative impact on GDP in 2020 would be somewhat larger than that of F1 (and somewhat smaller in 2013), but its benefit to consumption would also be larger. By 2020, the effect on coal use and annual emissions of CO_2, PM_{10}, SO_2, and NO_X compared to the base case would be substantial, with reductions ranging from 16% to 23%. Importantly, this scenario would result in sharp improvements in regional air quality, judged especially by concentrations of $PM_{2.5}$ and ozone. These in turn would translate into very substantial public health benefits, with an estimated 13,800 or as many as 83,200 premature deaths avoided from exposure to $PM_{2.5}$, depending on epidemiological assumptions. Another 5500 premature deaths from acute ozone exposure would also be avoided. In addition, the net ozone reductions would raise production of wheat, rice, and especially maize, with an estimated total economic value of 15.7 billion in constant 2007 yuan. Scenario F2 thus illustrates the large direct and indirect benefits that could be gained, at modest cost, if a carbon tax policy were implemented at rates similar to those discussed previously in Chinese policy-making circles.

Scenario F3, identical to F2 except with tax rates that accelerate more quickly to 100 yuan in 2020, naturally shows even larger effects by that year. The negative impact on GDP compared to the base case, −0.33%, would be sizable, though growth in consumption would also be large, +0.39%. Some might consider these economic impacts justified, however, given the enormous reductions in coal use and emissions and benefits to air quality, public health, and crop productivity that would result. Annual Chinese fossil CO_2 emissions would be hugely reduced under F3 compared to the base case, 30.8%, noting again that this is the only scenario analyzed that results in emissions falling over time from a peak in 2012. Emissions of domestic pollutants would also be sharply reduced, from 25% for PM_{10} to 34% for SO_2. This reduction would lead to 23,200 estimated avoided premature deaths from acute $PM_{2.5}$ exposures, or an enormous 139,000 avoided from chronic $PM_{2.5}$ exposures. Net reductions in ozone compared to the base case would lead to an additional 9900 premature deaths avoided, and 28.3 billion yuan of increased grain production. While comparing F3 to F2 indicates that a doubling of carbon tax rates would bring somewhat diminishing gains in terms of benefits, it also suggests that remarkable improvements in both carbon emissions and China's degraded air quality might be achievable at relatively manageable costs to its economy.

Scenario F4, identical again to F2 except that the tax revenues would instead be used to reduce existing taxes, illustrates an important option in carbon tax policy design. The effect on emissions compared to the base case (and therefore the benefits to air quality, health, and agriculture) would be effectively identical to those of F2, but the impacts on the economy of this alternative revenue recycling would differ significantly. The policy does not compensate households and thus does not raise consumption but would lead to higher retained earnings of enterprises, raising investment by 0.57% in 2020 compared to the base case. This higher capital stock would reduce distortions of the carbon tax, and could eventually leave GDP unaffected or even improved slightly over a business-as-usual case. In other countries, it has proven politically infeasible to use revenues of carbon pricing purely to offset more distortionary existing taxes in this way, because industries forced to pay most under new carbon pricing advocate forcefully to be compensated from the new revenues.[27] Scenario F4, however, suggests that if this sort of revenue-neutral policy design were politically feasible in China, it might offer the broad and substantial environmental benefits common to all our scenarios, but at costs to the economy that would be even smaller than in the other cases. The results that we have estimated, like all such modeling exercises, are driven by the methodological assumptions and parameter values. The estimated avoided damage depends on the emission factors, activity levels, and concentration-response coefficients, among many other factors, and these are subject to substantial uncertainty. The economic impacts depend on the assumptions regarding the base-case growth and elasticities of substitution, many of which have degrees of uncertainty that are difficult to quantify. One should keep this caution in mind and consult the other chapters for more details.

3.8 Conclusions

We draw two overarching conclusions about prospective carbon taxes in China.

First, any negative economic effects of taxes at the rates considered here would be modest at the aggregate level given the scale of recent developments in the Chinese economy that are already under way. In all but one of our cases (F4 in 2020), there would be small trade-offs between consumption, investment, and GDP in both the first and eighth year of introducing a carbon tax. For the most part, smaller reductions in investment could be gained by trading off increased consumption, and vice versa. This trade-off would mainly be driven by the choice of how the resulting tax revenues would be used: rebating them directly to consumers in lump sums (F1–F3, and F6) would increase consumption versus the base case, unsurprisingly, at the cost of slightly diminished GDP growth. Recycling the revenues to enterprises through a reduction of existing taxes (F4) would raise investment in the capital stock and diminish (or even reverse) the negative impact on GDP. Recycling through subsidies to vulnerable industries (F5) would diminish the biggest shocks and the impact on aggregate GDP.

In any case, nearly all these effects would be modest compared to the current and projected growth rates of aggregate GDP and other economic indicators. The estimated reductions in GDP in 2020 compared to the base case would be in the range of zero to three-tenths of a percent; that is, in the most aggressive case, F3, the annual growth rate would change from 8.07% during 2012–2020 in the base case to 8.03%. In our main scenario F2, the reduction in 2020 GDP is only one-tenth of a percent. While the size of these aggregate impacts depends on various technical assumptions of enterprise behavior in the economic model and assumptions in the base case, more pessimistic assumptions would raise these estimates only somewhat.[28] At the same time, gains to consumption in the near term under our consumer rebate scenarios would even be positive: from two- to four-tenths of a percent in 2020. In all cases the carbon tax would reduce exports by 2020, however, ranging from 0.7% (F1) to 2.1% (F3). This effect would be significant, enhancing a structural shift toward consumption-driven growth. This happens to align with the belief of many policy advisers that, ignoring carbon entirely, China's long-term economic prospects would be strengthened by a transition from export-led to consumption-driven growth.

Our second overall conclusion is that not only do carbon taxes efficiently limit China's carbon emissions, but they are also a powerful tool to address China's severely degraded domestic air quality, and associated damages, at the same time.

The fortuitous fact that the same fossil energy sources that emit carbon are also the primary sources of key air pollutants provides China with a remarkably convenient opportunity to align some of its most pressing domestic and global environmental aims.

All our tax policies would reduce cumulative CO_2 emissions from 2013 to 2020 compared to the base case, around 11–13% for most of our scenarios (F1–F2, F4-F6) and 20% under the higher tax rates of F3 (see figure 3.4). In annual terms, F1 would reduce 2020 CO_2 emissions by 12% compared to the base case, F2 and F4 by 19%, and F3 by 31%.

Such carbon benefits alone might justify incurring the economic impacts of the tax policies, and yet the policies would also bring ancillary reductions of PM_{10}, SO_2, and NO_X emissions at similar scales. These in turn would produce benefits to public health from reduced $PM_{2.5}$ and ozone concentrations that could be immense: from 19,300 avoided premature deaths to as many as 88,700 under F2 and F4 in 2020 alone, depending on epidemiological assumptions; and from 33,100 to as many as 148,900 under the higher tax rates of F3. Additional effects from net national reductions in ozone levels could boost grain production as much as 15.7 billion yuan under F2 and F4 in 2020, and 28.3 billion yuan under F3.

While our advanced atmospheric model allows us to trace the complex chemistry generating higher ozone levels and thus higher ozone risks to health in a few cities and to wheat production in several provinces due to the taxes, the increased acute ozone mortality in those regions would be more than offset by reductions in $PM_{2.5}$ mortality resulting from the same policy. Wheat farmers in the North China Plain would experience lower yields as a result, but they too would benefit from the lower health damages. We should note that these ozone increases could potentially be mitigated by a complementary policy targeting emissions of volatile organic compounds, the precursor pollutant that chapter 7 indicates governs the chemical formation of ozone in this region of China. The costs of such complementary policies could be spread across the entire population that benefits from the main carbon control policy, leaving no one worse off.

The debates over whether and how to implement a carbon reduction policy in wealthier countries have been contentious and acrimonious at times. In part this response may be due to the immediate and certain costs compared to the distant and uncertain benefits of climate risk mitigation. The pollution levels in the rich countries are comparatively low today, and the ancillary benefits of lower fossil fuel use are estimated at lower levels. The situation in China, and many other developing countries, however, is quite different. Here the air quality is often severely degraded, causing very high levels of health and other environmental damages. The Chinese government recognizes the urgent need to reduce air pollution, or at minimum to slow the growth of pollutant emissions.

Given this goal to reduce local emissions, which are mostly due to fossil fuel combustion, a natural question to consider is whether a policy that recognizes both local pollutants and carbon dioxide is preferable to policies dealing with these problems as though they are unrelated. Our consideration of carbon tax options here show that they can substantially reduce CO_2 and local pollutants at the same time, with a cost that is modest, especially compared to the reduced health damages.

This observation is true from a unilateral Chinese perspective, ignoring the benefits to the rest of the world of reducing its emissions of CO_2, SO_2, and other pollutants. The rest of the world derives substantial benefits from these reductions, however, including improvements in regional air quality across East Asia and well beyond, even across the Pacific Ocean. The reduction in climate risks for the world would be substantial given the large estimated tonnage of CO_2 reduced from the highest emitting country.

This effect indicates that the rest of the world should work with China to implement such emission reduction policies. There are many ways to cooperate internationally. We have already noted that establishment of a multilateral system of carbon prices, even if it took shape gradually, would reduce the effects on trade flows and ease the disruptions of implementing carbon prices in all countries. While China's availability as a source of offsets for carbon markets in other countries, such as under the Clean Development Mechanism of the Kyoto Protocol, appears to be declining for a variety of reasons, other modes of cooperation with wealthier nations are much discussed in the policy arena. These include technical assistance in the development of low-carbon policies, joint development of clean energy technologies and strategies, and scientific cooperation to build capacities and knowledge concerning the physical risks that underpin effective climate policy design. Given our results, we cannot but give our strongest encouragement for such coordinated international action.

Notes

1. As agreed at the 2012 UNFCCC meeting in Doha, Qatar, the protocol was extended to a second commitment period, but with a reduced number of nations agreeing to participate in binding emission constraints—representing only 15% of global GHG emissions.
2. The EU Emission Trading System (ETS) has notably encountered a number of well-documented problems as it has grown. These have included incentives for participating

nations to overstate their baseline emissions, which caused a sudden 60% collapse in the price of allowances in 2006 (Abboud 2008). There have been several forms of fraudulent trading, including the so-called carousel tax fraud believed to have cost the EU as much as 8 billion euros in 2008 and 2009 alone (Inman and Webb 2011). And in 2010 and 2011, thieves used computers to hack official carbon registries and steal as many as 2 million carbon allowances in the Czech Republic and Romania, with accounts in Austria also targeted (Gronewald and Fialka 2011). This theft led to a temporary suspension of ETS carbon trading to allow investigation and security enhancements, further disrupting and undermining confidence in the EU carbon market. As noted in the text, the price has more recently collapsed because of an oversupply of allowances due to economic recession.

3. Emission intensity is carbon emissions divided by GDP, in this case calculated for the locality. Use of this metric adds the considerable uncertainties of economic statistics and definitions to the uncertainties of emission estimation, complicating measurement of system compliance. Note also that a reduction in emission intensity can be consistent with a rise in emissions, if the economy is growing faster. Such a metric can thus incentivize GDP growth as much as emission reduction.

4. Estimates of both carbon emissions and GDP result from our research, and calculating carbon intensity is therefore straightforward. But a heavy emphasis on this value requires stronger trust in parameter assumptions than we think is warranted, in any modeling framework including our own. See also the paragraph and endnote that concludes the introduction to section 2.4.

5. This is not to say that we simply accept that all of those capacity targets and other energy plans will be realized. In fact we follow the capacity projections of the International Energy Agency, which makes expert judgments of the feasibility of the plans given China's recent energy trends, international market factors, and other considerations (IEA 2010).

6. Our base case includes improved energy efficiency and sharp expansions of nonfossil energy supply, as noted in the earlier discussion that includes endnote 4, and assumptions about other energy and pollution control policies described for each sector in sections C.1 and C.2. The base case does not assume any impact of the pilot carbon markets, as these are currently at experimental scales, they mostly lack conventional emission caps, and their effect by 2020 is highly uncertain.

7. This topic will be discussed further in section 3.4.1.

8. While this may seem like a rapid change in consumption, it is based on the intense discussions to rebalance the Chinese economy away from exports to domestic consumption. This rise in consumption parallels the assumption of a much smaller trade surplus. The consumption share fell from 46% in 2000 to 36% in 2007. That is, we saw rapid changes in the recent past, and thus it is reasonable to project a quick return to more typical patterns.

9. The carbon tax for 2006–2010 was set at 100 yuan per ton of carbon, which is approximately 27 yuan per ton of CO_2, in 2005 yuan.

10. Methodological note: In appendix A, this may be regarded as the variable $G\_transfer$ in equation (A.5). A scenario that adjusts lump sum revenue transfers to households on the basis of behavior, such as expenditures on energy, would be a good subject for future research.

11. Methodological note: This is the tax variable t^k in appendix A, equation (A.11).

12. See the end of section 2.2.2.

13. The projections are scaled to a rigorous estimate of CO_2 emissions for 2007 that is in fact calculated bottom-up, as will be introduced in section 3.4.1. In this case the bottom-up approach is to validate the base year estimate against instrumental measurements and to estimate uncertainty, but is not needed to judge the effects of the tax policies on carbon emissions.

14. Methodological note: The terms of trade may be regarded as the number of units of exportable goods that are needed to exchange for one unit of imports. That is, it is the price of imports divided by the price of exports. An appreciation of the terms of trade means that the country needs to export less for the same ton of imports. The terms of trade are represented by the variable e_t in appendix A, equations (A.28) and (A.29).

15. Of interest to those familiar with economic statistical systems, "industry" in this case refers to those defined at the six-digit NAICS classification level.

16. Methodological note: To do so, following appendix A.2 of Adkins et al. (2010), we first calculate the carbon content of the fossil fuel and electricity consumed by each of the 16 manufacturing industries. We then multiply it by the carbon price to get the yuan value of the cost increase due to higher energy prices. This procedure ignores the carbon content embodied in other intermediate inputs. The subsidy rate is then this value divided by the output value of the industry, as reported in the table.

17. Methodological note: In the model description in appendix A, we are changing the price of imports, PM_{it}^* in equation (A.28), and the price of exports, PE_{it}^* in equation (A.29).

18. See endnote 3.

19. Because our base year is 2007, the intensity back to 2005 is extrapolated using CO_2 emissions estimated by CDIAC (2010) from 2005–2007 and GDP from the *China Statistical Yearbook*.

20. National emissions are much easier for us to estimate for fossil CO_2 than for the other species, because the economic model directly projects energy use and CO_2 emissions after scaling to the aforementioned CO_2 inventory (Zhao, Nielsen, and McElroy 2012). The processes that emit the other species are far more diverse and complex, as amply shown in chapters 4–6. Although we could estimate bottom-up emissions of the other species for any given intervening year—2012, for instance, to spotlight how a carbon tax first imposed in 2013 would shift the emission trajectories—this step would require a large research effort that is otherwise unnecessary to draw our key conclusions about the 2013–2020 policy cases.

21. More precisely, these are denoted and defined as $\Delta NO_{x,F2/F4} = NO_{x,F2/F4} - NO_{x,base}$, $\Delta SO_{2,F2/F4} = SO_{2,F2/F4} - SO_{2,base}$, $\Delta O_{3,F2/F4} = O_{3,F2/F4} - O_{3,base}$, and $\Delta PM_{2.5,F2/F4} = PM_{2.5,F2/F4} - PM_{2.5,base}$.

22. East China is defined in these analyses as 100°E–120°E, 22°N–42°N.

23. The differences are defined similarly to those for F2F4, as $\Delta NO_{x,F3} = NO_{x,F3} - NO_{x,base}$, $\Delta SO_{2,F3} = SO_{2,F3} - SO_{2,base}$, $\Delta O_{3,F3} = O_{3,F3} - O_{3,base}$, and $\Delta PM_{2.5,F3} = PM_{2.5,F3} - PM_{2.5,base}$.

24. It may be reasonable to assume increasing stability of VOC in 2013–2020 compared to the growth in 2006–2010 (see figure 6.8 and section 6.3.3) given the attention VOCs are beginning to gain in air pollution control communities in China, and the model of VOC control in cities in more developed countries that have helped to drive down photochemical smog.

25. Comparing $\Delta O_{3,F2/F4}$ and $\Delta O_{3,F3}$, we find that the absolute value of the change in scenario F3, $\Delta O_{3,F3}$, is always larger than that of $\Delta O_{3,F2/F4}$, regardless of the sign of ΔO_3, because of the larger emission reductions of NO_X in the F3 case.

26. In principle it would be useful to compare our results to those of other carbon tax studies, but it is difficult to draw useful comparisons when research teams define tax policies differently and for different implementation periods.

27. This observation refers to actual or proposed use of revenues generated by auctions of carbon allowances in such programs as the EU Emission Trading System, the northeastern U.S. cap-and-trade program (the Regional Greenhouse Gas Initiative, RGGI), and the failed 2009 Waxman-Markey bill in the U.S. Congress. The EITE subsidies examined in scenario F5 represent an example of such compensation.

28. One of the key features driving costs is the ability to substitute low-carbon inputs for high-carbon ones, the elasticity of substitution in the economic model used. Our model assumes an elasticity of one for industry inputs. More inelastic parameters are used in other models that do estimate higher costs, but not extraordinarily high costs. For example, Adkins et al. (2010) estimate a -0.1% impact on U.S. GDP with a \$15/ton CO_2 price, and the Intertemporal General Equilibrium Model (IGEM) estimates that the Waxman-Markey plan would have reduced the growth rate in the United States from 2.67% to 2.63% (U.S. EPA 2009).

References

Abboud, L. 2008. Economist strikes gold in climate-change fight: Pollution market seen as test for U.S.; one "surreal" fall. March 13. *Wall Street Journal*, p. A1.

Adkins, L., R. Garbaccio, M. Ho, E. Moore and R. Morgenstern. 2010. The Impact on U.S. Industries of Carbon Prices with Output-Based Rebates over Multiple Time Frames. Resources for the Future, Discussion Paper 10-27, December.

Aldy, J., A. J. Krupnick, R. G. Newell, I. Parry, and W. A. Pizer. 2010. Designing climate mitigation policy. *Journal of Economic Literature* 48(4):203–234.

Bovenberg, A. L. 1999. Green tax reforms and the double dividend: An updated reader's guide. *International Tax and Public Finance* 6(3):421–443.

Carbon Dioxide Information Analysis Center (CDIAC). 2010. People's Republic of China fossil-fuel CO_2 emissions. Available at http://cdiac.ornl.gov/ftp/trends/emissions/prc.dat, last accessed February 13, 2012.

Energy Information Administration (EIA). 2009. *Annual Energy Outlook 2010*. December.

Fischer, C., and A. K. Fox. 2011. The Role of Trade and Competitiveness Measures in US Climate Policy. *American Economic Review* 101(3):258–262.

Gronewold, N., and J. J. Fialka. 2011. European Commission halts transfers of carbon emissions allowances until thefts are sorted out. January 20. *ClimateWire/New York Times*. Available at http://www.nytimes.com/cwire/2011/01/20/20climatewire-european-commission-halts-transfers-of-carbo-22394.html, last accessed on January 23, 2012.

International Energy Agency (IEA). 2010. *World Energy Outlook 2010*. Paris.

Inman, P., and T. Webb. 2011. Seven charged in carbon trading VAT fraud case. January 26. *The Guardian*. Available at http://www.guardian.co.uk/business/2010/jan/11/eu-carbon-trading-carousel-fraud, last accessed on January 23, 2012.

Lei, Y., Q. Zhang, C. P. Nielsen, and K. B. He. 2011. An inventory of primary air pollutants and CO_2 emissions from cement production in China, 1990–2020. *Atmospheric Environment* 45(1):147–154.

Liang, Q.-M., Y. Fan, and Y.-M. Wei. 2007. Carbon taxation policy in China: How to protect energy and trade-intensive sectors? *Journal of Policy Modeling* 29:311–333.

Lu, C. Y., Q. Tong, and X. M. Liu. 2010. The impacts of carbon tax and complementary policies on Chinese economy. *Energy Policy* 38:7278–7285.

National Bureau of Statistics (NBS). 2009. *Input-Output Tables of China 2007*. Beijing: China Statistics Press.

National Bureau of Statistics (NBS) and Ministry of Environmental Protection (MEP). 2010. *China Statistical Yearbook on Environment*. Beijing: China Statistical Press.

Newell, R., and W. Pizer. 2003. Regulating stock externalities under uncertainty. *Journal of Environmental Economics and Management* 45(2, Suppl.): 416–432.

Nordhaus, W. 2010. Carbon taxes to move towards fiscal sustainability. *The Economists' Voice* 7(3):1–5.

Parry, I., L. Goulder, and R. Williams. 1999. When can carbon abatement policies increase welfare? The fundamental role of distorted factor markets. *Journal of Environmental Economics and Management* 37:52–84.

Su, M., Z. H. Fu, W. Xu, Z. G. Wang, X. Li, and Q. Liang. 2009. Study on levying a carbon tax in China. Research Institute for Fiscal Science, Ministry of Finance, Beijing. In Chinese.

U.S. Environmental Protection Agency (U.S. EPA). 2009. *EPA Preliminary Analysis of the Waxman-Markey Discussion Draft: The American Clean Energy and Security Act of 2009 in the 111th Congress*. 20 April. Available at http://www.epa.gov/climatechange/economics/pdfs/WM-Analysis.pdf, last accessed on February 6, 2012.

———. 2010. *EPA Analysis of the American Power Act in the 111th Congress*. June. Available at http://www.epa.gov/climatechange/economics/economicanalyses.html, last accessed on December 20, 2011.

Wang, X., J. F. Li, and Y. X. Zhang. 2011. An analysis on the short-term sectoral competitiveness impact of carbon tax in China. *Energy Policy* 39:4144–4152.

Wei, T. 2012. Officials weighing green benefits of carbon taxation. January 6. *China Daily*. Available at www.chinadaily.com.cn/bizchina/2012-01/06/content_14391943.htm, last accessed on January 22, 2012.

Xinhua News Agency. 2012. China eyes nationwide emission trading programs in 2016–2020: Official. December 6. *Xinhuanet*. Available at http://news.xinhuanet.com/english/china/2012-12/06/c_132024197.htm, last accessed on December 8, 2012.

Yi, W.-J., L.-L. Zou, J. Guo, K. Wang, and Y.-M. Wei. 2011. How can China reach its CO_2 intensity reduction targets by 2020? A regional allocation based on equity and development. *Energy Policy* 39:2407–2415.

Zhao, Y., C. P. Nielsen, and M. B. McElroy. 2012. China's CO_2 emissions estimated from the bottom up: Recent trends, spatial distributions, and quantification of uncertainties. *Atmospheric Environment* 59:214–223.

II
Studies of the Assessment

4

Primary Air Pollutant Emissions of Coal-Fired Power Plants in China

Yu Zhao

4.1   Introduction

Coal is the primary energy source for Chinese power generation. As shown in figure 4.1, thermal power has maintained a large share of both electricity output and installed capacity (82% and 76%, respectively, for 2005). For many years, thermal power has been fueled predominantly by coal with very small amounts by oil or gas. During 2000–2007, the period in which the Chinese power sector developed fastest within the past 30 years, coal consumption by power plants increased from 560 to 1300 million tons (Mt) according to official statistics. As shown in figure 4.2, in recent years the share of total coal consumption by the power sector has remained around 50%. Investment in clean energy such as wind power and hydropower has also increased, but the overall Chinese power structure is still dominated by coal-fired generation, and this domination will continue in the foreseeable future.

As the largest coal-consuming sector, power generation has been considered the most important source of atmospheric pollutant emissions and thereby of regional air pollution. A series of studies has been conducted on Chinese emission inventories using top-down methods, that is, using industry-wide average characteristics and data. These studies indicate that the power sector accounts for 31–59% of national anthropogenic emissions of sulfur dioxide, SO_2 (Streets et al. 2003; van Aardenne et al. 2005; Ohara et al. 2007; Klimont et al. 2009; Zhang et al. 2009), 21–44% of nitrogen oxides, NO_X (Hao, Tian, and Lu 2002; Streets et al. 2003; van Aardenne et al. 2005; Ohara et al. 2007; Zhang et al. 2007a; Klimont et al. 2009; Zhang et al. 2009), and 9% of particulate matter, PM (Zhang et al. 2007b; Lei, Zhang, and Streets 2011).

In treating power generation as a single sector in a larger emission inventory framework, these studies have either applied uniform emission factors across the sector or referenced domestic emission standards to estimate emissions. They

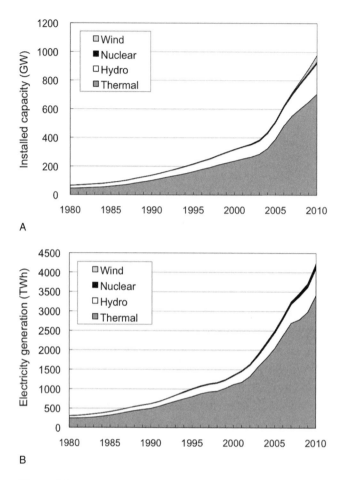

A

B

Figure 4.1
Chinese power sector development during 1980–2010. (A) Electricity output. (B) Installed capacity.

generally have ignored differences in technology and fuel characteristics among power units of different types, which have very different emission characteristics. Moreover, most studies have not reflected the rapid increase of coal consumption and electricity generation since the year 2000, and thus their results are less applicable for current policy analysis.

Under increasing environmental pressure, the Chinese government chose the coal-fired power sector as the most important target for abating atmospheric emissions during 2006–2010, the period of the 11th Five-Year Plan (11th FYP). Power plants thus faced more stringent environmental regulations related to siting and operations.

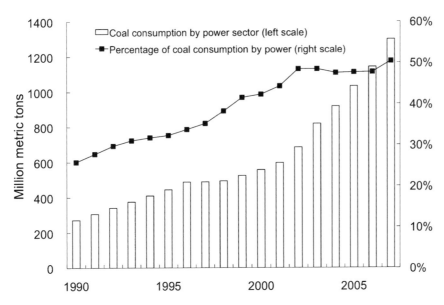

Figure 4.2
Coal consumption by the Chinese power sector during 1990–2007.

Two main measures were implemented: (1) all new thermal power units, as well as most existing ones, were required to install flue gas desulfurization (FGD) systems; and (2) most small units with low energy efficiency, estimated to total about 59 gigawatts (GW), were gradually shut down.

In order to provide a detailed picture of emissions for scientific research and policy analysis, this study uses an innovative unit-based method to explore current and future emissions of Chinese coal-fired power plants. We also consider the effects of currently implemented and prospective emission control policies. These results are integrated into the wider emission inventory study described in chapter 6, which in turn is applied in the integrated project's atmospheric simulations described in chapter 7.

4.2   Methods

4.2.1   Unit-Based Methodology

Our study covers 31 provinces and province-level administrative units of mainland China; Hong Kong, Macao, and Taiwan are not included. The pollutants examined in this chapter are SO_2, NO_X, and PM. In contrast to prior studies, emissions are estimated through a bottom-up methodology based on individual combustion units.

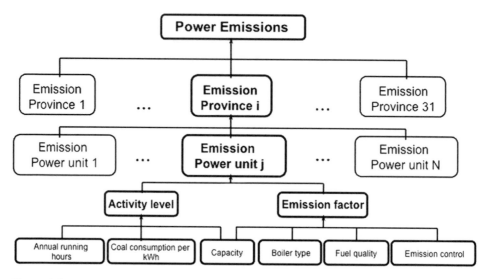

Figure 4.3
Flow chart of the unit-based bottom-up methodology.

(We are also of course interested in CO_2 generated by electric power plants, but as explained in chapter 6, we use a concurrent bottom-up study by the author and colleagues to estimate base year CO_2 emissions for all sectors [Zhao, Nielsen, and McElroy 2012]. The economic model of chapter 9 then scales CO_2 emissions under the policy scenarios to this base-year value.)

A database of the Chinese coal-fired power sector is established, in which all information related to emissions is compiled at the unit level, including geographical location, capacity, boiler type, initial year of operation, annual operating hours, fuel type, sulfur content, coal consumption per unit of electricity supply, and emission control technologies for SO_2, NO_X, and PM. A flow chart for this database is presented in figure 4.3. The database of coal-fired plants covers a total of 230 GW and 356 GW in 2000 and 2005, respectively. The data for 2001 to 2004 are derived using information on newly built and retired units in the corresponding years. For 2006 to 2010, the data for more than 350 GW of new power units are based on the national construction plans for power plants, including revisions as announced year by year through 2009.[1]

Annual emissions of each unit are calculated from unit-specific fuel consumption and emission factors and then aggregated to the regional level. Emissions of SO_2, NO_X, and PM from power plants in province i are given by

$$E_{SO_2,i} = \sum_j (AL_{j,i} \times EF_{SO_2,j,i}) = \sum_j AL_{j,i} \times S_{j,i} \times (1 - Sr) \times (1 - \eta) \times 2 \qquad (4.1)$$

$$E_{NOx,i} = \sum_k \sum_m \sum_n (AL_{i,k,m,n} \times EF_{NO_x,k,m,n}) \qquad (4.2)$$

$$E_{PM,y,i} = \sum_j (AL_{j,i} \times EF_{PM,i}) = \sum_k \sum_n AL_{i,k} \times A_i \times (1 - ar_k) \times f_{k,y} \times R_{k,n} \times (1 - \eta_{n,y}) \qquad (4.3)$$

where subscripts j, k, m, n, and y stand for power unit, boiler type, fuel type, emission control technology, and particulate size; EF is the NO_X emission factor; AL is the coal consumption; $R_{k,n}$ is the application rate of PM emission control technology n; η is the sulfur removal efficiency; $\eta_{n,y}$ is removal efficiency of control technology n for PM of size y; $S_{j,i}$ is the sulfur content of coal used in unit j; Sr is the sulfur retention of ash; A is the ash content of fuel; ar is the ratio of bottom ash; and f is the particulate mass fraction by size.

4.2.2   Scenarios

To analyze and compare the impacts of emission control polices, emissions of the Chinese coal-fired power sector in 2010 are estimated under these three scenarios: a base case, a carbon tax case, and an 11th FYP SO_2 control case.

The base case has the highest emissions of the three scenarios since the energy saving and emission control policies actually implemented during 2006–2010 are assumed not to have taken place. It is thus a hypothetical case designed to evaluate the impacts of policies of the 11th FYP that have in fact been implemented and to evaluate a counterfactual enactment of a carbon tax instead of the 11th FYP policies. In the base case, all the power plants in the national construction plans are assumed to have gone into operation, and the small, inefficient units slated for closure during 2006–2010 are assumed to have continued operating. The total installed capacity thus increases from 356 GW in 2005 to 710 GW by 2010. Regarding emission controls, installation of FGD systems is assumed to have occurred only in newly built coal-fired units, with no retrofitting of existing units as required in the 11th FYP. The national application rate of FGD still increases over time in the base case, from 14% (52 GW) in 2005 to 56% (406 GW) by 2010.

The 11th FYP case represents our best estimate of what actually happened in China from 2006 to 2010 and assumes full implementation of official measures on energy saving and emission abatement in the power sector. We estimate that approximately 59 GW of small, inefficient power units were shut down, yielding a total installed capacity of 651 GW. (Our estimates, made in 2009 at the outset of research, are extremely close to what the China Electricity Council reported in 2011.)[2] FGD systems were required not only for new power plants but also as retrofits of 125 GW of existing plants, and the application rate of FGD is estimated to have reached more than 80% of total thermal power capacity (531 GW) by 2010. This estimate compares well to an official estimate of 82.6% in 2010, reported after the 11th FYP had concluded.[3]

In the base case and 11th FYP case, the paths that we have described are used to directly estimate emissions of the coal-fired power sector, using methods described in the rest of this chapter. The multisector economic model described in chapter 9 is then used to generate paths of energy use in all other sectors of the economy. In the 11th FYP case, the costs of SO_2 controls are passed on through a rise in the price of electricity, which then affects the trajectories of the other sectors, capturing the indirect (or "general equilibrium") effects of these power sector policies through-out the rest of the economy. In both the base case and 11th FYP case, the energy consumption paths of all the nonpower sectors are translated into emissions as described in chapters 5 and 6, and combined with the power sector emissions esti-mated in this chapter for use in the atmospheric simulations of chapter 7.

The carbon tax case is a hypothetical scenario in which a tax of 100 yuan per ton of carbon (around 27 yuan per ton of CO_2) is applied each year from 2006 to 2010 on all fossil fuels. The rise in energy prices, especially of coal and coal-fired electricity, encourages energy efficiency, more use of low- and noncarbon energy, and structural economic changes toward less carbon-intensive production. These direct and indirect effects of the carbon tax on the trajectories of energy demand and supply of all sectors (including electric power generation) are first simulated using the multisector economic model described in chapter 9. The resulting path of reduced demand for electricity is met by the same power plant distribution of the base case, with annual running hours scaled down accordingly. Emissions are then estimated using methods described in the rest of this chapter. These power sector emissions are combined with emissions estimated for other sectors in chapters 5 and 6, also according to the energy projections of the economic model, for use in the air quality analyses of chapter 7.

4.2.3   Activity Levels

The coal consumption of each generating unit j is estimated according to its annual operating hours and coal consumption per unit of electricity supply:

$$AL_j = 1.4 \times U_j \times T_j \times Q_j \times 10^{-6} \tag{4.4}$$

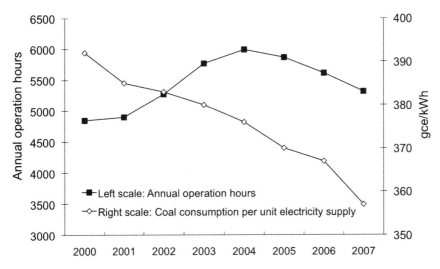

Figure 4.4
Average annual operation hours and coal consumption per unit of electricity supply of coal-fired power plants in China during 2000–2007. *Data source:* Editorial Board of China Electric Power Yearbook (2008).

where *AL* is the coal consumption (kt), *U* is the unit size (MW), *T* is the annual hours of operation, *Q* is the specific coal consumption per unit electricity supply (grams of coal equivalent per kilowatt hour, gce/kWh), and 1.4 converts grams of coal equivalent to grams of raw coal. For units lacking the *T* and *Q* parameters, national average values are applied. Data on coal quality are compiled at the unit level, based on investigation of power plants by the Ministry of Environmental Protection (MEP).

As shown in figure 4.4, the average annual operation hours of Chinese coal-fired power plants rose from 4900 to 5800 during 2000–2005. However, the average dropped quickly to 5300 hours in 2007, as new power plants alleviated the pressure on electricity supply. In this study, the annual average operation time is assumed to be 4800–5000 hours in the 11th FYP case for 2010. The national average coal consumption rates for these categories of generating units—smaller than 100 MW, 100–200 MW, 200–300 MW, 300–600 MW, and larger than 600 MW—are, respectively, 440, 379, 365, 335, and 326 gce/kWh (CEC 2007). With the rapid increase of large units, the national average declined from 392 to 357 gce/kWh during 2000–2007 (NBS 2007).

In this study, the national average value is calculated to be 350 gce/kWh in 2010 in the 11th FYP case, based on the distribution of units by size and corresponding

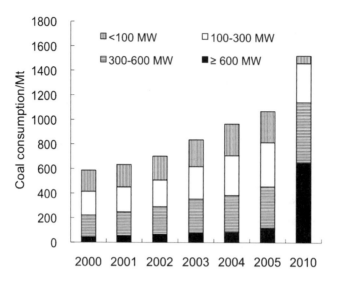

Figure 4.5
Coal consumption of power plants by unit size for 2000–2005 and 2010 (Mt).

coal consumption rates. For the base case in 2010, we assume that power generation would be the same as in the 11th FYP case, and we estimate slightly higher coal consumption reflecting the different capacity mix (i.e., including small, inefficient plants). For the carbon tax case, the results of the economic model indicate that there would be a 15% reduction in electricity demand, and we represent reduced electricity supply by scaling down operating hours over the capacity mix of the base case, calculating lower coal consumption accordingly.

National coal consumption of the power sector from 2000 to 2005 and projected for 2010 (under our 11th FYP case) is shown in figure 4.5 by size of generating units. In 2005 it was 1070 Mt, reaching 1521 Mt in 2010, an annual average growth rate of 7.3%. There are significant changes in the distribution by unit size; with the recent construction of large units, the coal consumption of those larger than 600 MW rose from 11.2% in 2005 to 42.8% in 2010, while the share of units smaller than 100 MW fell from 23.8% to 4.0%.

4.3   Improving Estimates of Emission Factors

The accuracy of a bottom-up emission inventory, together with the policy assessment that uses it, depends critically on the emission factors, that is, the parameters relating the quantity of pollutant released to the atmosphere with the level of activity associated with that release. In the current study, an "activity" mainly refers to burning coal of a particular quality in a thermal power plant employing particular combustion and emission control technologies. This chapter mainly focuses on the detailed research steps undertaken to develop the most accurate emission factors possible for the swiftly evolving Chinese electricity sector, and then applies them in section 4.4 to estimate emissions from the power sector under the several scenarios evaluated in this book.

To improve on existing emission factor estimates, field measurements of SO_2, NO_X, and PM emission characteristics were carried out in selected coal-fired power plants. Based on these test results and data from other recently published studies, an emission factor database for the Chinese coal-fired power sector is established, describing different emission levels by unit type. To better understand the uncertainty of using the database to estimate emissions, bootstrap and Monte Carlo simulations are applied to provide the variance and statistical distribution for given emission factors.

4.3.1   Field Measurements

In this section we summarize earlier field measurements conducted to support development of the emission inventory. Complete details on the procedures and instruments employed in the fieldwork are reported in Zhao (2008) and Zhao et al. (2010).

As shown in table 4.1, gaseous pollutant and PM emissions were measured for ten generating units at eight coal-fired power plants across China. This sample covers a range of boiler types, burner patterns, fuel quality, and emission control devices. As listed at the bottom of table 4.1, the boiler types are pulverized coal (PC), grate, and circulating fluidized bed (CFB) boilers. The control devices include flue gas desulfurization (FGD), low NO_X burner (LNB), electrostatic precipitator (ESP), selective catalytic reduction (SCR), and fabric filter (FF) systems. Burner pattern refers to the location of the burners and the direction of coal injection into the furnace of PC boilers. During the test period, all the power units were operating under normal conditions.

Instruments to measure pollutants were placed at both inlets and outlets of the emission control devices in order to obtain the emission levels and removal efficiencies of the devices; for example, PM emissions were measured before and after the dust collectors as well as after wet-FGD systems (if applicable). The sampling procedures followed the national regulations of the State Environmental Protection Administration (SEPA and CSBTS 1996).

Table 4.1

Operational parameters and gaseous pollutant concentrations (converted to the value at $O_2 = 6\%$) for the tested power plants

Plant Number	Unit Number	Size (MW)	Fuel	Sulfur Content (%)	Ash Content (%)	Boiler Type	Burner Pattern	Control Device	SO_2 (mg/m³)	NO_X (mg/m³)
1	2	200	Bituminous	1.33	16.2	PC	Tangential	LNB; ESP	3043	374
	3	200		1.33	16.2	PC	Tangential	ESP	2663	547
2	5	50	Lignite	0.15	21.7	PC	Wall	ESP	307	456
3	2	50	Lignite	0.18	21.7	PC	Wall	ESP	486	945
	4	100		0.18	21.7	PC	Wall	ESP	462	862
4	4	200	Bituminous	3.84	22.3	PC	Tangential	ESP; Wet-FGD	8228 (pre-FGD) 334 (post-FGD)	726
5	1	125	Bituminous	0.61	20.4	PC	Tangential	ESP	1379	792
6	2	29	Bituminous	0.77	20.4	Grate	—	Simple-FGD; Wet scrubber	1711 (pre-FGD) 1410 (post-FGD)	437
	8	58		0.77	20.4	CFB	—	ESP	1278	237
7	2	165	Anthracite	0.44	7.7	PC	W-fired	SCR; ESP; Wet-FGD	837 (pre-FGD) 47 (post-FGD)	351 (pre-SCR) 180 (post-SCR)
8	1	100	Bituminous	2.01	30.6	PC	Tangential	ESP; CFB-dry-FGD; FF	4593 (pre-FGD) 976 (post-FGD)	635

Notes: PC is pulverized (coal) combustion; CFB is circulating fluidized bed; FGD is flue gas desulfurization; LNB is low-NO_X burner; ESP is electrostatic precipitator; SCR is selective catalytic reduction; FF is fabric filter.

Gaseous pollutants (SO_2 and NO_X) as well as oxygen (O_2) were measured in the flue gas. For comparisons with other test results, the measured pollutant concentrations were converted to O_2-standardized values according to procedures set in the national regulations (SEPA and AQSIQ 2003). A filter drum was used to collect total PM, from which total PM emission concentrations were calculated.

A number of studies have used electrical or nonelectrical low-pressure impactors (ELPIs or LPIs) for real-time monitoring of particle size distribution from combustion sources, both in China (Yi et al. 2006; Gao 2006; Sui 2006) and abroad (Moisio et al. 1998; Lind et al. 2003; Lillieblad et al. 2004). In the fieldwork reported here, PM_{10} was sampled with gravimetric methods, using the combination of an ELPI and a two-stage dilution system described in Moisio (1999). This step yielded emission concentrations of PM for specific size categories.

The SO_2 and NO_X emission levels of measured power units are listed in the last two columns of table 4.1. The concentrations of combustion-generated SO_2 were very closely related to sulfur content of the coal burned ($R^2 = 0.994$). Compared with pulverized coal combustion (PC) boilers, the SO_2 emission from a circulating fluidized bed (CFB) boiler was around 25% lower. (This was measured at generating unit 8 of power plant 6; we refer to this as "plant 6 #8," and label other power units analogously.) Substantial variations in the effectiveness of SO_2 control were found for the tested FGD technologies. Wet FGD had the highest removal efficiency, above 90% (plant 4 #4 and plant 7 #2), followed by CFB-dry-FGD with about 80% efficiency (plant 8 #1). The removal efficiency of the tested simple FGD system (plant 6 #2) was merely 18%, even lower than that of the CFB boiler alone.

The emission levels of NO_X depend on a complex interaction of the coal quality, burner pattern, and emission control device, and more observations are needed to draw strong conclusions. During the tests, the lowest emission level was found for the CFB boiler (plant 6 #8) without further NO_X removal technologies. Selective catalytic reduction (SCR) reduced the NO_X level by 43% (plant 7 #2), while low-NO_X burners (LNB) reduced it by 27%.

The PM_{10} mass size distributions before and after dust collectors are shown in figure 4.6. It is widely reported that PM generated from coal combustion displays a bimodal size distribution; that is, a density graph with particle diameter on the horizontal axis has two peaks. There is a submicron mode in which particles are formed through vaporization, condensation, and nucleation of inorganic constituents, and another mode for the coarse particles in which they are formed through fragmentation and coalescence of surface ash droplets (Ylatalo and Hautanan 1998; Buhre et al. 2005; Yi et al. 2006). In this study of the 10 generating units,

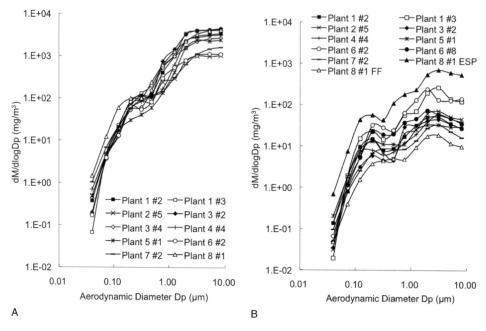

Figure 4.6
The PM$_{10}$ mass size distributions for the tested power units. Dp is aerodynamic diameter of particles; dM/dlog Dp is differential of PM concentration to Briggsian logarithm of Dp. (A) Pre–dust control. (B) Post–dust control.

submicron-mode peaks around 0.2–0.3 micron (µm) were observed in the flue gases before the dust collectors, while the coarse-mode peaks were missed because of the limit of the ELPI measurement range. In the flue gases after passing through the dust collectors, the bimodal size distribution was observed at all units, with peaks for submicron and coarse modes in the 0.2–0.3 and 2.0–3.0 micron ranges, respectively.

The calculated mass size fractions of PM$_{10}$ before and after dust collectors (and wet-FGD, if applicable) for each of the 10 generating units are shown in figure 4.7. Before the dust collectors, the fine-mode particles (PM$_{2.5}$) accounted for only 23–35% of the PM$_{10}$ mass, and the share of very fine PM$_{1.0}$ was less than 10% at all the tested units. After the dust collectors, the percentages of PM$_{2.5}$ and PM$_{1.0}$ increased to 38–60% and 14–28%, respectively, confirming that the removal efficiencies of dust collectors for finer particles were poorer than those for larger particles. The effect of dust collectors on the fine-particle share was smallest for the unit in which

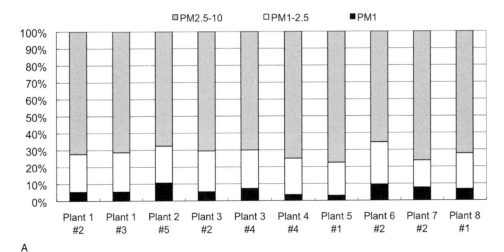

A

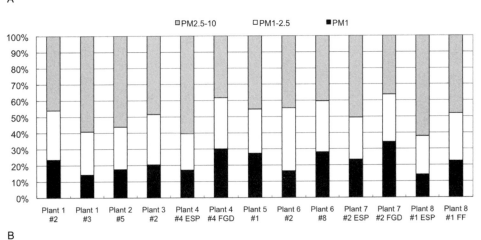

B

Figure 4.7
The PM_{10} composition by size for the tested power units. (A) Pre–dust control. (B) Post–dust control and desulfurization.

wet scrubbing was used instead of an ESP. A side effect of PM control was also found in wet-FGD technologies, particularly for large particles. After wet flue gas desulfurization, the percentages of $PM_{2.5}$ and $PM_{1.0}$ rose beyond 60% and 30%, respectively.

The removal efficiencies of different particle sizes can be calculated by comparing PM levels before and after the dust collectors. The measured removal efficiencies of

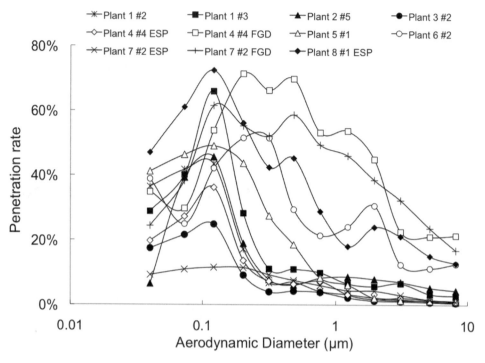

Figure 4.8
The PM$_{10}$ penetration by size for the tested dust collectors.

ESPs in this study (not including plant 8 #1, for reasons noted in the next paragraph) were in the 98.08–99.53% range for total PM, 93.25–98.78% for PM$_{10}$, and 90.88–97.86% for PM$_{2.5}$. These efficiencies are much higher than those of the tested wet scrubber. Since the ESP of plant 8 #1 applied only two electric fields (as opposed to three to five fields in typical Chinese ESPs), and was used for dust removal prior to the CFB-dry-FGD device, its removal efficiencies were lower than those of other ESPs, at 89.44%, 81.12%, and 74.27% for total PM, PM$_{10}$, and PM$_{2.5}$, respectively. When both ESP and wet-FGD are used (plant 4 #4 and plant 7 #2), the total PM removal rate could reach 99.8%.

Figure 4.8 shows the penetration of PM$_{10}$ through the dust collectors and wet FGDs. In accord with previous studies, the highest penetrations were found at the particle size range of 0.1–1.0 microns for all of the emission control devices, attributed to the effects of multiple PM capture mechanisms (Helble 2000; Yi et al. 2006). As mentioned earlier, the wet scrubber, wet-FGD, and ESP of plant 8 #1 had relatively poorer PM control effects and thus higher PM penetrations.

4.3.2 Emission Factor Database

The emissions factors of SO_2, NO_X, and PM from coal-fired power plants are calculated with equations (4.5) to (4.7), respectively; note that equations (4.1) to (4.3) multiply these values and activity levels to give emissions.

$$EF_{SO_2} = 10 \times S \times (1 - Sr) \times (1 - \eta) \times 2 \tag{4.5}$$

$$EF_{NO_X} = C_{NO_X} \times V / 1000 \tag{4.6}$$

$$EF_{PM,y} = 10 \times A \times (1 - ar) \times f_y \times (1 - \eta_y) \tag{4.7}$$

EF is the emission factor (kg/t); *S* is the sulfur content of fuel (%); *Sr* is the sulfur retention ratio of ash; η is the sulfur removal efficiency of the emission control technology (%); C_{NO_X} is the NO_X concentration in the flue gas (mg/m³); *V* is the volume of flue gas per unit of fuel consumption (m³/kg); *A* is the ash content of fuel (%); *ar* is the ratio of bottom ash to total ash; f_y is the particulate mass fraction of size *y*; and η_y is the removal efficiency for particulates of size *y*.

The flue gas volume per unit of fuel consumption (*V*) is affected mainly by the lower heating value of coals and is calculated separately for bituminous and lignite coal versus anthracite coal:

$$V = 1.04 \times Q_L / 4187 + 0.77 + 1.0161 \times (\alpha - 1) \times V_0 \tag{4.8}$$

$$V_0 = 0.251 \times Q_L / 1000 + 0.278 \text{ (bituminous and lignite)} \tag{4.9a}$$

$$V_0 = Q_L / 4140 + 0.606 \text{ (anthracite)} \tag{4.9b}$$

where Q_L is the lower heating value (kJ/kg), V_0 is the theoretical air volume (m³/kg), and α is the excess air coefficient; the value 1.4 was applied in this study according to the national emission standard of air pollutants for thermal power plants (SEPA and AQSIQ 2003).

The parameters of equations (4.5) to (4.9) were collected and classified for calculating emission factors according to the boiler/burner types, coal qualities, and emission control devices. Among all of the pollutants, the parameters for SO_2 are believed to be most certain (Streets et al. 2003), and thereby the measured data described in section 4.3.1 were sufficient to develop the database for these parameters.

Formation of NO_X during combustion is associated with the temperature and degree of oxygen enrichment, which can be significantly influenced by the unit capacity, burner, and fuel type. To better understand the NO_X emission characteristics, data for NO_X concentrations in flue gases were collected from many sources, including the field measurements in section 4.3.1 and other published results. The Chinese Journal Full-Text Database (CJFD) was thoroughly searched for published data on NO_X emissions from coal-fired power plants since 2000 for inclusion. This study also included field measurements by other research institutes referenced in additional literature (Tian 2003; Yi et al. 2006; CRAES 2006). The statistics of the NO_X concentrations by unit type obtained from all these sources are shown in figure 4.9. The heating values (Q_L) for different types of steam coals were taken from statistics of national coal quality (Jin 2001).

With regard to PM emission factors, field measurements of emission characteristics by size class have been conducted by several other domestic research teams in China (He and Wu 1999; Ge et al. 2001; Huang et al. 2003; J. Z. Liu et al. 2003; H. Xu et al. 2004; Gao 2006; Qi 2006; Sui 2006; Yi et al. 2006; Yao et al. 2007; H. Wang et al. 2008; P. Wang 2008). As listed in Zhao et al. (2008; 2010), ratios of bottom ash, particulate mass fractions by size, and size-specific removal efficiencies for dozens of different dust collectors were taken from those studies as well as from the results described in section 4.3.1. In addition to dust collectors, wet-FGD systems are increasingly common in China, and the benefits for PM control of this desulfurization technology should also be considered in the PM emission factor estimate.

The probability distributions of these parameters were calculated using a bootstrap simulation method (Frey and Zheng 2002). Statistical tests were used to fit observed data sets of each parameter to preliminary distributions of one of these forms: normal, lognormal, beta, gamma, or Weibull. Synthetic data sets of the same sample sizes as the original data sets were then generated using Monte Carlo methods from the assumed probability distribution by drawing random samples. This sampling was conducted 1000 times (i.e., 1000 synthetic data sets were generated for each parameter), and the mean value for each generated data set was calculated. The probability distribution for each parameter was finally determined by fitting those calculated mean values. To calculate the emission factor for each unit type, relevant parameters with corresponding statistical distributions were placed in a Monte Carlo framework, and 10,000 simulations were performed. This procedure yielded the mean value, distribution, and 95% confidence interval (CI) for each emission factor.

One example of the probability distribution bands is shown in figure 4.10, for the $PM_{2.5}$ emission factor of PC boilers with ESP. Through the bootstrap simulation, the mean values of the ash release ratio, $PM_{2.5}$ mass fraction of uncontrolled PM emissions, and removal efficiency of ESP for $PM_{2.5}$ were estimated to be 0.69%, 0.06%, and 92.31%, with beta, lognormal, and lognormal probability distributions, respectively. Based on these results, the uncontrolled and controlled $PM_{2.5}$ emission

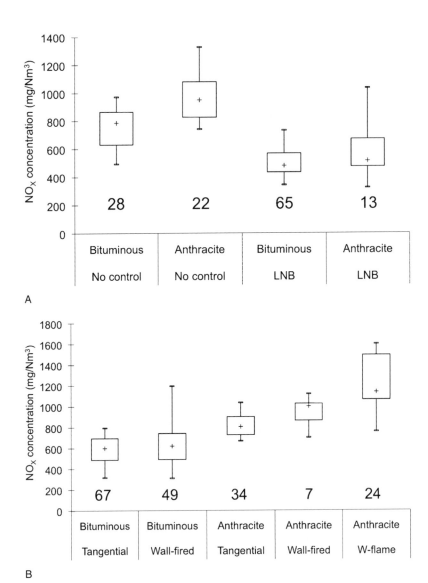

A

B

Figure 4.9
Surveyed NO$_X$ concentrations, 95, 75, 50, 25, and 5 percentiles, by unit type. The numbers under the concentration distributions are the sample sizes. (A) Units smaller than 300 MW. (B) Units larger than or equal to 300 MW, all of which have LNBs.

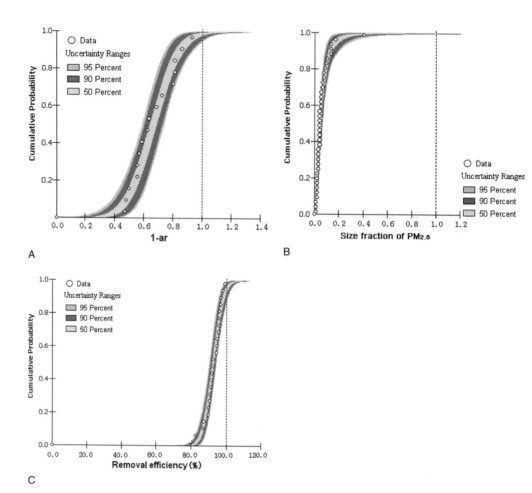

Figure 4.10
Probability distribution bands of three parameters of the emission factor for $PM_{2.5}$ from pulverized combustion boilers with electrostatic precipitators. (A) Ash release ratio $(1 - ar)$. (B) Mass fraction of $PM_{2.5}$ in uncontrolled PM emissions. (C) Removal efficiency of $PM_{2.5}$ by ESPs.

factors were calculated to be $0.41A$ (95% CI: $0.29A$–$0.59A$) and $0.032A$ (95% CI: $0.021A$–$0.046A$) respectively, as shown in figure 4.11, where A is the ash content.

The SO_2, NO_X, and PM emission factors of Chinese coal-fired power plants are summarized in table 4.2. For the unit types widely used in China, emission factors with 95% CI are provided in parentheses. For other rarely used unit types (e.g., grate boilers, circulating fluidized bed combustors, and those with wet scrubber dust

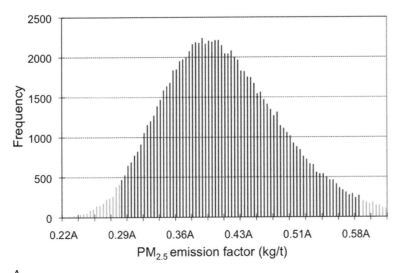

A

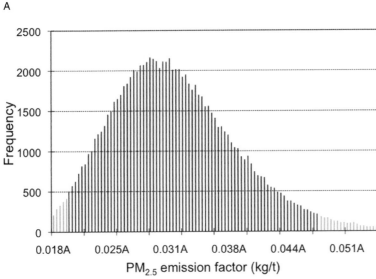

B

Figure 4.11
Probability distribution of $PM_{2.5}$ emission factor for pulverized combustion boilers with electrostatic precipitators. A is the ash content of coal (in percents). The gray bars are out of the 95% CI. (A) Uncontrolled. (B) Controlled.

Table 4.2

Emission factor database for Chinese coal-fired power plants (kg/t). The numbers in parentheses indicate the 95% CIs. For those without 95% CIs, considerable uncertainty can be expected because of small sample sizes, and these emission factors should be regarded as tentative.

Sulfur Dioxide

Boiler	Coal	Uncontrolled	Controlled		
			Wet-FGD	Dry-FGD	Simple FGD
PC and grate boiler	Bituminous and anthracite	18.0S (17.2S–18.5S)	0.9S	3.6S	15.0S
	Lignite	15.0S	N/A	N/A	N/A
CFBC	All types	13.0S*	N/A	N/A	N/A

Nitrogen Oxides

Boiler	Capacity	Coal	Control	Burner	Emission Factor
PC and grate boiler	<300 MW	Bituminous and lignite	No	All types	6.1 (5.3–7.1)
		Anthracite	No	All types	9.0 (8.1–9.9)
		Bituminous and lignite	LNB	All types	4.0 (3.5–4.6)
		Anthracite	LNB	All types	5.5 (4.3–6.8)
	≥300 MW	Bituminous and lignite	LNB	Tangential	4.7 (4.1–5.4)
		Bituminous and lignite	LNB	Wall-fired	5.2 (4.4–6.1)
		Anthracite	LNB	Tangential	7.6 (7.1–8.1)
		Anthracite	LNB	Wall-fired	8.6 (7.4–9.9)
		Anthracite	LNB	W-flame	11.2 (9.9–12.5)
CFBC	All	All types	No	CFBC	1.5

Note: In all cases, S is the sulfur content, in percent, of the coal as fired. Because of low sulfur content (usually less than 0.2%), FGD systems are not generally installed when burning lignite.

*Result from a single field test of a unit burning bituminous. It is generally believed that the SO_2 emission of CFBC systems is closely related to both the sulfur content and the calcium-to-sulfur ratio.

collectors), values are given tentatively, without uncertainty estimates, because of limited available survey data.

The uncontrolled SO_2 emission factor for PC boilers combusting bituminous or anthracite coal, the most common unit type and fuels, was estimated to be 18.0S kg/t, where S is the sulfur content of the coal. This represents a sulfur retention ratio of roughly 0.10, which is lower than the value of 0.15 conventionally used in previous official estimates. Acknowledging the uncertainty, wet-FGD (e.g.,

Table 4.2 (continued)

Particulate Matter

				Controlled		
		Uncontrolled	ESP	Wet Scrubber	FF	ESP + wet-FGD
PC	$PM_{2.5}$	$0.4A^1$ (0.3A–0.5A)	0.032A (0.021A– 0.046A)	0.135A	0.0019A	0.0147A (0.0092A– 0.0225A)
	PM_{10}	1.5A (1.1A–1.9A)	0.065A (0.039A– 0.092A)	0.291A	0.0034A	0.0210A (0.0129A– 0.0317A)
	PM	6.9A (5.8A–7.9A)	0.094A (0.065A– 0.132A)	0.479A	0.0042A	0.0231A (0.0142A– 0.0348A)
Grate	$PM_{2.5}$	0.10A	0.008A	0.032A	N/A	N/A
	PM_{10}	0.26A	0.012A	0.054A	N/A	N/A
	PM	1.50A	0.019A	0.098A	N/A	N/A
CFB	$PM_{2.5}$	0.45A	0.034A	N/A	N/A	N/A
	PM_{10}	1.54A	0.067A	N/A	N/A	N/A
	PM	4.80A	0.085A	N/A	N/A	N/A

Note: In all cases A is the ash content, in percent, of the coal as fired.

technology using limestone and/or gypsum sorbents), currently installed in over 50% of Chinese coal-fired units, indicates satisfactory SO_2 removal efficiency (95%). The control effects of other FGD systems that are mostly applied in relatively small units were poorer. Regarding the sulfur content (S), the data were obtained unit by unit in surveys by the Ministry of Environmental Protection (MEP). Figure 4.12 shows the sulfur content distribution by province in 2005. Relatively high sulfur content was found in southwestern China, while the northeast had the lowest values. The national average sulfur content of coal used by the power sector was 1.01% in 2005.

Because China has been implementing a policy retiring small units and requiring newly built units to equal or exceed 300 MW in capacity, this size is used as a threshold between small and large units for classifying NO_X emission factors. At higher combustion temperatures, coal with low volatile matter content (e.g., anthracite) reliably generates higher NO_X emissions. In this study, the NO_X emission factors were generally 30–40% lower for units burning bituminous and lignite coals than those fired by anthracite. Regarding burner types, wall-fired boilers produced

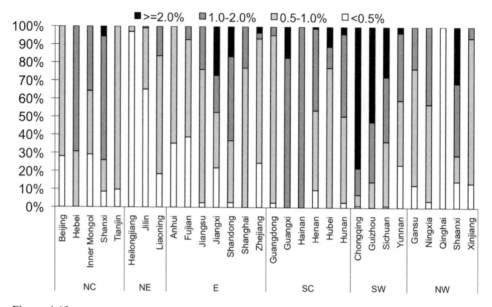

Figure 4.12
Percentage of coal consumption with different sulfur contents by region. NC is north-central; NE is northeast; E is east; SC is south-central; SW is southwest; and NW is northwest.

10–15% higher emissions than tangentially fired boilers, while W-flame boilers, which are mainly designed to burn anthracite, had the highest emission factor, 11.2 kg/t. Low NO_X burners (LNB) are widely applied in large units, and their NO_X control efficiency was estimated to be 30–40%. Since there is only incipient use of SCR in China and few measurements exist, its control effect was not evaluated in this study.

Regarding PM emission factors, PC boilers with ESP and wet-FGD systems are the most widely used configurations. The share of fine particles ($PM_{2.5}$) of the total uncontrolled emissions was estimated to be 5.8%, and increased to 34.0% and 63.6% with addition of ESPs and ESP combined with wet-FGD systems, respectively. Currently the market share of fabric filter systems (FF), a dust collector with higher PM removal efficiency than ESP, has been rising and is believed to have reached 10% in 2010. The controlled PM emission factor of units using FF was calculated to be 0.0042A, 82% lower than that of ESP combined with wet-FGD, indicating the huge benefits of this technology; however, we should note the large uncertainty surrounding this estimate.

The ash content of coal consumed in each region was calculated from the ash content of mined coal and a coal transportation flow matrix by CCTA (2003) and

Jiang (2004). The national average of ash content in coal for 2005 is estimated at 21.7%.

4.3.3   Discussion

The objective of the research described so far has been to establish an integrated emission factor database for the development of emission inventories of Chinese coal-fired power plants, which is a key link in the integrated research of this volume to evaluate Chinese national emission control policies. Before describing the effects of policy on emissions, let us point out some key implications of the work just described.

We first note some differences between the results of this study and AP-42 (U.S. EPA 1999), the widely used emission factor database for the United States.[4] For SO_2, the emission factors for bituminous and lignite combustion in AP-42 are $19.0S$ and $17.5S$ kg/t, respectively, higher than those obtained here ($18.0S$ and $15.0S$ kg/t). The value for anthracite in AP-42 is even higher, reaching $19.5S$ kg/t.

The comparison of NO_X emission factors is similar, as shown in figure 4.13A where the solid squares represent the estimates from this study and the open circles represent AP-42. The emission factors for four pulverized coal technology–fuel combinations and one circulating fluidized-bed combustion (CFBC) system are compared; the PC units represented here are those larger than 300 MW. The emission factors in AP-42 are 4–19% higher than the mean values of the corresponding unit types in the current assessment, but within the 95% CI for three of the PC types. For tangentially fired burners combusting anthracite, the AP-42 value is beyond the upper limit of the 95% CI of this study.

We also computed the emission factors in terms of emission per unit of energy input (kg/GJ), and these are shown in figure 4.13B. Here, our estimates are much closer to the AP-42 factors. The emission factor for wall-fired boilers burning anthracite in AP-42 is even lower than that obtained in this study. Therefore, the difference between the AP-42 and this study's NO_X emission factors (kg of NO_X per ton of coal) can be largely explained by the difference in coal heating values between the two countries. The average heating values for bituminous and anthracite in the United States are around 24.4 and 27.9 MJ/kg, respectively, more than 10% higher than those in China (U.S. EPA 1999; Jin 2001).

Regarding PM, we compare the emission factors that are normalized to the total PM levels for all of the uncontrolled and controlled technology types as shown in figure 4.13C. The factors are given for $PM_{2.5}$, PM_{10}, and PM. For uncontrolled PM, the AP-42 emission factors of the different size fractions are 25–28% lower than

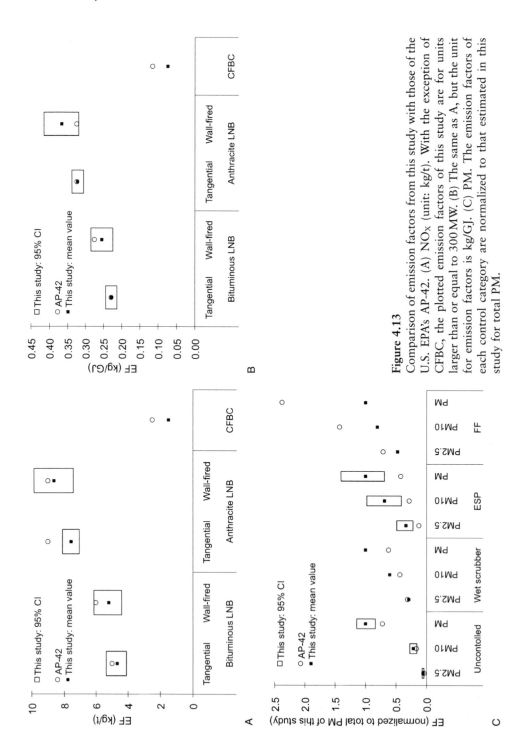

Figure 4.13

Comparison of emission factors from this study with those of the U.S. EPA's AP-42. (A) NO$_x$ (unit: kg/t). With the exception of CFBC, the plotted emission factors of this study are for units larger than or equal to 300 MW. (B) The same as A, but the unit for emission factors is kg/GJ. (C) PM. The emission factors of each control category are normalized to that estimated in this study for total PM.

those found in this study. For the wet scrubber–controlled and ESP-controlled groups, the AP-42 factors are much lower—respectively 28–38% and 58–62% lower—than ours. (Wet scrubber–controlled $PM_{2.5}$, however, is 7.1% higher.) This difference implies that the removal efficiencies of wet scrubbers and ESPs in the United States are higher than those in China. However, the AP-42 emission factors for FF control of PM are significantly higher than those of this study, although this is a far less conclusive comparison because of our small sample size of only two field tests.

We next discuss the contributions of the various parameters to the variance of the emission factors. The quantitative uncertainties of the emission factors were calculated by dividing the differences between the upper and lower limits of the 95% CIs by the mean values. The contributions of different parameters to the emission factor variance were provided through Monte Carlo simulations described in section 4.3.2. Ignoring the sulfur content variation, the uncertainty for uncontrolled SO_2 from PC boilers was estimated to be 7% in this study.

For NO_X, the uncertainties with the variance contributions of the various parameters are shown in figure 4.14A. The uncertainties of the NO_X emission factors for different unit types are represented by the black squares and varied from 14% to 45%. The shaded area represents the contribution of the uncertainty of NO_X concentration, and the contribution of the uncertainty of fuel heating value is the unshaded area. The emission factor variances for only three out of the listed nine unit types were dominated by the uncertainty of fuel heating value, while the remaining six were dominated by the uncertainty of NO_X concentrations, particularly for units combusting anthracite. In order to reduce the uncertainty, more field measurements of NO_X emission levels should thus be conducted at power plants using anthracite.

As shown by the black squares in figure 4.14B, the uncertainties of PM emission factors generally increased from coarse to fine particles. The emission factors for units with ESP and wet-FGD systems had uncertainties over 80% for all the size fractions, much higher than those for SO_2 and NO_X. There are four sources of uncertainty for the PM factors: the ash release ratio, ESP removal efficiency, mass fraction by size, and FGD removal efficiency. The uncertainty of the ash release ratio contributed more than that of mass fraction by size (78% versus 22%) to the variance of the uncontrolled emission factor for large particles (>10 microns), but the opposite was true for small particles (8% versus 92% for $PM_{2.5}$). For postcontrol emission factors, the contributions to variance of the ash release ratio decreased substantially, to less than 10%. Electrostatic precipitator removal efficiencies played

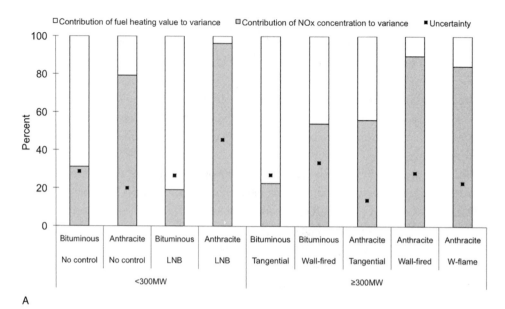

A

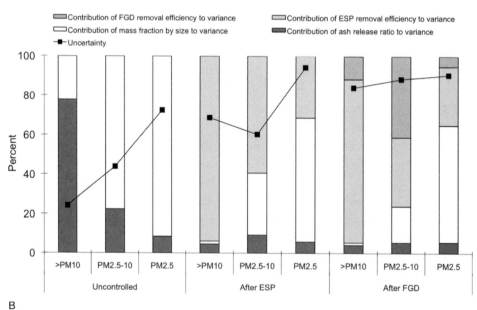

B

Figure 4.14
Uncertainty analyses of emission factors. (A), NO$_X$. (B) PM.

very important roles for the variances of all size fractions, particularly for large particles (more than 80% for particles larger than 10 microns), while the mass fraction uncertainty contributed around 60% to the variance of the $PM_{2.5}$ emission factor. To reduce the uncertainties of PM emission factors, studies of mass fractions of fine particles and ESP removal efficiencies are still needed.

The limitations of this research on emission factors should be acknowledged. Besides boiler types, fuel qualities, and emission control devices, emission levels can also be influenced by the operating parameters of power units like boiler load, coal fineness, oxygen enrichment, and dust-dislodging rapping cycles of the collector (Bi and Chen 2004; Yi et al. 2006; X. W. Liu et al. 2006). Although all the emissions data in this study were obtained during normal working conditions, those operating conditions were hardly identical. The testing instruments used by various studies were also different and thus subject to different sensitivity levels. Accordingly, the uncertainties of emission factors could possibly be even larger than estimated here.

At present, the penetration of advanced technologies into the Chinese power sector is increasing quickly, including ultra-supercritical boilers (600–1000 MW) with wet-FGD, SCR, and FF systems. To date, few field measurements have been taken and published on these units. Moreover, the removal efficiencies of FGD and SCR can vary considerably depending on the quantities of sorbent used.[5] Therefore, the emission factors obtained for these units should be applied with caution.

4.4 Results

4.4.1 Emissions from 2000 to 2005

The emissions of SO_2, NO_X, and PM from coal-fired power plants from 2000 to 2005 are calculated using unit-based activity data and the emission factors described earlier. The national totals, as well as the regional contributions, are shown in figure 4.15.

From 2000 to 2005, the period when the power sector in China developed the fastest in the past 30 years, SO_2 emissions increased by 50% nationwide and, notably, by 80% in the northwest and 60% in south-central China. Sulfur dioxide emissions from coal-fired power plants in 2005 are estimated to have been 16,097 kt, approximately 53% of total national emissions. Five provinces with the largest coal consumption (Shandong, Henan, Hebei, Shanxi, and Jiangsu) emitted more than 1000 kt of SO_2 each, followed by Guizhou and Sichuan, where high sulfur coal was commonly used. The SO_2 emissions of power plants in Beijing, Hainan, and Qinghai were less than 100 kt. Emissions from units less than 300 MW, which accounted for

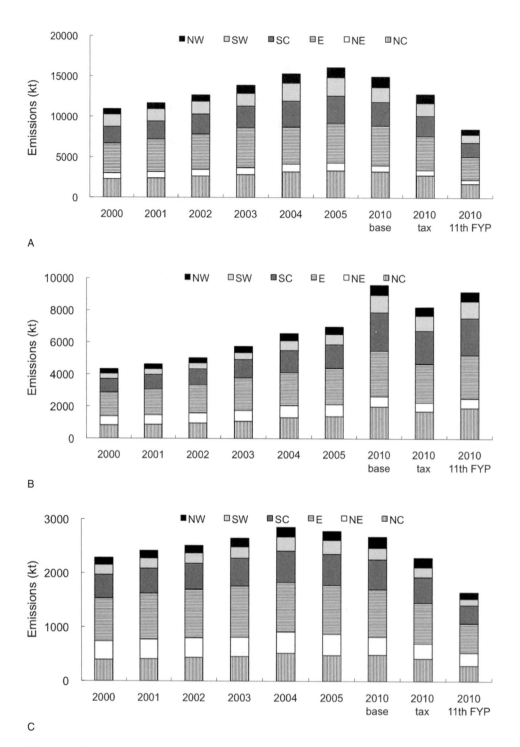

Figure 4.15
Emissions of coal-fired power plants from 2000 to 2005 and 2010. (A) SO₂. (B) NOₓ. (C) PM.

half of the total capacity in 2005, reached 10,124 kt, accounting for 63% of the total emissions.

Rising even faster than SO_2, national NO_X emissions increased 60% from 2000 to 2005. Southwestern and south-central China were the areas with the biggest increases, by 90% and 70%, respectively. The NO_X emissions from coal-fired power plants in 2005 are estimated to have reached 6965 kt, approximately 36% of total national emissions. Jiangsu, Shandong, and Henan each emitted over 500 kt of NO_X, while Hainan and Qinghai generated less than 20 kt. The generating units with low NO_X burners (LNBs), which accounted for 62% of the total capacity and consumed 53% of coal in the power sector in 2005, emitted 3255 kt of NO_X, or 47% of total emissions. This figure reflects the relatively modest emission control benefits of LNBs. As for the effects of fuel quality on emissions, anthracite causes higher NO_X emission because of the relatively low content of volatile matter compared with bituminous and other coals. In 2005 the share of anthracite and "lean" coals (those with volatile matter content of 10–20%) was about 20%, and the units using those coals emitted 2230 kt NO_X, 32% of the total emitted nationally from the power sector.

Primary PM emissions by the power sector rose 20% from 2000 to 2004, to 2848 kt, with the southwest and northwest regions having the biggest growth (40%). In 2005 emissions dropped slightly to 2774 kt, less than 10% of total national emissions. Henan, Jiangsu, and Shandong each emitted more than 200 kt in 2005. Emissions of PM_{10} and $PM_{2.5}$ in the same year were 1842 and 994 kt, that is, 66% and 36% of the total PM mass, respectively. Large differences of PM size characteristics were found between emissions from pulverized coal boilers and from grate boilers. The ratios of $PM_{2.5}$ and PM_{10} to TSP emission from grate boilers were 65% and 89%, respectively, much higher than those from pulverized coal boilers, at 31% and 63%, respectively. The main reason is that the wet scrubbers and cyclones commonly applied in grate boiler plants have much lower removal efficiency of fine particles than the ESPs used in PC boiler plants.

We compared our estimations with other studies for year 2000, as well as other years where available. These studies include the official NBS (2006), S. X. Wang (2001), Streets el al. (2003), Tian (2003), Zhu, Wang, and Zheng (2004), van Aardenne et al. (2005), Ohara et al. (2007), Zhang et al. (2007a, 2007b), and Klimont et al. (2009). As shown in figure 4.16A, the SO_2 emissions from the coal-fired power sector in 2000 have been estimated in seven studies to be in the range 7,100–13,330 kt. Our result (marked by open diamonds in the figure) is 10,950 kt, which is higher than other studies except for the EDGAR data set (van Aardenne

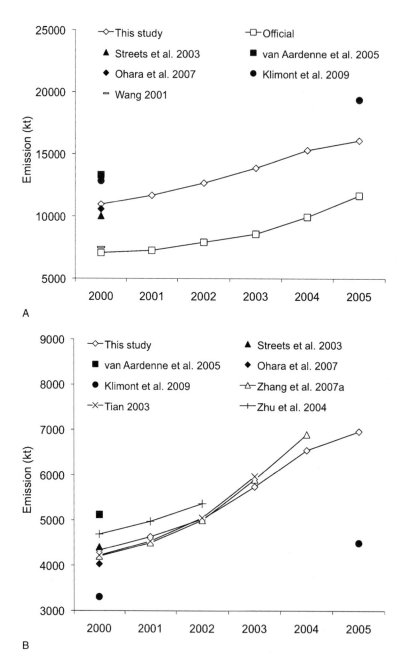

Figure 4.16
Emission comparisons with other studies. (A) SO$_2$. (B) NO$_X$. (C) PM.

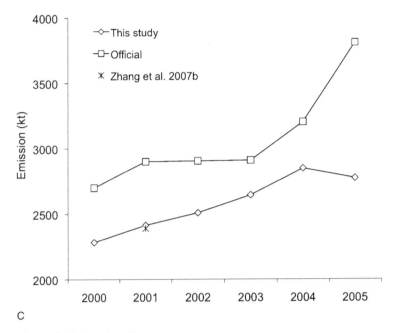

C

Figure 4.16 (continued)

et al. 2005) and Klimont et al. (2009). In 2005, our estimate was 38% higher than the official estimate (NBS 2006). The reasons for our higher estimates are as follows: (1) the activity level estimated with our unit-specific method is somewhat higher than the national statistics, as was the case with Akimoto et al. (2006); and (2) the sulfur retention used in the official estimation was usually 0.15, higher than the value we determined previously.

A similar situation is also found for NO_X. As shown in figure 4.16B, NO_X emissions from the power sector are estimated by the different sources to be between 2430 and 5120 kt in 2000. Our result is 4340 kt, which is lower than EDGAR and Zhu, Wang, and Zheng (2004), but higher than other studies. The estimate by Zhang et al. (2007a) in 2003–2004 is slightly higher than that of this study, attributed mainly to the application of relatively high emission factors for small units in their work.

Regarding PM, Zhang et al. (2007b) estimate the emissions from the Chinese power sector at 2390 kt in 2001, consistent with the result of this study, 2413 kt. Lei et al. (2011) estimated such emissions in 2000 at a very similar level, 2320 kt, but estimated those in 2005 to be 3090 kt, higher than our value of 2774 kt. The official estimate by MEP, however, was consistently higher than ours and these other estimates during 2000–2005. This discrepancy may result from the different levels of removal efficiency of dust collectors (including the PM control benefits from wet-FGD) in the estimates.

4.4.2   Emissions in 2010

The estimated emissions of the power sector in 2010 under the three scenarios assessed in this volume are illustrated in the last three columns of figures 4.15A and C. In the base case, although coal consumption by power plants would have sharply increased by 42% from 2005 to 2010, SO_2 emissions would have slightly decreased by 7% to 14,964 kt, attributed mainly to the installation of FGD equipment on newly built units. Units less than 300 MW would have continued contributing strongly to SO_2 emissions, accounting for 52% of the power sector total. As for regional growth, emissions in north-central, northeastern, south-central, and southwestern China would have decreased by 4–18%, while those in the northwest would have increased by 4% and those in the east would have been roughly unchanged. Shandong, Shanxi, Henan, Hebei, and Jiangsu provinces would each have had emissions exceeding 1000 kt.

In the carbon tax case, a 15% reduction in coal consumption projected in the power sector by the economic analyses of chapter 9 translates into a reduction of estimated SO_2 emissions to 12,800 kt in 2010, which is assumed to occur evenly as a 15% reduction at all power plants.

In the 11th FYP policy case (our estimate of what actually happened), with coal consumption assumed to be equal to that of the base case, the total SO_2 emissions from the power sector in 2010 are estimated to have reached only 8470 kt, a reduction of 47% compared to 2005. This outcome represents enormous benefits from the government emission control measures over the five years. As a result of the retirement of small units, the estimated share of SO_2 emissions from units smaller than 300 MW decreased to 45% from 63% in 2005. Emissions in Hainan, Sichuan, and Hunan provinces declined by over 60%.

In contrast to SO_2, no strengthened control measures for NO_X had been implemented in recent years prior to 2010 except for an amendment to the emission standard in 2003. Advanced NO_X control technologies, such as SCR, are not believed to have been widely deployed by 2010; the penetration rate of SCR in the thermal power sector was estimated to have been lower than 5% in 2008 according to the Ministry of Environmental Protection. Thus NO_X emissions are believed to have continued rising, and would have reached 9581 kt in the 2010 base case. Units with LNBs, increasing to 70% of the total capacity, would have emitted 64% of this total. Emissions of NO_X in southwest and south-central China would have increased by over 60%, followed by north-central, east, and northwest China, by 25–44%. The emissions of Henan, Shandong, Shanxi, Guangdong, and Jiangsu would have each exceeded 600 kt. Shanxi and Henan would have had the largest incremental emission gain in the base case, that is, 329 and 305 kt, respectively. Emissions of NO_X in Shanghai and Beijing, however, would have declined as a result of the deployment of SCR systems.

In the carbon tax case, the national emissions of NO_X by the power sector would have only reached 8195 kt in 2010, a sizable reduction versus the base case total of 9581 kt, as a result of reduced coal consumption. The effect on NO_X emissions of the 11th FYP policy case, by contrast, was considerably less, with total emissions in 2010 estimated to be 9159 kt, a mere 4% reduction from the base case. The emissions from units with LNBs reached 6697 kt, approximately 73% of the total emissions by the power sector, compared to 64% in the base case. This shows that phasing out small, inefficient power units (whatever its benefits both economically and on emissions of other pollutants) is not an effective approach to NO_X control. The benefit of LNBs in the power sector, additionally, is limited. It appears that without wider deployment of SCR systems, NO_X emissions will continue to rise as long as coal-fired power generation continues to expand in China.

Although no strengthened controls of PM were implemented from 2005 to 2010, in the base case emissions from the power sector would have decreased slightly from 2774 kt in 2005 to 2671 kt in 2010. In the carbon tax and 11th FYP policy cases, emissions would have declined (or did decline) even further, to 2285 and 1645 kt, respectively. These are reductions of 18% and 40% compared to 2005.

Significant benefits in PM emission control are clearly indicated for both policy options, but especially for the 11th FYP measures (even though they target SO_2, not PM). There are two main reasons for this: (1) widely installed wet-FGD systems have a considerable side benefit in control of PM, particularly larger particles, by operating similarly to wet scrubbers and with comparable removal efficiency; and (2) many small units employing dust collectors with relatively low removal efficiency were replaced by large units employing high-efficiency dust collectors such as ESP or FF systems. Importantly for health implications, however, the estimated share of PM_{10} and $PM_{2.5}$ in total PM increased to 39% and 70%, respectively, attributed mainly to the poor removal efficiencies of FGD on finer particles. Regarding regional differences, estimated PM emissions in the southwest decreased by 54% from 2005 to 2010 in the 11th FYP case, followed by south-central and north-central regions, declining 41% and 40%, respectively. Jiangsu, Shandong, Henan, Heilongjiang,

Zhejiang, and Inner Mongolia each had estimated PM emissions in 2010 exceeding 100 kt in the 11th FYP policy case.

4.4.3   Spatial Allocation

To better understand regional distributions and ultimately to support atmospheric simulations, the plant-level emissions estimated in this chapter are located in a 0.5° by 0.67° grid covering east Asia matched to the nested grid of the GEOS-Chem atmospheric model used in chapter 7.

Gridded coal-fired power emissions of SO_2, NO_X, and PM in 2005 and the three 2010 scenarios are shown in the 12 maps in figure 4.17. It can be seen from the figure that the eastern region has the highest emission intensity for all species, followed by the south-central and north-central regions. These regions, which cover only 35% of China's territory but 69% of the national population and 78% of GDP in 2005, accounted for more than 70% of the national power emissions of SO_2, NO_X, and PM in 2005 and in all scenarios in 2010. This spatial distribution illustrates that population density and concentration of economic activity largely determine power plant location and emissions. Even though the development of western China is emphasized under current national policies, the spatial distribution of power plant construction and emissions is not expected to change significantly in the near future.

4.5   Summary and Conclusions

A generating unit-based method is developed to estimate the SO_2, NO_X, and PM emissions from coal-fired power plants in China, based on detailed information of boiler type, fuel quality, emission control technology, and geographical location. The main findings follow.

The emission factor of uncontrolled SO_2 from pulverized coal boilers burning bituminous or anthracite coal is estimated to be 18.0S kg/t, with a 95% CI of 17.2S–18.5S (where S is the sulfur content). Nitrogen oxide emission factors for PC boilers range from 4.0 to 11.2 kg/t, with uncertainties of 14%–45% for different unit types. The emission factors for uncontrolled $PM_{2.5}$, PM_{10}, and total PM emitted by PC boilers are estimated to be 0.4A, 1.5A, and 6.9A kg/t, respectively, with 95% CIs of 0.3A–0.5A, 1.1A–1.9A, and 5.8A–7.9A (where A is the ash content). The analogous PM values for emissions with electrostatic precipitator (ESP) controls are 0.032A (95% CI: 0.021A–0.046A), 0.065A (0.039A–0.092A), and 0.094A

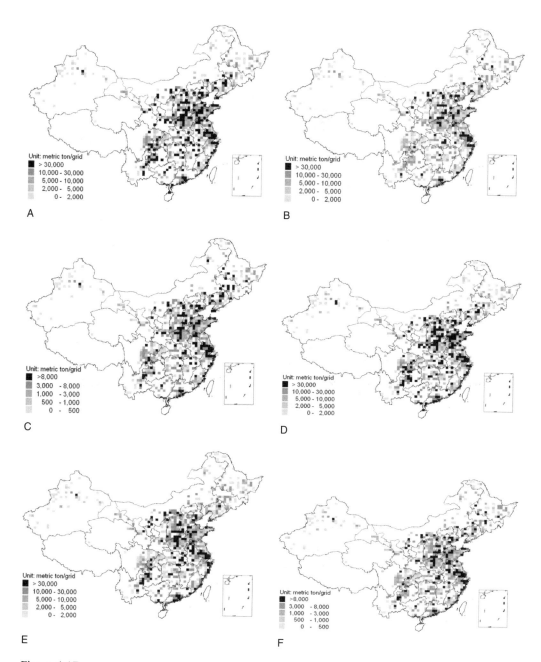

Figure 4.17
Gridded emissions by the power sector in 2005 and 2010 under the three scenarios. (A) SO₂
2005. (B) NO_X 2005. (C) PM 2005. (D) SO₂ base case 2010. (E) NO_X base case 2010.
(F) PM base case 2010.   (*Figure continued*)

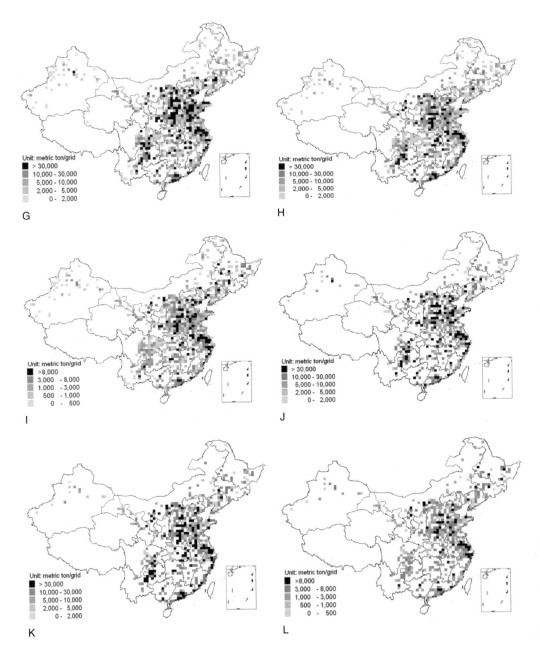

Figure 4.17 (continued)
(G) SO$_2$ carbon tax case 2010. (H) NO$_X$ carbon tax case 2010. (I) PM carbon tax case 2010.
(J) SO$_2$ 11th FYP policy case 2010. (K) NO$_X$ 11th FYP policy case 2010. (L) PM 11th FYP
policy case 2010.

(0.065A–0.132A) kg/t; and for those with both ESP and wet-FGD systems 0.015A (0.009A–0.022A), 0.021A (0.013A–0.032A), and 0.023A (0.014A–0.035A).

From 2000 to 2005, the period when the power sector developed the fastest in the past 30 years, SO_2, NO_X, and PM emissions of coal-fired power plants increased by 50%, 70%, and 20%, respectively. In 2005, the estimated SO_2, NO_X, and PM emissions were 16,097, 6965, and 2774 kt, respectively. East, south-central, and north-central China accounted for more than 70% of the total emissions, illustrating the role of population density and economic concentration on determining the location of Chinese power plants and associated emissions.

Emissions are estimated for 2010 under a base case and two policy cases, a carbon tax and the 11th FYP SO_2 controls. In the hypothetical base case, the SO_2, NO_X, and PM emissions by the power sector in 2010 would have reached 14,964, 9581, and 2671 kt, or 93%, 138%, and 96% of the 2005 levels, respectively. In the hypothetical carbon tax case, emissions of all species from power plants would have declined 15% because of reduced coal consumption. Wide deployment of FGD and retirement of small units under the 11th FYP produced sizable benefits in SO_2 and PM emission control in the power sector, reducing estimated emissions to 8470 and 1645 kt, respectively, or –43% and –38% compared to the base case. In contrast, estimated NO_X emissions reached 9159 kt, merely 4% lower than the base case (or 32% higher than in 2005).

With regard to future research, we note considerable uncertainties in emission factors of units with advanced control technologies because of the limited deployment of such systems to date in China and a corresponding lack of field data. More measurements of such units are needed to improve the emission factor database.

Concerning future policy, these assessments spotlight that NO_X emissions from power plants not only tracked growth of coal consumption by the power sector during the period of the 11th FYP but also have clearly become a major new challenge in Chinese air pollution control. To date, deployment of advanced NO_X-control systems like SCR is just incipient in China. That the 12th FYP includes a 10% NO_X emission reduction target is an important step in addressing this escalating challenge. The strong success of the 11th FYP SO_2 controls in the power sector assessed here might provide guidance for development and implementation of specific policies and emissions standards to achieve the new NO_X reduction targets. Whatever the policy mechanism chosen, the effort to reduce China's NO_X emissions will need to address not only power plants but also other large industrial point sources, and especially China's cement kilns, the subject of the next chapter.

Acknowledgments

Portions of this chapter are adapted from two articles published in the journal *Atmospheric Environment* (Zhao et al. 2010; Zhao et al. 2008); permission to reproduce sections of these articles was kindly granted by the editors of this journal. This chapter was developed from research originally funded by the Hi-Tech Research and Development Program (No. 2006AA06A305) and the National Basic Research Program of China (No. 2005 CB422206). Further development of the research for the purposes of this book was supported by the China Sustainable Energy Program of the Energy Foundation and the U.S. National Science Foundation (Grant ATM-1019134).

Notes

1. These data come from internal sources including personal communications with government officials.

2. The China Electricity Council reported that 62 GW of small plants were closed in the 11th FYP, compared to our estimate of 59 GW. It reported 707 GW of installed *thermal* power capacity by the end of 2010, which, accounting for the gas- and oil-fired share and the several additional GW of retired units, amounts to ~650 GW of coal-fired units, compared to our 651 GW. See http://tj.cec.org.cn/tongji/niandushuju/2011–02–23/44236.html, last accessed on November 27, 2012. Employing the CEC numbers with benefit of hindsight would alter our final results, but only very slightly.

3. See endnote 1 in chapter 2 and http://news.xinhuanet.com/english2010/china/2011–08/29/c_131081860.htm, last accessed October 10, 2012.

4. We make comparisons especially to U.S. EPA's AP-42 because it is the most comprehensive source of such data in the world and is the default source that researchers cite routinely to place new results from other countries in comparative context. However, there is a substantial and constantly improving scientific literature on emission factors, as indicated by the citations in this chapter and in the next two chapters. We recommend that fellow researchers consult this literature for comparisons to other results for China and other countries, as well as to keep abreast of new findings.

5. Assumptions about removal efficiencies of FGD are important for this and other assessments of China's SO_2 controls, but our plant-by-plant database does not include these data, and all other studies face a similar challenge: finding defensibly generalizable estimates. Steinfeld, Lester, and Cunningham (2009) and Y. Xu (2011) provide helpful evidence, but both have deficiencies for our purposes, specifically sampling biases toward plants in a couple of provinces that are not obviously representative of the national industry. We started with an internal national survey conducted by the Chinese MEP early in the 11th FYP (2007) and adjusted it upward to 85% for 2010 based on evidence in early versions of Y. Xu (2011) of operations that were improving as a result of rising effluent fees, the price premium for desulfurized electricity, and rising penalties and inspections.

References

Akimoto, H., T. Ohara, J. Kurokawa, and N. Horii. 2006. Verification of energy consumption in China during 1996–2003 by using satellite observational data. *Atmospheric Environment* 40(40):7663–7667.

Bi, Y. S., and G. H. Chen. 2004. Countermeasures and suggestions for controlling the NO_X emission of utility boilers. *Electric Power* 37(6):37–41. In Chinese.

Buhre, B. J. P., J. T. Hinkley, R. P. Gupta, T. F. Wall, and P. F. Nelson. 2005. Submicron ash formation from coal combustion. *Fuel* 84(10):1206–1214.

China Coal Transportation Association (CCTA). 2003. Statistical data on inter-province coal transportation in China. Internal information. In Chinese.

China Electricity Council (CEC). 2007. National power industry statistics, 2006. Internal information. In Chinese.

Chinese Research Academy of Environmental Science (CRAES). 2006. *Study on the Methodology of Checking Air Pollutant Emission Quantity*. 2003BA614A-05. In Chinese.

Editorial Board of China Electric Power Yearbook. 2008. *China Electric Power Yearbook, 2008*. Beijing: China Electric Power Press. In Chinese.

Frey, H. C., and J. Y. Zheng. 2002. Quantification of variability and uncertainty in air pollutant emission inventories: Method and case study for utility NO_X emissions. *Journal of the Air & Waste Management Association* 52(9):1083–1095.

Gao, X. P. 2006. Study on mineral matter vaporization and submicron particle formation during coal combustion. Master's thesis, Huazhong University of Science and Technology, Wuhan, China. In Chinese.

Ge, S., Z. P. Bai, W. L. Liu, T. Zhu, T. J. Wang, S. Qing, and J. F. Zhang. 2001. Boiler briquette coal versus raw coal: Part I—Stack gas emissions. *Journal of the Air & Waste Management Association* 51(4):524–533.

Hao, J. M., H. Z. Tian, and Y. Q. Lu. 2002. Emission inventories of NO_X from commercial energy consumption in China, 1995–1998. *Environmental Science & Technology* 36(4): 552–560.

He, L. G., and Y. H. Wu. 1999. Experimental study on fly ash particulate size distribution of electrostatic precipitator in power plants. *Electric Power Environmental Protection* 15(4): 4–7. In Chinese.

Helble, J. J. 2000. A model for the air emissions of trace metallic elements from coal combustors equipped with electrostatic precipitators. *Fuel Processing Technology* 63(2–3): 125–147.

Huang, W., Z. Y. Luo, H. Xu, T. Wang, P. Wang, M. X. Fang, X. Gao, and K. F. Cen. 2003. Research on distribution of aerosol and PAHs in power plant stack. *Energy Engineering* 2:29–32. In Chinese.

Jiang, J. K. 2004. Preliminary studies on emission and control of atmosphere mercury in China. Master's thesis, Tsinghua University, Beijing, China. In Chinese.

Jin, W. Q. 2001. *Power Plant Boiler*. Beijing: China Waterpower Press.

Klimont, Z., J. Cofala, J. Xing, W. Wei, C. Zhang, S. Wang, K. J. Jiang, P. Bhandari, R. Mathur, P. Purohit, P. Rafaj, A. Chambers, and M. Amann. 2009. Projections of SO_2,

NO$_X$ and carbonaceous aerosols emissions in Asia. *Tellus Series B—Chemical and Physical Meteorology* 61(4):602–617.

Lei, Y., Q. Zhang, K. B. He, and D. G. Streets. 2011. Primary anthropogenic aerosol emission trends for China, 1990–2005. *Atmospheric Chemistry and Physics* 11:931–954.

Lillieblad, L., A. Szpila, M. Strand, J. Pagels, K. Rupar-Gadd, A. Gudmundsson, E. Swietlicki, M. Bohgard, and M. Sanati. 2004. Boiler operation influence on the emissions of submicrometer-sized particles and polycyclic aromatic hydrocarbons from biomass-fired grate boilers. *Energy & Fuels* 18(2):410–417.

Lind, T., J. Hokkinen, J. K. Jokiniemi, S. Saarikoski, and R. Hillamo. 2003. Electrostatic precipitator collection efficiency and trace element emissions from co-combustion of biomass and recovered fuel in fluidized-bed combustion. *Environmental Science & Technology* 37(12):2842–2846.

Liu, J. Z., H. Y. Fan, J. H. Zhou, X. Y. Cao, and K. F. Cen. 2003. Experimental studies on the emissions of PM$_{10}$ and PM$_{2.5}$ from coal-fired boiler. *Proceedings of the CSEE* 23(1):145–149. In Chinese.

Liu, X. W., M. H. Xu, D. X. Yu, Y. Yu, X. P. Gao, and Q. Cao. 2006. Research on formation and emission of inhalable particulate matters at different oxygen content during coal combustion. *Proceedings of the CSEE* 26(15):46–50. In Chinese.

Moisio, M. 1999. Real time size distribution measurement of combustion aerosols. PhD diss., Tampere University of Technology, Tampere, Finland.

Moisio, M., A. Laitinen, J. Hautanen, and J. Keskinen. 1998. Fine particle size distributions of seven different combustion power plants. *Journal of Aerosol Science* 29(Suppl. 1): S459–S460.

National Bureau of Statistics (NBS). 2006. *China Environment Statistical Yearbook, 2006.* Beijing: China Statistics Press.

———. 2007. *China Energy Statistical Yearbook, 2006.* Beijing: China Statistics Press.

Ohara, T., H. Akimoto, J. Kurokawa, N. Horii, K. Yamaji, X. Yan, and T. Hayasaka. 2007. An Asian emission inventory of anthropogenic emission sources for the period 1980–2020. *Atmospheric Chemistry and Physics* 7(16):4419–4444.

Qi, L. Q. 2006. Study on the electrostatic precipitability and escape mechanism of the fine particulate from electrostatic precipitator of coal-fired boilers. PhD diss., North China Electric Power University, Beijing, China. In Chinese.

State Environmental Protection Administration (SEPA) and General Administration of Quality Supervision, Inspection and Quarantine of the People's Republic of China (AQSIQ). 2003. Emission standard of air pollutants for thermal power plants. National standard no. GB13223-2003. In Chinese.

State Environmental Protection Administration (SEPA) and China State Bureau of Technology Supervision (CSBTS). 1996. The determination of particulates and sampling methods of gaseous pollutants emitted from exhaust gas of stationary source. National standard no. GB/T16157-1996. In Chinese.

Steinfeld, E. S., R. K. Lester, and E. A. Cunningham. 2009. Greener plants, grayer skies? A report from the front lines of China's energy sector. *Energy Policy* 37:1809–1824.

Streets, D. G., T. C. Bond, G. R. Carmichael, S. D. Fernandes, Q. Fu, D. He, Z. Klimont, S. M. Nelson, N. Y. Tsai, M. Q. Wang, J. H. Woo, and K. F. Yarber. 2003. An inventory of gaseous and primary aerosol emissions in Asia in the year 2000. *Journal of Geophysical Research—Atmospheres* 108(D21), Art. No. 8809.

Sui, J. C. 2006. Study on submicron particle formation and emission during coal combustion. PhD diss., Huazhong University of Science and Technology, Wuhan, China. In Chinese.

Tian, H. Z. 2003. Studies on present and future emissions of nitrogen oxides and its comprehensive control policies in China. PhD diss., Tsinghua University, Beijing, China. In Chinese.

U.S. Environmental Protection Agency (USEPA). 1999. Emissions Factors & AP 42: Compilation of Air Pollutant Emission Factors. Available at http://www.epa.gov/ttnchie1/ap42/, last accessed March 25, 2013.

van Aardenne, J. A., F. Dentener, J. G. J. Olivier, and J. A. H. W. Peters. 2005. The EDGAR 3.2 Fast Track 2000 dataset (32FT2000). Available at http://www.mnp.nl/edgar/, last accessed March 25, 2013.

Wang, H., Q. Song, Q. Yao, and C. H. Chen. 2008. Experimental study on removal effect of wet flue gas desulphurization system on fine particles from a coal-fired power plant. *Proceedings of the CSEE* 28(5):1–7. In Chinese.

Wang, P. 2008. Study on PM$_{2.5}$ emission and control in coal combustion boiler of power plants. PhD diss., Zhejiang University, Hangzhou, China. In Chinese.

Wang, S. X. 2001. Technical options and planning on control of sulfur dioxide emitted from coal-fired power plants in China. PhD diss., Tsinghua University, Beijing, China. In Chinese.

Xu, H., Z. Y. Luo, T. Wang, P. Wang, X. Gao, Z. L. Shi, and K. F. Cen. 2004. Studies on the characteristics of aerosol and trace metals emitted from a CFB coal-fired power plant. *Acta Scientiae Circumstantiae* 24(3):515–519. In Chinese.

Xu, Y. 2011. Improvements in the operation of SO$_2$ scrubbers in China's power plants. *Environmental Science and Technology* 45:380–385.

Yao, Q., L. S. Chen, Z. W. Chen, M. R. Wei, and H. Li. 2007. Discussion on control technology of boiler flue gas PM$_{10}$ emission in coal-fired power plants. *Electric Power Environmental Protection* 23(1):52–54. In Chinese.

Yi, H. H., J. M. Hao, L. Duan, X. H. Li, and X. M. Guo. 2006. Characteristics of inhalable particulate matter concentration and size distribution from power plants in China. *Journal of the Air and Waste Management Association* 56(9):1243–1251.

Ylatalo, S. I., and J. Hautanen. 1998. Electrostatic precipitator penetration function for pulverized coal combustion. *Aerosol Science and Technology* 29(1):17–30.

Zhang, Q., D. G. Streets, G. R. Carmichael, K. B. He, H. Huo, A. Kannari, Z. Klimont, I. S. Park, S. Reddy, J. S. Fu, D. Chen, L. Duan, Y. Lei, L.T. Wang, and Z. L. Yao. 2009. Asian emissions in 2006 for the NASA INTEX-B mission. *Atmospheric Chemistry and Physics* 9(14):5131–5153.

Zhang, Q., D. G. Streets, K. He, Y. Wang, A. Richter, J. P. Burrows, I. Uno, C. J. Jang, D. Chen, Z. Yao, and Y. Lei. 2007a. NO$_X$ emission trends for China, 1995–2004: The view from the ground and the view from space. *Journal of Geophysical Research—Atmospheres* 112(D22), Art. No. D22306.

Zhang, Q., D. G. Streets, K. B. He, and Z. Klimont. 2007b. Major components of China's anthropogenic primary particulate emissions. *Environmental Research Letters* 2(4), Art. No. 045027.

Zhao, Y. 2008. Study on air pollutant emission of coal-fired power plants in China and its environmental impacts. PhD diss., Tsinghua University, Beijing, China. In Chinese.

Zhao, Y., C. P. Nielsen, and M. B. McElroy. 2012. China's CO_2 emissions estimated from the bottom up: Recent trends, spatial distributions, and quantification of uncertainties. *Atmospheric Environment* 59:214–223.

Zhao, Y., S. X. Wang, L. Duan, Y. Lei, P. F. Cao, and J. M. Hao. 2008. Primary air pollutant emissions of coal-fired power plants in China: Current status and future prediction. *Atmospheric Environment* 42(36):8442–8452.

Zhao, Y., S. X. Wang, C. P. Nielsen, X. H. Li, and J. M. Hao. 2010. Establishment of a database of emission factors for atmospheric pollutant emissions from Chinese coal-fired power plants. *Atmospheric Environment* 44(12):1515–1523.

Zhu, F. H., S. Wang, and Y. F. Zheng. 2004. NO_x emitting current situation and forecast from thermal power plants and countermeasures. *Energy Environmental Protection* 18(1): 1–5. In Chinese.

5

Primary Air Pollutants and CO_2 Emissions from Cement Production in China

Yu Lei, Qiang Zhang, Chris P. Nielsen, and Kebin He

5.1 Introduction

China is the largest cement-producing and cement-consuming country in the world today. Cement production in China was 1.39 billion metric tons in 2008 (CMIIT 2009), which accounted for 50% of the world's production (USGS 2009). Enormous quantities of air pollutants are emitted from cement production, including sulfur dioxide (SO_2), nitrogen oxides (NO_X), carbon monoxide (CO), and total particulate matter (PM) of all sizes,[1] resulting in significant regional and global environmental problems. Accordingly, the cement industry is a primary source of air pollution in China. For example, it is the largest source of PM emissions, accounting for 40% from all industrial sources[2] in 2000 according to official statistics (CEYEC 2001) and 21% of total national PM emissions in 2005 (Y. Lei, Zhang, and He 2011). Cement production also releases large amounts of carbon dioxide (CO_2) from both fuel combustion and the chemical process producing clinker, in which calcium carbonate ($CaCO_3$) is calcined and reacted with silica-bearing minerals. (Clinker is the solid material produced by a cement kiln that is then milled with gypsum and other additives into cement.) According to the two official national greenhouse gas inventories issued to date (SDRC 2004, NDRC 2012), the cement industry share of process (noncombustion) emissions of CO_2 from Chinese industrial sources rose from 57% in 1994 to 72% in 2005.

There are two main kiln types in China: shaft kilns and rotary kilns. With their higher productivity and efficiency, rotary kilns have dominated the cement industry in western countries since the middle of the 20th century. Starting in the 1980s in China, however, small but easy-to-construct shaft kilns were built all over the country to meet the rapidly increasing demands of the construction industry. By the mid-1990s, they accounted for 80% of production (Q. Z. Lei 2004). The extremely rapid increase in the number of shaft kilns resulted in poor operating practices within the Chinese cement industry. There were more than 7000 cement plants in China in 1997 (Zhou 2003), most of them small and producing high emissions. At the end of the 1990s, China began to restrict construction of new shaft kilns and instead promoted precalciner kilns, which are the most advanced rotary cement kilns. Consequently, the production from precalciner kilns increased very rapidly, and by 2008 they accounted for more than 60% of production (CMIIT 2009).

Since China's cement industry is an important source of many key air pollutant types, systematic and reliable estimation of its emissions is essential for atmospheric modeling and air pollution policy making. Existing emission inventories for China usually treat the cement industry as part of the industrial sector, simply estimating its emissions based on coal consumption (Streets et al. 2003; Ohara et al. 2007). Our previous studies have estimated cement industry emissions of specific air pollutants (Y. Lei et al. 2008) or for a specific year (Zhang et al. 2009), but the historical trends of emissions from China's cement industry have not previously been estimated. In this chapter, we adapt our previously published study (Y. Lei, Zhang, Nielsen, et al. 2011; see the acknowledgments) that developed a historical emission inventory of major air pollutants from China's cement industry for the period 1990–2008 to explore the effects of recent regulations and technologies on these emissions. We then use this emission inventory to estimate emissions from cement for 2010 under the base case and two policy cases (SO_2 control and a carbon tax) introduced in the preceding chapters.

5.2   Methodology and Data

5.2.1   Bottom-up Methodology

The emission inventory developed here includes four gaseous air pollutants (CO_2, SO_2, NO_X, and CO) and PM in three different size ranges: $PM_{2.5}$ (particulates with diameter less than 2.5 microns), PM_{10} (those less than 10 microns), and TSP (total suspended particulates). The kiln is the primary source of most air pollutants in cement production. In this study, kilns are classified into three groups: shaft kilns, precalciner kilns, and other (non-precalciner) rotary kilns.

The burning of fuel in the kilns is the sole source of SO_2, NO_X, and CO emissions. Almost all cement kilns in China use coal as fuel. The emissions of pollutant k in year j, $E_{j,k}$, is given by summing over the emissions from all provinces (i) and types of kilns (m):

$$E_{j,k} = \sum_{i,m} E_{i,j,k,m} = \sum_{i,m} CC_{i,j,m} \times EF_{k,m} \qquad (5.1)$$

where $E_{i,j,k,m}$ is the emission of k from all kilns of type m in province i; this is given by the coal consumption ($CC_{i,j,m}$) and the emission factor for pollutant k from kiln m ($EF_{k,m}$).

As noted previously, CO_2 is produced in cement making not only by fuel combustion but also by process emissions, in this case the calcination of carbonates such as limestone. Theoretically, process CO_2 emissions are determined by the amount and chemical composition of the carbonates consumed during clinker production, but it is impractical to collect the latter information for cement production across the entire country. An alternative estimation approach, recommended by the Intergovernmental Panel on Climate Change (IPCC 2006), is therefore used (as discussed in section 5.2.3). Total CO_2 emissions from the cement industry in year j ($E_j^{CO_2}$) is the sum over all kiln types (m) and provinces (i), and the CO_2 from all kilns of type m in province i ($E_{i,j,m}^{CO_2}$) is the sum of the combustion and process sources:

$$E_j^{CO_2} = \sum_{i,m} E_{i,j,m}^{CO_2} = \sum_{i,m} CC_{i,j,m} \times EF_m + \sum_i CP_{i,j} \times EFC \qquad (5.2)$$

where the first term on the right-hand side gives the CO_2 emissions from coal combustion as the product of the coal consumed ($CC_{i,j,m}$) and the emission factor for kiln type m (EF_m) (using the same approach as for the other gaseous pollutants in equation 5.1). In the second term, process CO_2 is given as the product of clinker production (CP) and the emission factor of CO_2 from calcination during clinker production (EFC).

Estimating PM emissions is more complicated because PM is released from several processes other than the operation of kilns, such as milling of raw material and clinker. Moreover, usage of PM control technology is another important parameter effecting final emissions. Emissions of PM of size y in year j ($E_{j,y}^{PM}$) are the sum over provinces (i) and types of kiln or processes other than clinker making (m). Considering the complexity of cement-producing processes and PM control options, the emissions of PM of size y from kilns using process type m in province i ($E_{i,j,m,y}$) are given as the product of the cement produced ($P_{i,j,m,}$) and the final emission factor ($ef_{j,m,y}$):

$$E_{j,y}^{PM} = \sum_{i,m} E_{i,j,m,y} = \sum_{i,m} P_{i,j,m} \times ef_{j,m,y} \qquad (5.3)$$

$$ef_{j,m,y} = EF_m \times F_{m,y} \times \sum_n C_{m,n,j} \times (1 - \eta_{n,y}) \qquad (5.4)$$

The final emission factor for PM of size y in year j for kiln per process type m ($ef_{j,m,y}$) is given as the product of three terms: (1) the unabated emission factor for process m prior to the effect of PM control devices (EF_m); (2) the fraction of all PM from process m that is in size range y ($F_{m,y}$, where $\Sigma_y F_{m,y} = 1$); and (3) the fraction removed $\Sigma_n C_{m,n,j} \times (1 - \eta_{n,y})$. The variable $C_{m,n,j}$ is the penetration rate of the PM control technology n with $\Sigma_n C_n = 1$, and $\eta_{n,y}$ is the PM_y removal efficiency of the control technology. The three size categories of PM identified are those smaller than 2.5 microns ($PM_{2.5}$), those between 2.5 microns and 10 microns ($PM_{2.5-10}$), and those larger than 10 microns ($PM_{>10}$).

5.2.2   Activity Rates

As shown in equations (5.1) through (5.4), the quantity of coal consumption, clinker production, and cement production by kiln type are required for estimating different kinds of air pollutants. However, there are no detailed statistical data for coal consumption or clinker production by kiln type and province. Instead we estimate these by using the typical energy efficiency and raw mix proportion data of the Chinese cement industry, as follows.

Annual cement production by province from 1990 onward is available from the *China Statistical Yearbook* (NBS 1991–2011). Cement production from different types of kilns is estimated based on the historical capacity of precalciner kilns and other rotary kilns (Kong 2005; Q. Z. Lei 2004; Zeng 2004). As shown in figure 5.1, there have been two periods of especially rapid development in China's cement industry. First, from 1990 to 1995, shaft kilns were the largest contributor to the increase in cement production. From 2002 to 2007, cement production from precalciner kilns increased dramatically, exceeding production from shaft kilns in 2006.

In general, cement is produced by mixing auxiliary materials with milled clinker. In China, the mass proportion of clinker in cement is in the range of 0.701 to 0.738 (Zhou 2003), and so for this study we assume a proportion of 0.72. The energy efficiency of production improves with the application of energy-saving technologies and use of high-capacity production lines. China's Cleaner Production Standard recommends 120 g (grams) of coal equivalent (gce) per kilogram of clinker production as the basic level for clean production of cement (MEP 2009). This level represents the current average energy efficiency of precalciner kilns. From this, the historical energy efficiency of different kiln types is interpolated with modifications based on practical experiences from the cement industry (Zhou 2003).

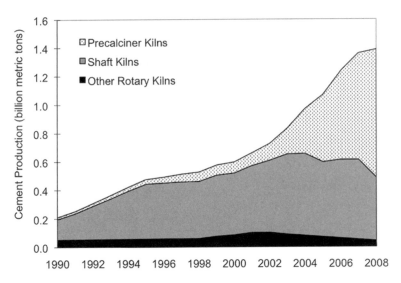

Figure 5.1
Cement production in China from different types of kilns, 1990 to 2008. *Note:* 1 billion
metric tons equals 1000 Tg.

5.2.3   Emission Factors
Sulfur Dioxide, Nitrogen Oxides, and Carbon Monoxide   The manufacture of
clinker generates SO_2, NO_X, and CO. SO_2 mainly comes from the oxidation of sulfur
in the coal used. In precalciner kilns, approximately 70% of this SO_2 is absorbed
by reaction with calcium oxide (CaO) (H. Q. Liu 2006), while much less is absorbed
in other rotary kilns and in shaft kilns (Su, Gao, and Ye 1998). Utilization of baghouse filters, as required with the new precalciner kilns, further reduces SO_2 emissions by absorption in the PM filtered at the surface of the bags. Assuming that SO_2
absorption is 80% for the entire precalciner kiln process and 30% for other types
of kilns, we estimate SO_2 emissions by province using a mass balance approach and
using the average sulfur content of coal in each province derived by B. J. Liu (1998).
    The dominant mechanism of NO_X generation in cement kilns is thermal forma-
tion. As thermal formation is highly dependent on temperature and oxygen availability, the quantity of NO_X emissions varies a lot with the operational conditions
of the kilns. In general, the operating temperature of rotary kilns is higher than
shaft kilns, and coupled with more automatic and stable ventilation, the high temperature results in higher NO_X emission factors compared to shaft kilns. Zhang
et al. (2007) estimated NO_X emission factors from different types of Chinese kilns
based on local test results. The average NO_X emission factors of the cement industry

Table 5.1
Emission factors of SO_2, NO_X, and CO from cement kilns (in grams per kilogram of coal combusted in kilns)

	SO_2*	NO_X	CO
Precalciner kilns	2.9	15.3	17.8
Other rotary kilns	12.3	18.5	17.8
Shaft kilns	12.3	1.7	155.7
1990 average	11.6	7.3	108.1
1995 average	11.5	4.9	127.8
2000 average	11.4	6.1	116.9
2005 average	8.8	8.7	88.0
2008 average	6.8	10.8	64.3

*These are national average SO_2 emission factors weighted by provincial coal consumption. Different SO_2 emission factors are applied for different provinces according to sulfur content of coal from that province.

as a whole are calculated from these kiln-specific factors and the historical proportion of different kiln types; the results are given in table 5.1. We can see the steady rise in the national average NO_X emission factor from 4.9 in 1995 to 10.8 in 2008 with the rising share of rotary kilns.

Carbon monoxide is formed in relatively low-temperature and reducing atmospheres. Shaft kilns can generate a lot of CO because of their relatively low operational temperature and variable operational conditions. Streets et al. (2006) estimated the CO emission factors for rotary and shaft kilns based on local emission tests conducted in China. The weighted average CO emission factors for each kiln type are calculated and listed in table 5.1. The national average CO emission factor has fallen from 128 g/kg of coal in 1995 to 64 in 2008.

Carbon Dioxide   As shown in equation (5.2), calcination of carbonates from raw materials and combustion of coal are the two main sources of CO_2 emissions. The IPCC has recommended that CO_2 emissions from raw materials can be estimated according to the mass proportion of quicklime (CaO) and some minor minerals in clinker when the chemical composition of the raw materials is not available (IPCC 2006). Following this approach, Cui and Liu (2008) estimated a CO_2 emission factor from the typical chemical composition of clinker in China. The estimate of the CO_2 emission factor from material calcination (*EFC* in equation 5.2) is 0.5482 kg for each kilogram of clinker, including CO_2 from decomposition of carbonates and oxidation of organic materials.

Table 5.2
Unabated PM emission factors from three types of kiln and other processes
(in grams per kilogram of cement)

	Total PM	PM$_{2.5}$	PM$_{2.5-10}$	PM$_{>10}$	Emission Factor Range*
Precalciner kilns	105	18.9	25.2	60.9	58.2–317.9
Other rotary kilns	98	14.0	21.0	63.0	24–330
Shaft kilns	30	3.3	6.0	20.7	13.4–91.2
Other processes	140	9.5	23.8	107.7	63–235

*Sources of the emission factor range are U.S. EPA (1995), Jiao (2007), and SEPA (1996a).

The emission factor of CO$_2$ from coal combustion is determined by the quality of coal and the efficiency of combustion. Based on global average coal characteristics, the IPCC has recommended CO$_2$ emission factors for industrial stationary coal combustion in the range of 1.83 to 2.40 kg per kilogram of coal (IPCC 2006). Cui and Liu (2008) estimated the CO$_2$ emission factor to be 1.94 kg per kilogram of coal for coal consumption in the Chinese cement industry, assuming an oxidation rate of carbon of 98%.

Particulate Matter   As shown in equation (5.4), the emission factor of PM is affected by the type of kiln or processes other than clinker making, as well as by the effectiveness of PM emission control devices. Our previous study (Y. Lei et al. 2008) reviewed prior research and calculated the unabated PM emission factors for different types of kilns and processes (EF_m in equation 5.4). Assuming that all cement plants have the same unabated PM emission factors for all processes except for clinker making, we summarize in table 5.2 the unabated PM emission factors for the three PM size ranges, and for three types of kilns and for other processes. For total PM, the emission factors range from 30 g of PM per kilogram of cement for shaft kilns to 105 for precalciner kilns.

PM control devices can reduce PM emissions by 10% to 99.9%, depending on the type of control technology employed and the size distribution of PM in the raw flue gas (Y. Lei et al. 2008). Since the PM removal efficiency of a specific technology is relatively constant across kiln types, the penetration rate of the different technologies ($C_{m,n,j}$) is the key parameter that governs the actual PM emission factors of the cement industry as a whole. Although more efficient PM control technologies come with higher investment and operational costs, improving emission standards is driving the promotion of these technologies in the industry. At the time of research there have been three editions of air pollutant emission standards for the cement industry published by the government: in 1985, 1996, and 2004. The emission standard for PM concentration in kiln flue gas dropped from 800 to 50 mg/m³ in these 20 years (SEPA 1985, 1996b, 2004). With the enforcement of each new standard, a PM control technology with a lower efficiency has generally been replaced by one with a higher efficiency. For example, cyclones were used in almost all shaft kiln plants and in most rotary kiln plants before 1985, whereas baghouse filters are the major PM control devices in most new cement plants.

Based on the PM concentration requirements of the three successive emission standards, the penetration rates of the different PM control technologies in newly built cement plants are estimated for four periods: before 1985, 1985–1996, 1997–2004, and after 2004. Although the emission standards published in 1996 and 2004 allow three to ten years for existing plants to reduce their PM emission rates, a reduction is not likely to be significant in existing plants as there are few measures to enforce the revised standards. In calculating the emission factors here, we assume that all plants retrofit their whole production line every 15 years, and in doing so meet the present standards for new plants. We calculate the penetration rates of PM control technologies across the cement industry for the period 1990–2008, and then estimate the corresponding PM emission factors, as shown in figure 5.2. Over the 18-year period, the overall emission factor of TSP from the cement industry dropped by 88%, from 27.93 g/kg to 3.31 g/kg (the black circles in figure 5.2). The reduction

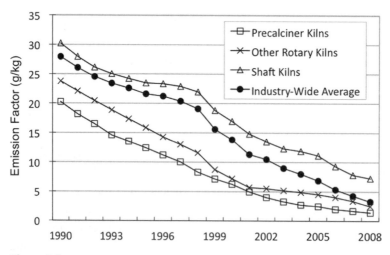

Figure 5.2
PM emission factors (in grams per kilogram of cement) for all cement-producing processes in China's cement industry, 1990 to 2008.

in the PM_{10} and $PM_{2.5}$ emission factors is also significant: 18.08 g/kg to 2.53 g/kg and 10.65 g/kg to 1.61 g/kg, respectively.

5.3   Results for Historical Emissions from 1990 to 2008

5.3.1   Total Emissions

Using the methods, activity levels, and emission factors just described, we estimated the emissions of PM, SO_2, NO_X, CO, and CO_2. Figure 5.3 shows the estimated emissions from China's cement industry for the period 1990–2008.

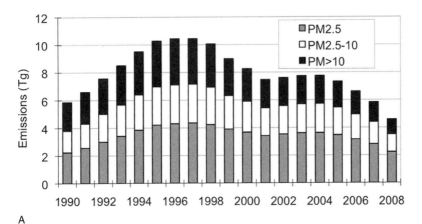

A

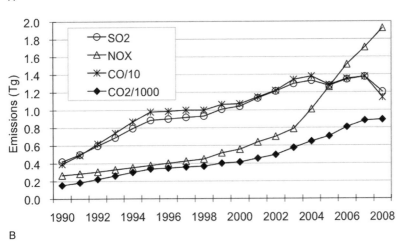

B

Figure 5.3
Emissions of PM, SO_2, NO_x, CO, and CO_2 from China's cement industry, 1990 to 2008. (A) $PM_{2.5}$, $PM_{2.5-10}$, $PM_{>10}$. (B) SO_2, NO_x, CO, and CO_2.

Emissions of PM rose rapidly from 1990 to 1995, when cement production grew at an average annual rate of 17.8%. In the second half of the 1990s, the expansion of China's cement industry slowed and the new emissions standard released in 1996 promoted the application of electrostatic precipitators (ESPs) in shaft kilns, resulting in an industry-wide decrease in PM emissions. After 2000, although the average annual growth in cement production was greater than 12%, PM emissions gradually decreased as a result of the replacement of shaft kilns by precalciner kilns and the application of high-performance PM removal technology, especially after 2004. Over the whole period, PM emissions reached a peak in 1997, with 4.36 Tg (teragrams, equivalent to million metric tons)[3] of $PM_{2.5}$, 7.16 Tg of PM_{10}, and 10.44 Tg of TSP.

The emissions of CO_2 increased from 153 Tg in 1990 to 892 Tg in 2008, consistent with the continuously rising cement production. (Note that the units in figure 5.3 are scaled to fit them in the same chart; for CO_2, the values in the figure must be multiplied by 1000.) The proportion of CO_2 emissions from fuel combustion compared to that from calcination of carbonates decreased from 46.0% in 1990 to 38.6% in 2008, representing improved energy efficiency in the cement industry. We calculate CO_2 emissions from material calcination to be 5% higher than the estimate of the Chinese government (SDRC 2004). A possible reason is that our study accounts not only for the major source of CO_2 by calcination of $CaCO_3$, but also two relatively minor sources: calcination of magnesium carbonate ($MgCO_3$) and oxidation of organic carbon in raw materials. In our study, CO_2 emissions for 1990–2006 are systematically 20% lower than the estimates made by the widely cited Carbon Dioxide Information Analysis Center of Oak Ridge National Laboratory (Boden et al. 2009), which assumed a higher clinker to cement ratio (83–100%) on a global basis (Boden et al. 1995). In this work, we use the value of 72%, derived from Chinese investigation of cement-making practice (Cui and Liu 2008).

Emissions of SO_2 from the cement industry rose from 0.42 Tg in 1990 to 1.39 Tg in 2007, accounting for approximately 5% of China's total SO_2 emissions, then dropped to 1.21 Tg in 2008 (a reduction of 12.6%). The decline in SO_2 emissions in 2008 is attributed to two factors. First, the global economic recession suppressed the construction industry and saw the annual rate of increase in cement production drop from 10% in the previous year to 2% in 2008. Second, the nationwide replacement of shaft kilns by precalciner kilns from 2007 to 2008 led to a 20% reduction of cement production from shaft kilns, which emit several times more SO_2 per mass unit of cement.

The trend observed for CO emissions is very similar to that of SO_2, with a sharp fall in 2008 (note that the CO values in figure 5.3 must by multiplied by 10). In contrast, NO_X is the only pollutant with emissions that increased over the entire period of analysis. During the 1990s, NO_X emissions doubled, and the 2000 emissions were in turn tripled by 2008. With the recent rapid expansion of precalciner kilns in China, the annual increase in NO_X emissions from the cement industry between 2003 and 2008 was over 220 Gg (gigagrams, equivalent to 1000 metric tons or kilotons) per year. This accounts for about 20% of the incremental NO_X emissions seen for China as a whole according to the INTEX-B emission inventory (Zhang et al. 2009). As awareness grows of China's increasing NO_X emissions and their consequences for ozone pollution (see chapter 7) and acidification (Zhao et al. 2009), policies to combat photochemical smog and acid rain pollution will inevitably have to specifically address the cement industry.

To understand the contribution of the cement industry to China's emissions of air pollutants, we compare our results with estimates of all anthropogenic emissions from other studies. As shown in table 5.3, the cement industry is a major source of PM in China, contributing more than a quarter of $PM_{2.5}$ and PM_{10} in 2005 (the latest year in which all sources and all pollutant types are estimated). The contribution of the cement industry to total PM emissions in China, however, can be expected to drop with the progressive implementation of stricter emission standards. In 2005, the percentage contribution of SO_2 and CO emissions from the cement industry was approximately 5.1% and 7.7%, respectively. The contribution of NO_X

Table 5.3
Estimated emissions (in teragrams) of air pollutants and carbon dioxide from China's cement industry, 1990 to 2008, and cement share of emissions (percentage) from all anthropogenic sources in 2005

Year	$PM_{2.5}$	PM_{10}	TSP	SO_2	NO_X	CO	CO_2
1990	2.23	3.79	5.86	0.42	0.27	3.93	153.33
1995	4.21	6.97	10.28	0.89	0.38	9.81	336.74
2000	3.68	5.90	8.25	1.04	0.56	10.70	413.31
2005	3.48	5.47	7.33	1.29	1.26	12.83	704.83
2008	2.23	3.52	4.59	1.21	1.92	11.41	892.14
All sources in 2005*	12.9	18.8	34.3	25.5	19.8	167	5626
Cement share in 2005 (percent)	26.9	29.0	21.4	5.1	6.4	7.7	12.5

Sources: $PM_{2.5}$, PM_{10}, and TSP data from Lei, Zhang, and He (2011); SO_2 from SEPA (2007); NO_X and CO from Zhang et al. (2009); and CO_2 from Boden et al. (2009).

from cement was 6.4% in 2005; in the subsequent three years to 2008, however, the swift rise in the industry's NO_X emissions from 1.26 to 1.92 Tg likely drove up this share. CO_2 emissions from cement production accounted for about one eighth of all Chinese anthropogenic CO_2 emissions, with approximately 40% from fuel combustion and 60% from the chemical process of calcination, indicating that China's cement industry is a significant contributor of greenhouse gases (GHGs).

5.3.2   Spatial Distribution of Emissions

The spatial distribution of emissions changes yearly. Using 2005 as a base year, cement production data from 5294 plants were collected from the China Cement Association, including the capacity of 612 clinker production lines installed with precalciner kilns. These plants accounted for almost all precalciner kilns and more than 95% of cement production in China in that year. The location of these plants and production lines is determined at the county level from cement plant registration information (CCA 2006). Thus emissions of $PM_{2.5}$, SO_2, and NO_X from the cement industry in 2005 were mapped onto an 18′ by 18′ grid for China, as shown in figure 5.4.

In different ways, the distribution of $PM_{2.5}$, SO_2, and NO_X emissions reflect regional operational differences and kiln combinations within China, a point illustrated by the following examples. The grid cells indicating high $PM_{2.5}$ emissions in figure 5.4A show a greater emission rate in the North China Plain. The provinces of Shandong, Hebei, and Henan account for 31.5% of total cement $PM_{2.5}$ emissions. By consuming coal with much higher sulfur content than other provinces, Sichuan in the southwest has the second-highest provincial emissions of SO_2, after Shandong, as shown in figure 5.4B. Shandong and provinces farther south including Zhejiang, Jiangsu, and Anhui are the largest contributors to cement NO_X emissions, shown in figure 5.4C, because a number of clinker-producing centers use large precalciner kilns in these provinces.

As discussed earlier, precalciner kilns emit less PM and SO_2, while shaft kilns emit less NO_X. The different emission characteristics result in regional variations in emissions of the different pollutants. Nonetheless, the highest $PM_{2.5}$, SO_2, and NO_X emissions are all located in the same grid cell in Shandong province, where the city of Zaozhuang is found. Further analysis shows that the production of cement from Zaozhuang is much higher than that of other cities, thus increasing air pollution and lowering local air quality. According to monitoring of daily ambient air quality,[4] increases in annual number of nonattainment days in Zaozhuang from 2004 to 2009 were higher than in any other city in China.

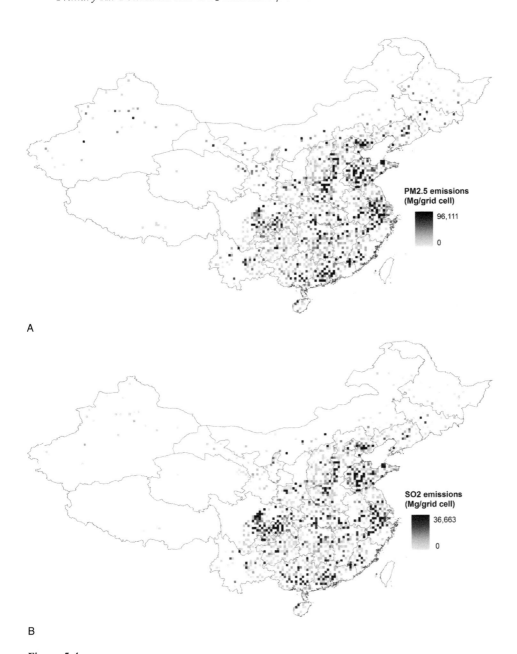

A

B

Figure 5.4

Emissions in 2005 of PM$_{2.5}$, SO$_2$, and NO$_x$ from the cement industry in China. (A) PM$_{2.5}$. (B) SO$_2$. (C) NO$_X$. *Notes:* Emissions are plotted on an 18′ by 18′ grid. One megagram (Mg) equals 1 kiloton or one one-millionth of 1 Tg.   *(Figure continued)*

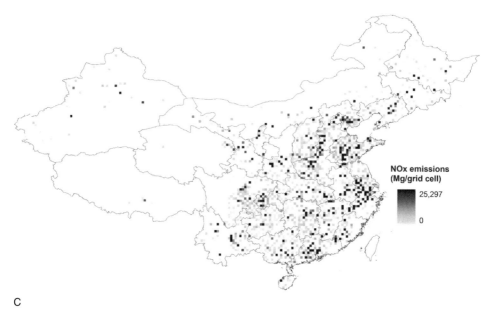

C

Figure 5.4 (continued)

5.4   Emission Estimates for 2010

Increases in cement production are placing new burdens on China's environment. Since emissions from China's national cement industry contribute significant amounts of several air pollutants, accurate emission projections are necessary to inform Chinese national strategies on air pollution control and GHG mitigation. To date, there have been few such projections for China's cement industry because of a lack of reliable emission factors. In this study, the total production of cement for 2010 is estimated for three scenarios (base, carbon tax, and 11th FYP SO_2 control cases) by the economic model presented in chapter 9, and emissions are estimated thereafter.

5.4.1   Cement Production

Massive urbanization and expansion of transportation networks within China have driven a great increase in the cement industry in the past two decades. Cement production increased in the final years of the 11th FYP (2006–2010) due to the 4 trillion yuan economic stimulus package of the Chinese government, which emphasized infrastructure construction. But as annual per capita cement production has

Table 5.4
Estimated cement production and energy consumption in 2010 (in teragrams)

Scenario	Cement Production			Coal Use
	Precalciner Kilns	Shaft Kilns	Other Rotary Kilns	
Base case	1146	357	25	191
Carbon tax	990	305	25	165
11th FYP SO₂ control	1159	361	25	193

now exceeded 1000 kg, there may be relatively little room for further increases in cement production. In 2006 the Chinese government published a specific plan for development of the cement industry, which projected that China's annual cement production would reach 1.2 billion metric tons in 2010 and 1.3 billion metric tons in 2020 (NDRC 2006). However, this plan underestimated the demands for cement from widespread and large-scale construction. The Chinese government therefore reconsidered the potential of the cement market, and in 2009 estimated that the peak cement requirement would be approximately 1.6 billion metric tons per year in 2020 (NDRC 2009).

Official statistics indicate that more than 97% of the cement produced in China is used by domestic consumers (NBS 2008), and thus Chinese cement production is effectively determined by domestic demand. Precalciner kilns are proliferating quickly in China. The government set a target to increase the share of precalciner kilns to 70% in 2010, in terms of cement production (NDRC 2006). But as small shaft kilns were closed to help meet the 20% energy intensity reduction target of the 11th FYP, the share of precalciner kilns had already exceeded 70% in 2009. In this study, we estimate the ratio to have reached 75% in 2010. As shown in table 5.4, cement production is estimated to have exceeded 1.5 billion metric tons under the SO₂ control scenario, as it also would have under the base case. For comparison, the cement production for 2010 as reported in official statistics was 1.9 billion metric tons (NBS 2011); this difference likely reflects the higher-than-expected infrastructure development across the nation. The estimated cement production is approximately 0.2 billion metric tons less under the carbon tax than the base case and SO₂ control case, however, and the consumption of coal is accordingly smaller.

5.4.2 Emission Factors
Deployment of advanced emission control technologies is driven by the requirements set out in emission control policies. In China, emission standards are the

Table 5.5

Emission factors for PM and gaseous pollutants from cement kilns in 2010
(in grams per kilogram)

Scenario	Cement				Coal Combusted		
	$PM_{2.5}$	PM_{10}	TSP	CO_2	SO_2	NO_X	CO
Base case	0.80	1.27	1.56	634	6.4	11.2	59.9
Carbon tax	0.77	1.22	1.51	632	6.4	11.3	59.3
11th FYP SO_2 control	0.80	1.27	1.56	634	6.4	11.2	59.9

major regulatory basis of emission control in the cement industry. As mentioned earlier, as of 2010 the latest emission standard for the cement industry was published in 2004. Based on the requirements of this standard, we estimated the emission factors in 2010 applying the same methodology presented in section 5.2.

PM emissions have been the major focus of previous emission standards. As of 2010, all new cement plants were required to meet an emission standard of 50 mg of PM per cubic meter of flue gas (SEPA 2004), which equates to a PM emission factor for the whole production process of approximately 1 g per kilogram of cement (CRAES 2003). As for SO_2, the standards required that the emission factor should not exceed 0.43 g per kilogram of cement. Although some old, small plants located in suburban and rural areas cannot meet the standards for PM and SO_2, they are gradually being replaced by newer and cleaner plants with baghouse filters installed.

The emission standard for NO_X as of 2010 was lax. Although average NO_X emission factors increased with the spread of precalciner kilns, most cement plants could meet the current emission standard without additional abatement measures. There were no emission standards for CO and CO_2. Therefore, the emission factors of NO_X, CO, and CO_2 were determined by the increasing penetration rate of precalciner kilns, as discussed in section 5.2.3.

The emission factors of PM and gaseous pollutants in 2010 are listed in table 5.5. Note the different emission factors in the carbon tax scenario. This is because imposition of a carbon tax leads to less cement produced by old kilns.

5.4.3   Emissions

Despite the estimated increase in cement production, the emission trends of air pollutants are estimated to have declined during the 11th FYP period, except for NO_X and CO_2. Table 5.6 lists the ratio of projected emissions in 2010 to those in 2005.

Table 5.6
Ratio of emissions of PM and gaseous pollutants in 2010 compared to 2005

Scenario	PM$_{2.5}$	PM$_{10}$	TSP	CO$_2$	SO$_2$	NO$_X$	CO
Base case	0.378	0.385	0.350	1.375	0.827	1.677	0.761
Carbon tax	0.314	0.317	0.292	1.184	0.715	1.456	0.652
11th FYP SO$_2$ control	0.383	0.385	0.354	1.390	0.836	1.694	0.770

PM emissions are estimated to have decreased sharply from 2006 to 2010. This decrease is attributed to (1) the retirement of old cement kilns with a total production capacity of 148 million metric tons by 2010, as required by the National Development and Reform Commission (NDRC),[5] and (2) the imposition of stricter emission standards on existing cement plants under the previously noted 2004 ruling that were scheduled to take effect by the end of 2010.

SO$_2$ emissions also decreased as shaft kilns were gradually closed, with estimated annual emissions of approximately 1.2 Tg by 2010 under the base case and SO$_2$ control scenarios. Nitrogen oxide emissions are estimated to have reached 2.2 Tg in 2010 under the base case and SO$_2$ control policy scenarios, becoming the third-largest sector contributor to NO$_X$ pollution after power generation and vehicles. In the carbon tax scenario, emissions of pollutants would have fallen by 14%, mainly attributed to lower production of cement.

5.5   Conclusions

The cement industry plays an important role in emissions of many air pollutants in China. This study estimates the emissions of major air pollutants from cement production based on information on the development of production technologies and tightening emission standards in China's cement industry. Our analysis shows that with the replacement of old shaft kilns by precalciner kilns, there is an opportunity to reduce PM emissions through the implementation of stricter emission standards and promotion of high-performance PM control technologies. Compared with the historical high (4.36 Tg of PM$_{2.5}$, 7.16 Tg of PM$_{10}$, and 10.44 Tg of TSP) in 1997, PM emissions from the cement industry may have declined by 75% by 2010, despite a tripling of cement production by this year. Other air pollutants such as CO and SO$_2$ have also declined as shaft kilns have been gradually retired. However, promotion of precalciner kilns and a rapid increase in cement production has led to greatly increased NO$_X$ emissions.

Accounting for approximately one-eighth of total CO_2 emissions, China's cement industry also plays an important role in the processes implicated in global climate change. With the estimated successful control of PM, SO_2, and CO emissions, abatement of NO_X and CO_2 emissions should be of increasing concern because of their large effect on local air quality and global climate, as well as the significant role that the cement industry plays in their emissions.

Acknowledgments

Substantial portions of this chapter have been previously published in the journal *Atmospheric Environment*, vol. 45, no. 1, by Y. Lei, Q. Zhang, C. P. Nielsen, and K. B. He, "An inventory of primary air pollutants and CO_2 emissions from cement production in China, 1990–2020," pages 147–154, Copyright 2011, adapted with permission of Elsevier. This chapter was developed from research originally supported by China's National Basic Research Program (2005CB422201) and China's National High Technology Research and Development Program (2006AA06A305). Further development for the purposes of this book was supported by the China Sustainable Energy Program of the Energy Foundation.

Notes

1. "PM" in this usage is equivalent to total suspended particulates (TSP), but in the form of emissions rather than an ambient concentration.

2. In official statistics, "industry" (gōng yè) refers to mining, primary materials production, manufacturing, and utilities supplying power, heat, gas, and water.

3. While these two units are equivalent, Tg is often used in the literature for emissions while Mt is often used as a unit of energy or industrial production. This book uses both units, depending on the context.

4. Available at http://datacenter.mep.gov.cn/report/air_quality/air_dairy_year.jsp, last accessed July 2, 2011.

5. Available at http://www.chinacements.com/cementFile/2007227154719411.doc, last accessed June 27, 2011.

References

Boden, T. A., G. Marland, and R. J. Andres. 1995. Estimates of global, regional, and national annual CO_2 emissions from fossil-fuel burning, hydraulic cement production, and gas flaring: 1950–1992. ORNL/CDIAC-90, NDP-30/R6. Oak Ridge National Laboratory, U.S. Department of Energy, Oak Ridge, Tennessee.

————. 2009. Global, regional, and national fossil-fuel CO₂ emissions. Carbon Dioxide Information Analysis Center, Oak Ridge National Laboratory, U.S. Department of Energy, Oak Ridge, Tenn., U.S.A., doi:10.3334/CDIAC/00001

China Cement Association (CCA). 2006. *Yellow Book of Licensed Cement Manufacturers in China*. Beijing: China Building Material Industry Publishing House. In Chinese.

China Environment Yearbook Editorial Committee (CEYEC). 2001. *China Environment Yearbook 2001*. Beijing: China Environment Yearbook Press. In Chinese. In Chinese.

Chinese Ministry of Industry and Information Technology (CMIIT). 2009. *Operation Status of Major Industry Sectors in 2008: Material Industry*. Beijing: CMIIT. In Chinese.

Chinese Research Academy of Environmental Sciences (CRAES). 2003. Description on developing emission standard of air pollutants for cement industry. Internal report. In Chinese.

Cui, S. P., and W. Liu. 2008. Analysis of CO₂ emission mitigation potential in cement producing processes. *China Cement* 4:57–59. In Chinese.

Intergovernmental Panel on Climate Change (IPCC), 2006. 2006 IPCC Guidelines for National Greenhouse Gas Inventories. Available at www.ipcc-nggip.iges.or.jp/public/2006gl/index.html, last accessed June 30, 2011.

Jiao, Y. D. 2007. *Air Pollution Control in Cement Industry*. Beijing: Chemical Industry Press. In Chinese.

Kong, X. Z. 2005. Scale and technical equipments status of Chinese cement industry. *China Building Materials* 5:34–38. In Chinese.

Lei, Q. Z. 2004. The developing space of precalciner kilns in China. *Henan Building Materials* 4:3–6. In Chinese.

Lei, Y., K. B. He, Q. Zhang, and Z. Y. Liu. 2008. Technology-based emission inventory of particulate matters (PM) from cement industry. *Environmental Science* 29:2366–2371. In Chinese with abstract in English.

Lei, Y., Q. Zhang, and K. B. He. 2011. Primary anthropogenic aerosol emission trends for China, 1990–2005. *Atmospheric Chemistry and Physics* 11:931–954.

Lei, Y., Q. Zhang, C. P. Nielsen and K. B. He. 2011. An inventory of primary air pollutants and CO₂ emissions from cement production in China, 1990–2020. *Atmospheric Environment* 45:147–154.

Liu, B.J. 1998. Life cycle assessment on controlling sulfur from coal. PhD diss., Tsinghua University, Beijing, China. In Chinese with abstract in English.

Liu, H. Q. 2006. Control of SO₂ from cement kiln systems. *China Cement* 11:74–77. In Chinese.

Ministry of Environmental Protection (MEP). 2009. *HJ 467-2009 Cleaner Production Standard: Cement Industry*. Beijing: China Environmental Science Press. In Chinese.

National Bureau of Statistics (NBS). 1991–2011. *China Statistical Yearbook (1991–2011 editions)*. Beijing: China Statistics Press.

National Development and Reform Commission (NDRC). 2006. Specific Plan on Cement Industry Development. Available at www.sdpc.gov.cn/zcfb/zcfbtz/tz2006/t20061019_89080.htm, last accessed June 27, 2011. In Chinese.

———. 2009. Some suggestions to restrain redundant capacity and projects and lead healthy development for some industry. Available at www.chinanews.com.cn/cj/news/2009/09-30 /1894456.shtml, last accessed June 27, 2011. In Chinese.

———. 2012. The Second National Communication on Climate Change of The People's Republic of China. Report to the United Nations Framework Convention on Climate Change. Available at http://unfccc.int/national_reports/non-annex_i_natcom/items/2979.php, last accessed March 25, 2013.

Ohara, T., H. Akimoto, K. Kurokawa, N. Horii, K. Yamaji, X. Yan, et al. 2007. An Asian emission inventory of anthropogenic emission sources for the period 1980–2020. *Atmospheric Chemistry and Physics* 7:4419–4444.

State Development and Reform Commission (SDRC). 2004. The People's Republic of China Initial National Communication on Climate Change. Report to the United Nations Framework Convention on Climate Change. Available at http://unfccc.int/national_reports/non-annex_i_natcom/items/2979.php, last accessed March 25, 2013.

State Environmental Protection Administration (SEPA). 1985. *GB 4915-85: Emission Standard of Pollutants for Cement Industry*. Beijing: China Environmental Science Press. In Chinese.

———. 1996a. *Handbook of Industrial Pollution Emission Factors*. Beijing: China Environmental Science Press. In Chinese.

———. 1996b. *GB 4915-1996 Emission Standard of Air Pollutants for Cement Plant*. Beijing: China Environmental Science Press. In Chinese.

———. 2004. *GB 4915-2004 Emission Standard of Air Pollutants for Cement Industry*. Beijing: China Environmental Science Press. In Chinese.

———. 2007. Report On the State of the Environment in China 2005. Available at http://english.mep.gov.cn/standards_reports/soe/soe2005/200708/t20070827_108448.htm, last accessed June 27, 2011.

Streets, D. G., T. C. Bond, G. R. Carmichael, S. D. Fernandes, Q. Fu, D. He, Z. Klimont, S. M. Nelson, N. Y. Tsai, M. Q. Wang, J.-H. Woo, and K. F. Yarber. 2003. An inventory of gaseous and primary aerosol emissions in Asia in the year 2000. *Journal of Geophysical Research* 108(D21), 8809, doi:10.1029/2002JD003093.

Streets, D. G., Q. Zhang, L. T. Wang, K. B. He, J. M. Hao, Y. Wu, Y. H. Tang, and G. R. Carmichael. 2006. Revisiting China's CO emissions after the Transport and Chemical Evolution over the Pacific (TRACE-P) mission: Synthesis of inventories, atmospheric modeling, and observations. *Journal of Geophysical Research* 111, D14306, doi:10.1029/2006JD007118.

Su, D. G., D. H. Gao, and H. M. Ye. 1998. Pollution of hazardous gases from cement kilns and its control. *Chongqing Environmental Science* 20:20–23. In Chinese.

United States Environmental Protection Agency (USEPA). 1995. AP-42, Fifth Edition, Compilation of Air Pollutant Emission Factors, vol. 1: Stationary Point and Area Sources. Chapter 11.6, Mineral Products Industry, Portland Cement Manufacturing. Available at http://www.epa.gov/ttn/chief/ap42/ch11/index.html, last accessed June 27, 2011.

United States Geological Survey (USGS). 2009. *Mineral Commodity Summaries 2009*. Washington DC: United States Government Printing Office.

Zeng, X. M. 2004. Review of 100 years development of Chinese cement industry. *China Cement* 11:35–39. In Chinese.

Zhang, Q., D. G. Streets, G. R. Carmichael, K. B. He, H. Huo, A. Kannari, Z. Klimont, I. S. Park, S. Reddy, J. S. Fu, D. Chen, L. Duan, Y. Lei, L. T. Wang, and Z. L. Yao. 2009. Asian emissions in 2006 for the NASA INTEX-B mission. *Atmospheric Chemistry and Physics* 9:5131–5153.

Zhang, Q., D. G. Streets, K. B. He, Y. X. Wang, A. Richter, J. P. Burrows, et al. 2007. NO_X emission trends for China, 1995–2004: The view from the ground and the view from space, *Journal of Geophysical Research* 112, D22306, doi:10.1029/2007JD008684.

Zhao, Y., L. Duan, J. Xing, T. Larssen, C. P. Nielsen, and J. M. Hao. 2009. Soil acidification in China: Is controlling SO_2 emissions enough? *Environmental Science and Technology* 43(21):8021–8026.

Zhou, D. D. 2003. *China's Sustainable Energy Scenarios in 2020*. Beijing: China Environmental Science Press. In Chinese.

6

An Anthropogenic Emission Inventory of Primary Air Pollutants in China for 2005 and 2010

Yu Zhao, Wei Wei, and Yu Lei

6.1 Introduction

A regional emission inventory is a fundamental input into air quality modeling. Without accurate emission profiles of all relevant chemical species located in a sufficiently fine spatial distribution it would not be possible to simulate atmospheric concentrations and fluxes accurately—that is, to conduct simulations that can be confirmed by observations and can be used for policy analysis.

In recent years a series of studies have developed emission inventories for Asian countries including China, with most of them choosing 2000 as the base year (Streets and Waldhoff 2000; Klimont et al. 2001; Streets et al. 2003; Ohara et al. 2007; Q. Zhang et al. 2009). Streets et al. (2003) is among the most cited inventories of Chinese emissions, and these researchers evaluated sulfur dioxide (SO_2), nitrogen oxides (NO_X), carbon monoxide (CO), non-methane volatile organic compounds (NMVOC), black carbon (BC), organic carbon (OC), ammonia (NH_3), and methane (CH_4) for two major studies, the Asian Pacific Regional Aerosol Characterization Experiment (ACE-Asia) and the Transport and Chemical Evolution over the Pacific (TRACE-P) experiment. Most of those inventories, however, are not informed by the most recent local emission field tests or energy statistics. Akimoto et al. (2006) compared the 1996–2003 trends of energy consumption to column concentrations of nitrogen dioxide (NO_2) measured from satellites (Irie et al. 2005; Richter et al. 2005) and found underestimates of emissions derived from energy statistics from both the International Energy Agency (IEA) and the National Bureau of Statistics (NBS) of China during that period. Y. X. Wang et al. (2004) applied TRACE-P aircraft measurements and ground station observations in conjunction with model simulations run in inverse mode, and indicated that the bottom-up NO_X emissions estimate should be raised by 47%, with the largest adjustment in central China.

They also estimated that the emission adjustment for east China should be about 33% and implied that more uncertainties existed in the emissions from biomass burning and microbial sources than fuel combustion (Y. X. Wang et al. 2007).

Since 2000, Chinese economic growth has been higher than that projected in government plans, and this has been accompanied by a dramatic increase in total energy consumption at an annual average growth rate of 10% according to official statistics (versus 4% for 1980–2000). Pollutant emissions from China have been rapidly growing accordingly. To limit the degradation of the atmospheric environment, China's government set compulsory targets for energy saving and emission reduction during 2006–2010, the period of the 11th Five-Year Plan (11th FYP). Specifically, the national energy intensity (energy consumption per unit GDP) and emissions of SO_2 should be reduced by 20% and 10%, respectively, between 2005 and 2010. To realize these targets, the 11th FYP required early retirement of 50 GW of inefficient, small thermal power units, and required that flue gas desulfurization (FGD) systems be installed at most power plants by 2010.

More recently, under pressure to abate carbon emissions, the government has begun exploring the implications of implementing a carbon tax on fossil fuels at some point. We seek to evaluate and compare how these 11th FYP and carbon control policies improve regional air quality and reduce health and agricultural risks. To do so, we must first quantify the emissions of relevant atmospheric pollutants in 2005 and 2010 for a no-policy base case and different policy scenarios for use in atmospheric simulation. With that primary objective, this study developed a comprehensive and detailed emission inventory differentiated by sectors and regions of China, incorporating results from chapters 4 and 5 for the power and cement sectors and taking into account the latest domestic information on emission factors and activity levels for other sectors. Our estimates of this inventory are applicable for policy analyses like those in this volume, as well as for more purely scientific research.

6.2   Methods and Sources of Data

6.2.1   Methodology

The geographical domain of this study covers the 31 provinces and province-level autonomous regions and municipalities in mainland China. Hong Kong, Macao, and Taiwan were not included. Emissions of sulfur dioxide (SO_2), nitrogen oxides (NO_X), particulate matter (PM) in several size classes, black carbon (BC), organic carbon (OC), non-methane volatile organic compounds (NMVOC), and ammonia (NH_3) were estimated at the provincial level and then aggregated to give the national total.[1]

For brevity, we do not describe in complete detail here the methodologies for estimating the emissions of so many species from so many processes; instead we point out the references for those interested in further information. In general, the emissions of each species by sector and region were calculated on the basis of activity levels (the energy-consumption or industrial-production level), the unabated emission factors, and the removal efficiencies of applicable emission control technologies. The unabated emission factors (EFs) are expressed as the mass of emitted pollutant per unit fuel combusted, or per unit industrial production, prior to any emission control.

For anthropogenic emissions of SO_2, NO_X, and PM, the emission sources were classified as either combustion or noncombustion sources, concentrating on five major sectors as categorized in the official statistics: (1) energy supply (thermal power plants, heating supply, coking plants, oil refineries and gas works); (2) industry (industrial boilers and kilns, and processes like production of cement, lime, and chemicals, and smelting of iron, steel, and nonferrous metals); (3) transportation (on-road vehicles, railway transportation, construction machinery, agricultural vehicles, and inland shipping); (4) domestic (i.e., commercial and residential) uses; and (5) open burning of biomass. (Note that the sectors differentiated by the economic model of chapter 9 are more numerous, including many subdivisions of "industry.")

Based on the availability of detailed source information, coal-fired power plants, cement plants, and iron and steel smelting plants were treated as large point sources in this study. For each large point source (LPS) we first determine the location, energy consumption, production level, and production technology, and then estimate the emissions from those sources plant by plant. The other sectors were treated as area sources, including both mobile and stationary source types.

Ammonia emissions come mainly from agricultural activities, in particular the raising of livestock and use of fertilizer. Relatively minor emissions that have also been quantified come from stationary sources (including the production of ammonia and nitric acid), transportation, and open burning of biomass.

The main sources of NMVOC emissions are solvent production and use; transportation; oil and gas mining, storage and transport; fuel combustion; industrial processes including coking, iron and steel smelting, and chemical manufacturing; and waste treatment. The detailed classification of the sources of NMVOCs is described in Wei et al. (2008).

Emissions of species *i* from stationary combustion sources in province *j* were calculated from activity levels, emission factors, and pollutant removal efficiency:

$$E_{i,j} = \sum_k \sum_m \sum_n AL_{j,k,m,n} \times EF_{i,j,k,m} \times R_{i,j,k,m,n} \times (1 - \eta_{i,n})$$ (6.1)

where *k*, *m*, and *n* are indexes for the sector, fuel type, and emission control technology, respectively; *AL* is the activity level; *EF* is the unabated emission factor; *R* is the application rate of emission control technology *n*, where the set of technologies includes a "no-control" case such that the sum of *R* over *n* is 1; and $\eta_{i,n}$ is the removal efficiency of technology *n*. The emission factors are developed for each source type from a wide sample of measurements that we made ourselves and from the literature. The emission control technologies of primary concern are those for SO_2 and PM in power generation and manufacturing.

The level of SO_2 emissions from combusting fuel *m* is closely related to the sulfur content ($S_{j,m}$), and thus the emissions from province *j* are calculated as the product of the activity level, sulfur content, and the rate of use of control technology *n*, adjusting for the sulfur retention rate ($Sr_{k,m}$) and removal efficiency (η_n), and summing over all sectors, fuel types, and technologies:

$$E_{SO_2,j} = \sum_k \sum_m \sum_n AL_{j,k,m,n} \times S_{j,m} \times (1 - Sr_{k,m}) \times R_{j,k,m,n} \times (1 - \eta_n) \times 2$$ (6.2)

Sr is the sulfur retention in ash (%) and the 2 at the end is the coefficient converting sulfur (32 g/mol) to SO_2 (64 g/mol).

Similarly, the level of PM emissions of size *y* from coal combustion (i.e., *m* = coal) in province *j* is given by the activity level, ash content ($A_{j,m}$), and rate of application of control technology *n*:

$$EF_{PM,y,j,m} = \sum_k \sum_n AL_{j,k,m,n} \times A_{j,m} \times (1 - ar_{k,m}) \times R_{j,k,m,n} \times f_{k,m,y} \times (1 - \eta_{n,y})$$ (6.3)

where *ar* is the ratio of bottom ash to total ash (%), $f_{k,m,y}$ is the particulate mass fraction of size *y*, and $\eta_{n,y}$ is the removal efficiency for PM(*y*) using technology *n*.

As detailed in chapter 5, emissions from cement production were characterized at two stages, clinker firing (including material breaking, material and coal grinding, firing, and cooling) and cement processing (mainly grinding clinker into cement). Clinker firing can emit both gaseous pollutants and PM, while cement processing only produces PM. The total emissions of pollutant *i* from cement plant *p* is the sum of these two stages:

$$E_{p,i} = \sum_t (AK_{p,t} \cdot EF_{i,t}) + AC_p \cdot ef$$ (6.4)

where t indexes the technologies of clinker firing, $AK_{p,t}$ is the amount of clinker produced using technology t, AC is the quantity of cement produced, $EF_{i,t}$ is the emission factor for pollutant i for the clinker firing, and ef is the PM emission factor during cement processing.

Emissions from the iron and steel industry were considered from sintering, iron smelting, and steel smelting (including the two main steelmaking technologies, electric furnaces and basic oxygen converters). The total emissions of pollutant i from plant p is the product of the quantity produced using technology t ($AIS_{p,t}$) and the emission factor ($EF_{i,t}$), summed over all technologies:

$$E_{p,i} = \sum_t AIS_{p,t} \cdot EF_{i,t} \tag{6.5}$$

For other noncombustion stationary sources, the emissions of different species and sectors were simply calculated as a product of industrial or agricultural (livestock and fertilizer) production and the corresponding emission factors.

For emissions from on-road sources in the transport sector, vehicles were categorized into nine types: light-duty gasoline vehicles (LDGV), light-duty gasoline trucks (LDGT), light-duty diesel vehicles (LDDV), light-duty diesel trucks (LDDT), heavy-duty gasoline vehicles (HDGV), heavy-duty gasoline trucks (HDGT), heavy-duty diesel vehicles (HDDV), heavy-duty diesel trucks (HDDT), and motorcycles (MC). Emissions of pollutant i in province j from on-road vehicle sources are given by the mileage and fuel consumption per mile for each vehicle type, summed over all types:

$$E_{i,j} = \sum_t (VP_{j,t} \times VMT_{j,t} \times FE_{j,t}) \times EF_{i,j,t} \tag{6.6}$$

where t is the vehicle type, VP is the vehicle population, VMT is the annual average vehicle miles traveled, $FE_{j,t}$ is the average on-road fuel economy of vehicle type t (consumed fuel divided by miles traveled), and $EF_{i,j,t}$ is the average emission factor (emissions of i per unit fuel). The vehicle miles traveled, $VMT_{j,t}$, was estimated from passenger traffic and freight traffic volume, hours traveled, and vehicle speed.

A flow chart summarizing the estimation of emissions from all these different sources is given in figure 6.1.

6.2.2 Activity Levels in 2005

The data on activity levels in the base year of 2005 were mainly obtained from statistics published by a variety of government agencies in China. In those instances when activity data were unavailable for 2005, they were extrapolated from statistical data from previous years.

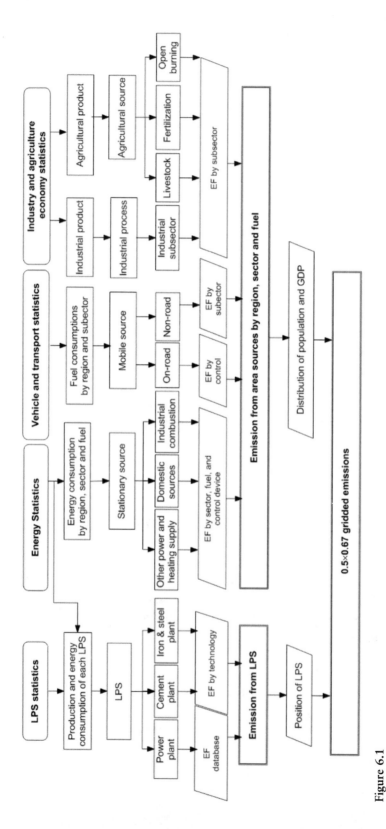

Figure 6.1
Flow chart of the estimation of emissions by source type.

Consumption of fossil fuels and biofuels by province, sector, and fuel type in 2005 was obtained from the *China Energy Statistical Yearbook* (NBS 2006a) and is listed in table 6.1. In this study, the activity levels of combustion-related stationary sources including heating supply, industrial (manufacturing) activity, and domestic (commercial and residential) use were derived from the statistical yearbooks. For the two energy-converting industries, oil refineries and coking plants, their fuel consumptions were derived from the total consumption of oil and coke, respectively, also provided by the energy statistical yearbook. The percentages of open-burned biomass by province were estimated in a national survey by Tsinghua University using questionnaires (Li 2007).

Activity levels of large point sources were estimated for each plant. As described in chapter 4, coal consumption in the power sector was estimated at the unit level by applying information on annual operation hours and specific coal consumption per unit of electricity supply from surveys by the Ministry of Environmental Protection (MEP).

As shown in equation (6.4) and described in chapter 5, the activity levels in the cement industry include clinker firing and cement production. The data for cement production by plant were taken from the China Cement Association. With regard to the clinker assessment, a transfer flow model recently developed by Lei (2008) was applied.

The production of iron and raw steel by plant was obtained from the China Iron and Steel Association (unpublished). To estimate the technologies used, plants producing both pig iron and raw steel were assumed to use basic oxygen converters, while those producing only steel were assumed to use electric furnaces.

For estimating noncombustion emissions from industry, the production of the main processing sectors was obtained from the *China Industry Economy Statistical Yearbook* (NBS 2006b) and the *China Chemical Industry Yearbook* (CPCIA 2006). Solvent consumption, which is a significant source of NMVOC, was derived from various publications (CNLIC 2006; CNCIA 2006; Xie 2005; Shang, Zhang, and Shen 2006). The agricultural sector is the most important source of NH_3, and the amounts of livestock production and fertilizer use were obtained from the *China Animal Industry Yearbook* (MOA 2006) and the *China Rural Statistical Yearbook* (NBS 2006c).

Activity levels for the transportation sector were estimated using a different approach, since emission factors differ considerably across vehicle types. The transportation fuel consumption was calculated as a product of the number of a given vehicle type, average annual mileage traveled, and fuel economy. Vehicle numbers

Table 6.1

Energy consumption by region and fuel in China 2005 (in petajoules)

Region	Power Plants			Industry			Transport		Domestic			Total
	Coal	Oil	Other	Coal	Oil	Other	Oil	Other	Coal	Biofuel	Other	
Anhui	787	0	42	658	25	30	189	3	196	419	24	2373
Beijing	197	6	23	190	20	137	175	5	135	25	167	1080
Chongqing	222	2	5	378	12	73	78	5	95	309	34	1213
Fujian	476	4	0	411	100	106	129	2	84	61	85	1458
Gansu	474	1	1	306	7	34	75	12	128	121	13	1172
Guangdong	1282	391	7	546	395	174	507	0	40	317	349	4008
Guangxi	145	1	0	231	37	255	131	2	8	675	79	1564
Guizhou	564	0	0	388	15	36	68	2	327	451	27	1878
Hainan	38	1	17	18	8	38	36	1	2	52	13	224
Hebei	1330	3	55	2437	77	274	394	12	383	397	62	5424
Heilongjiang	552	1	7	238	80	46	128	18	21	356	163	1610
Henan	1712	3	15	886	69	167	337	4	275	402	48	3918
Hubei	724	3	2	757	87	119	171	20	137	346	29	2395
Hunan	429	2	0	812	38	99	165	5	262	322	54	2188
Inner Mongolia	1104	0	25	393	33	128	116	33	257	193	74	2356
Jiangsu	1963	17	40	1203	147	197	338	10	52	465	104	4536
Jiangxi	274	1	0	281	15	112	105	0	73	212	31	1104
Jilin	379	2	18	325	43	79	98	18	126	210	97	1395
Liaoning	727	5	20	600	84	474	230	15	128	272	84	2639
Ningxia	261	0	0	119	4	2	30	2	30	40	8	496
Qinghai	12	0	4	47	2	7	21	2	19	21	51	186
Shaanxi	706	2	4	219	1	9	105	5	196	168	83	1498
Shandong	1894	0	0	2100	163	56	496	23	231	429	222	5614
Shanghai	697	25	115	326	78	91	383	3	36	0	158	1912
Shanxi	1404	0	48	648	41	320	165	15	271	122	36	3070
Sichuan	654	1	8	598	19	63	192	2	127	895	118	2677
Tianjin	314	0	12	249	27	7	346	4	42	15	85	1101
Tibet	0	0	0	4	0	0	12	0	0	0	0	16
Xinjiang	213	1	24	186	45	122	78	11	124	128	55	987
Yunnan	277	1	22	524	12	96	139	4	96	252	19	1442
Zhejiang	1089	76	2	692	170	39	308	1	48	85	146	2656

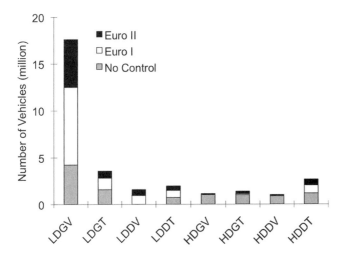

Figure 6.2
The number of vehicles by control level and type in China 2005.

by type were obtained from the *China Automotive Industry Yearbook* (CAAM 2006). In 2005, new vehicles were required nationwide to meet Euro II emission standards, while older vehicles met Euro I standards or no standards. The estimated fleet composition of these three emission standards in 2005 for each vehicle type is shown in figure 6.2. The most numerous type is LDGV, and just under half of them met the Euro I standard and a quarter met the Euro II standard. None of the HDDV meet any standard.

The average annual mileage and fuel economy by vehicle type were obtained from K. B. He et al. (2005). The activity levels of off-road transport were calculated as the product of the number of passengers or amount of freight and the fuel consumption per unit of such traffic. These data were taken mainly from the *China Statistical Yearbook for Regional Economy* (NBS 2006d).

6.2.3 Activity Levels in 2010

To analyze and compare the effects of the emission control polices introduced in prior chapters, we estimated emissions in 2010 for each of the three scenarios (two policy cases and the base case).[2] The projection of activity levels under the different scenarios depends most on the database of coal-fired electric power plants described in chapter 4 and on the general equilibrium model of the economy described in chapter 9; we used a slightly different sequence of steps to analyze the different policies, as described in discussion of figure 2.1.

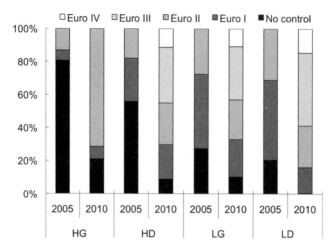

Figure 6.3
The share of vehicles by control level and type in 2005 and 2010. HG, heavy-duty gasoline vehicles; HD, heavy-duty diesel; LG, light-duty gasoline; LD, light-duty diesel.

The base and policy cases are fully introduced in sections 2.2 and 4.2.2. As a quick reminder to the reader, the 11th FYP case is closest to the actual energy and emission path from 2006 to 2010; that case assumed implementation of official measures to save energy and abate emissions in the power sector. The carbon tax case is a hypothetical scenario in which a tax of 100 yuan per ton of carbon was applied economy-wide to fossil fuels for the years 2006 through 2010; this would have encouraged increased energy efficiency, more use of noncarbon energy sources, and a structural change in the economy toward less carbon-intensive sectors. The base case is the most pessimistic scenario, in which the energy saving and emission control policies of the 2006–2010 11th FYP are assumed not to have taken place. It is thus, like the carbon tax, also hypothetical.

In addition to the modeled carbon tax and 11th FYP policies, we assumed that an emission standard for new on-road vehicles was implemented on schedule in all scenarios. The share of vehicles by control level and type under this assumption was estimated for 2010 and is shown in figure 6.3.

The total primary energy consumption by sector in 2005 is illustrated in figure 6.4, as well as the consumption under the different scenarios in 2010. ("Primary" refers only to the combustion of fossil fuels and does not include hydropower.) In 2005, power and industry accounted for around 70% of national primary energy consumption, followed by domestic (residential and commercial) use (~20%) and

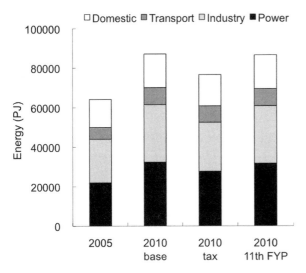

Figure 6.4
Primary energy consumption by major sectors in 2005 and 2010 under the base case and two policy cases.

transportation (~10%). In the base and 11th FYP scenarios, the annual rates of increase of energy consumption are estimated at 8%, 6%, 7%, and 4% for power, industry, transport, and domestic sectors, respectively.[3] In the carbon tax scenario, however, the rates of increase are only 5%, 2%, and 2% for power, industry, and domestic use, respectively; energy consumption by transport still grows by 7% given the relative small tax on liquid fuels.

6.2.4 Emission Factors

Emission factors were determined for each of the species, sectors, and fuel types from an array of both published studies and unpublished field measurements. Those obtained through studies carried out in China were considered the most appropriate and used first for emission estimation. In some cases where local information was lacking, however, foreign sources were consulted such as the AP-42 (USEPA 1999), the EMEP/CORINAIR Atmospheric Emission Inventory Guidebook (EEA 2005), and the scientific literature (Kato and Akimoto 1992; Streets et al. 2003; Ohara et al. 2007). Table 6.2 summarizes the mean emission factors used in this study by sectors and fuels.

The sulfur content of fuels is a key parameter determining the SO_2 emission factor (see equation 6.2). For SO_2 from coal combustion, an interprovince transfer matrix

Table 6.2
Mean emission factors for energy consumption sources (in grams per gigajoule)

	Power Plants			Industry			Transport	Domestic			Open Burning
	Coal	Oil	Other	Coal	Oil	Other	Oil	Coal	Biofuel	Other	
SO_2	783.8	249.1	13.7	416.5	108.2	22.1	71.3	554.2	91.4	41.0	42.6
NO_x	339.1	191.7	71.6	232.4	200.3	88.6	678.3	97.5	96.6	155.1	234.7
TSP	131.8	28.7	4.0	332.8	25.0	56.9	59.0	426.5	477.2	14.0	603.0
PM_{10}	87.2	21.1	4.0	134.3	23.1	54.8	57.5	167.1	477.1	13.8	578.9
$PM_{2.5}$	46.8	16.1	4.0	71.7	19.0	53.2	55.8	93.9	462.2	13.7	560.8
BC	2.3	1.6	0.4	14.0	3.1	10.3	23.9	29.9	75.8	2.9	84.1
OC	0.4	0.7	1.2	2.8	0.8	39.8	18.0	17.1	293.3	1.5	319.7

based on CCTA (2003) and NBS (2001) was developed by S. X. Wang (2001) and Jiang (2004) to quantify the coal flows by type and sector throughout the country. The sulfur content of coal consumed by region was then calculated from the mined coal quality and the coal flows. The national average sulfur content of raw coal was estimated at 1.01% in 2005. For liquid fuels, the sulfur content was determined according to the national standards, which are 0.28%, 0.22%, and 0.05% for heavy oil, medium distillates (e.g., crude oil and diesel), and gasoline, respectively.

Hao et al. (2002) developed a NO_X emission factor database for Chinese fuel combustion sources, mainly based on field surveys and measurements. Q. Zhang et al. (2007a) summarized domestic tests and the international literature for various sources of NO_X emissions, suggesting average emission factors by sector and technology. The results of these two studies were applied in this work, with a few modifications based on more recent field measurements in different provinces by Tsinghua University that are not yet published.

Unabated PM emission factors from various sources were calculated as the product of the fuel ash content and the ratio of fly ash. The ash content of coal consumed in different regions was calculated using the coal quality surveyed by MEP and the transfer matrix mentioned earlier. The national average ash content of raw coal was 21.7% in 2005. The fractions of particles under 2.5 microns ($PM_{2.5}$) and under 10 microns (PM_{10}) in diameter were mainly derived from Q. Zhang (2005).

Regarding black carbon (BC) and organic carbon (OC) emission factors, Q. Zhang (2005) indicated ratios of BC and OC to $PM_{2.5}$. Cao, Zhang, and Zheng (2006) summarized relevant studies and generated an emission factor database for Chinese BC and OC emissions, which was applied in this work. The removal efficiencies of different dust collectors were based on domestic field tests (SEPA 1996) and recent measurements by studies at Tsinghua University (Yi et al. 2006, 2008; Zhao et al. 2010).

For non-methane volatile organic compounds (NMVOC), Wei et al. (2008) systematically compiled emission factors by species. NH_3 emission factors derived in two European studies were applied in the current study (Bouwman et al. 1997; Klimont et al. 2001), because of the lack of domestic estimates.

Specific sectors merit additional explanation. For coal-fired power plants, an integrated emission factor database was established based on field tests and data surveys, along with considering the effects of unit capacity, boiler technology, fuel quality, and emission control devices on emission levels. The complete details are described in chapter 4 and Zhao et al. (2010). For two more types of large point sources, cement plants and iron and steel plants, Lei (2008) thoroughly summarized the emission characteristics of each working process and technology, generating an emission factor database for all species. Full details on the cement sector, to illustrate another key industry, are presented in chapter 5 and Lei, Zhang, Nielsen, et al. (2011).

Emission factors for on-road vehicles by type and control level (Euro II, Euro I, or no standards) were obtained from the emission standards for each type in each year. Emission factors from other industry processes were mainly taken from the *Handbook of Industrial Pollution Emission Rates* (SEPA 1996), with supplementary data from AP-42 and EMEP/CORINAIR, noted previously. For biofuel use and open burning of biomass, the results of a series of field measurements of emission factors conducted by Li, Wang, et al. (2007) and Li, Duan, et al. (2007) were applied in this study.

Except for the power and transport sectors, the emission factors of all other sectors were assumed to have remained unchanged during 2005–2010, because there was no improvement of emission standards or regulations affecting them over this period.

6.3 Results and Discussion

6.3.1 National, Regional, and Sector Emissions in 2005

The emissions of SO_2, NO_X, NMVOC, PM, BC, OC, and NH_3 by region and sector are summarized in this section. Provincial emissions for each species are listed in table 6.3, and the sector distribution is plotted in figure 6.5.

National emissions of carbon dioxide (CO_2) are also reported in this section, but without regional disaggregation because the location of sources is irrelevant to the global effects of this greenhouse gas.

Sulfur Dioxide The total SO_2 emissions in 2005 were estimated to be 29.4 Tg (teragrams),[4] which is 5.0 Tg higher than the total in the official statistics (SEPA 2006). Compared to emissions in 2000 estimated by Streets et al. (2003) and SEPA (2001), 20.4 and 20.0 Tg, respectively, the SO_2 emissions in 2005 were 44–47% higher. Shandong and Hebei, with very high fossil fuel consumption, and Sichuan, where extremely high-sulfur coal is widely used, were estimated to have the highest emissions of the provinces, exceeding 2.0 Tg each. Other high-emission provinces included Hubei, Jiangsu, Shanxi, Guizhou, and Henan. Coal combustion is the most important source of SO_2 by fuel type, accounting for 83% of total emissions. Of

Table 6.3
Emissions by region 2005 (in gigagrams)

Region	SO$_2$	NO$_X$	PM$_{10}$	PM$_{2.5}$	BC	OC	NMVOC	NH$_3$
Anhui	688	688	731	537	80	186	951	883
Beijing	224	302	154	96	15	16	319	93
Chongqing	804	272	367	275	41	108	376	324
Fujian	429	419	349	228	25	38	452	369
Gansu	305	289	248	181	25	53	243	252
Guangdong	1174	1270	934	616	59	136	1819	790
Guangxi	905	403	680	519	70	223	930	624
Guizhou	1635	414	575	407	69	159	429	371
Hainan	57	79	65	48	6	18	123	133
Hebei	2054	1256	1342	912	111	199	1103	1090
Heilongjiang	277	620	513	400	62	156	633	421
Henan	1630	1128	1380	888	121	237	1203	1693
Hubei	2012	741	766	535	71	153	704	809
Hunan	966	506	739	505	68	134	602	795
Inner Mongolia	1087	715	512	367	58	107	439	488
Jiangsu	1680	1357	1072	757	94	217	1783	1082
Jiangxi	548	339	497	311	37	93	383	436
Jilin	363	411	461	334	50	104	438	482
Liaoning	1016	802	715	502	67	123	789	522
Ningxia	372	151	116	81	10	19	97	89
Qinghai	20	46	56	41	4	9	50	86
Shaanxi	958	353	414	265	41	81	347	365
Shandong	2969	1634	1829	1213	139	256	1840	1592
Shanghai	675	763	180	115	18	14	406	72
Shanxi	1673	841	788	531	108	141	500	264
Sichuan	2415	846	1082	832	113	325	1573	1120
Tianjin	359	554	139	90	17	14	216	98
Tibet	1	10	7	5	0	1	12	46
Xinjiang	266	271	244	180	28	59	350	345
Yunnan	431	312	482	343	50	104	430	558
Zhejiang	1431	967	628	390	40	53	1039	352
Total	29,426	18,759	18,066	12,506	1698	3535	20,580	16,645

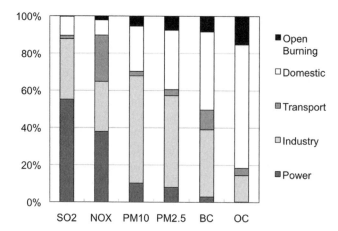

Figure 6.5
Sector distributions of SO_2, NO_X, PM, BC, and OC emissions in 2005.

the four sectors identified, coal-fired power plants emitted the most SO_2, accounting for 54%, followed by industry combustion with 23%. As noted earlier, MEP investigation revealed that only 14% of coal-fired power plants had desulfurization systems installed in 2005; hence the targeting of thermal power generation for SO_2 control in the 11th FYP.

Nitrogen Oxides Emissions of NO_X in 2005 were estimated to total 18.8 Tg, approximately 49–68% higher than the 11.2–12.6 Tg estimated for 2000 by various studies (Streets et al. 2003; Ohara et al. 2007; Tian 2003; Q. Zhang et al. 2007a). Q. Zhang et al. (2007a) evaluated the average annual growth rate of NO_X emissions at 6.1% between 1995 and 2004, with a total of 18.6 Tg in 2004, a result close to our estimate. Jiangsu, Shandong, Guangdong, Hebei, and Henan emitted more than 1.0 Tg each. Power plants were the largest sector emission source of NO_X (37%), followed by industry (27%, including both combustion and process emissions, and transportation (25%). By fuel type, NO_X emissions from coal and oil combustion were 57% and 24%, respectively.

Particulate Matter Total primary PM emissions were estimated to be 32.3 Tg in 2005, 18% higher than the estimate for 2001 by Q. Zhang et al. (2007b). Our team's GEOS-Chem model of chapter 7 requires primary PM_{10} and $PM_{2.5}$ rather than primary PM, and emissions of those species were estimated at 18.1 and 12.5 Tg, respectively.

Of the four sectors, industry was the greatest source, accounting for 57% and 49% of PM_{10} and $PM_{2.5}$ emissions, respectively. Emissions from cement production reached 4.9 and 3.1 Tg for PM_{10} and $PM_{2.5}$, respectively, contributing about half of all industrial emissions. Besides industry, domestic (residential and commercial) sources were another big contributor, accounting for 24% and 32% of PM_{10} and $PM_{2.5}$ emissions. Despite high fuel consumption, power generation had a relatively low share, only 10%, especially because of wide application of electrostatic precipitators (ESP). Workshops and small boilers with inefficient technologies were major sources of industrial dust, control of which will be a great challenge for China. The share of total PM emissions from on-road transport was relatively low but should nevertheless be regarded as serious, because of the high fraction of fine particles and the fact that they are emitted at ground level and mainly in populated areas, all of which raises the risks to human health.

Black Carbon   Black carbon emissions were estimated to total 1.70 Tg, but with considerable uncertainty about both emission factors and activity levels. Given the strong interest in the effects of BC on radiative forcing, many studies have been carried out to evaluate Chinese emissions applying emission factors from abroad. The results for 2000 vary from 1.05 Tg (Streets et al. 2003) to 1.50 Tg (Cao, Zhang, and Zheng 2006). These studies do not include industrial process emissions, however, which may be an important source. Q. Zhang et al. (2007b) estimated BC emissions in 2001 at 1.71 Tg, including 0.46 Tg in process emissions.

In the current study, the provinces of Shandong, Henan, Sichuan, Hebei, and Shanxi emitted more than 100 Gg each, mainly because of their high consumption of biofuels. Overall, domestic fuel combustion was the greatest source of BC (42%), followed by industrial sectors (36%) and transportation (11%, of which more than 75% was from nonroad sources like construction machinery). The contributions to total BC emissions by fuel type were coal, 26%; biomass, 37%; and oil, 9%.

Organic Carbon   Total OC emissions were estimated at 3.54 Tg, with great uncertainties similar to those of BC. A series of studies (Ohara et al. 2007; Cao, Zhang, and Zheng 2006; Streets et al. 2003) indicated that OC emissions in 2000, not counting industrial process sources, were between 2.56 and 4.24 Tg. Taking process emissions into account, Q. Zhang et al. (2007b) calculated OC emissions for 2001 at 3.58 Tg, even lower than the estimate by Cao, Zhang, and Zheng (2006).

In the current study, Sichuan had extremely high OC emissions at 325 Gg, followed by the provinces of Shandong, Henan, and Guangxi. For the sector distribution, combustion for domestic use accounted for 66%, followed by open burning (15%) and industry (14%). The contributions by type of fuel were coal, 4%; biomass, 68%; and oil, 3%.

Non-Methane Volatile Organic Compounds   Emissions of NMVOC in China were estimated to total 20.6 Tg in 2005, 18% and 32% higher than the results for 2000 by Streets et al. (2003) and Klimont et al. (2002), respectively, and close to the estimate for 2001 by L. T. Wang (2006). Shandong, Guangdong, and Jiangsu led the provinces in such emissions.

With transportation and the chemical industry developing swiftly in the recent years, NMVOC emissions from petroleum products, natural gas, and solvent use are growing rapidly. Of all sectors, biofuel combustion and road transportation were the largest contributors, accounting for 29% and 23%, respectively, followed by solvent use (17%), industry (15%), and waste treatment (6%). The speciation of total NMVOC, which is important to the chemistry driving some of the air quality results of chapter 7, will be discussed in section 6.3.2.

Ammonia   National NH_3 emissions in 2005 were 16.6 Tg, 22% higher than the estimate for 2000 by Streets et al. (2003). Henan, Shandong, Sichuan, Hebei, and Jiangsu, provinces with strong agricultural production, produced the most. Fertilizer use and livestock were the most important sources, accounting for 51% and 43% of total emissions, respectively.

Carbon Dioxide   For CO_2, the most critical greenhouse gas, an updated emission inventory framework is applied to estimate the national emission at 5516 Tg from fossil fuel use in 2005 (Zhao, Nielsen, and McElroy 2012).[5]

6.3.2 Speciation of NMVOC

As described in Wei et al. (2008), the source profiles of NMVOC were mainly derived from the U.S. EPA SPECIATE database, supplemented by other sources. These provided information on emissions from vehicles (Hwa et al. 2002; Schauer et al. 2002); gasoline and diesel storage, loading, and unloading (Cheng et al. 2003); coal-fired domestic stoves, combustion of agricultural residues, natural gas, coal gas, and liquid petroleum gas (J. Zhang et al. 2000; Andreae and Merlet 2001); and coking (Q. He et al. 2005).

Figure 6.6 summarizes speciation of total NMVOC in 2005 by combining 40 species into six main categories. The contribution from alkanes is 24% (ranging

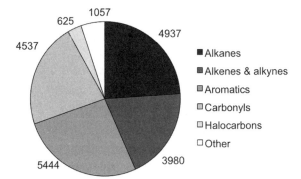

Figure 6.6
The NMVOC emissions by species (in gigagrams).

from 11% to 38% in different provinces), followed by alkenes and alkynes 19% (9–29%), aromatics 26% (21–35%), carbonyls 22% (10–29%), halocarbons 3% (1–6%), and others 5% (2–10%).

Emissions of toxic NMVOC were also estimated by Wei et al. (2008) because of their hazardous impacts on human health; these include n-hexane, 1,3-butadiene, styrene, and benzene. The national emissions of toxic NMVOC were estimated to be 5.8 Tg, approximately 28% of the total NMVOC. Biofuel combustion was the largest emission source (27%), followed by transportation (26%) and solvent use (19%). Across the provinces, the fraction of toxic NMVOC generally varied between 24% and 30%, with the exception of Shanxi, which has many large coking plants resulting in a high 37% toxic share of total NMVOC.

In the GEOS-Chem atmospheric model of chapter 7, emissions of specific species of NMVOC are required, including C_2H_6, C_3H_8, other alkanes, C_2H_4, other alkenes, CH_2O, C_2H_4O, other aldehydes, and ketone. In all, those species accounted for about 60% of total NMVOC emissions. Emissions of those species by six sectors in 2005 are illustrated in figure 6.7. Biofuel combustion accounted for nearly all of acetaldehyde (C_2H_4O) emissions and more than half of alkene and ketone emissions. Transport accounted for nearly all of the formaldehyde (CH_2O) emissions and more than half of ethane (C_2H_6) and ethylene (C_2H_4) emissions. Solvent use contributed nearly one-third of ketone emissions. Industry contributed nearly half and one-third of propane (C_3H_8) and other alkane emissions, respectively.

6.3.3 Emissions in 2010

The emissions in the three different scenarios by sector in 2010 are shown in figures 6.8A–G for each of the seven air pollution species.

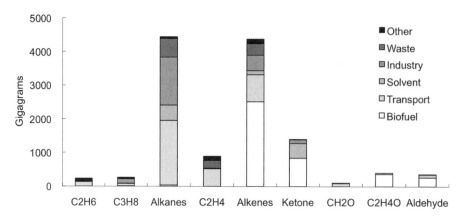

Figure 6.7
Sector emissions of NMVOC species used in GEOS-Chem (in gigagrams).

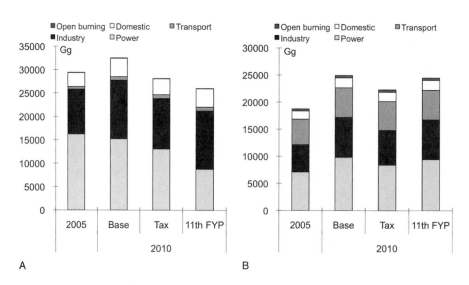

Figure 6.8
Emissions in 2005 and 2010 by scenario and sector: (A) SO_2, (B) NO_X, (C) PM_{10}, (D) $PM_{2.5}$, (E) BC, (F) OC, (G) NMVOC.

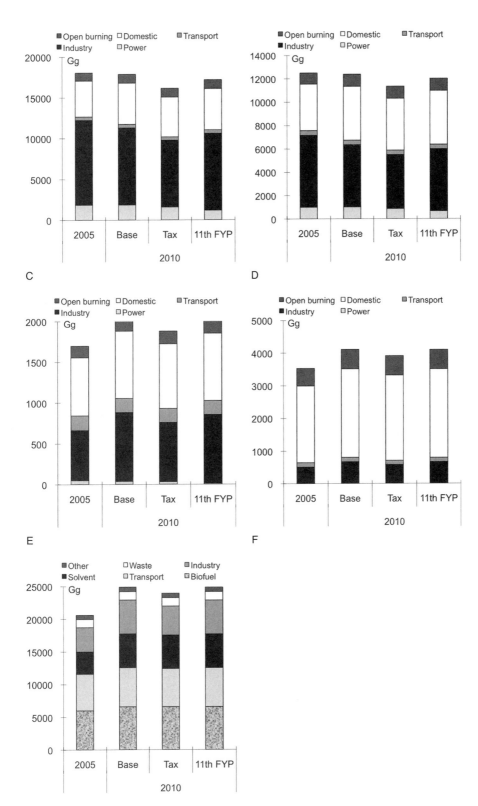

C

D

E

F

G   Figure 6.8 (continued)

For the first six species, we identified five sectors: open burning, electric power generation, industry, domestic (residential and commercial), and transportation. The emissions in 2005 are also given in the first bar in each figure.

In the base case, national SO_2 and NO_X emissions in 2010 would be 10% and 29% higher than in 2005, respectively; the much slower growth of SO_2 is due to a partial FGD requirement, while there is little NO_X control in this period of rapid growth in energy consumption. The NO_X emissions of the three leading sources—that is, power, industry, and transport—would rise by 38%, 41%, and 9%, respectively. Emissions of BC and OC would increase by 11% and 6%, mainly because of the growth of industry (except cement) and domestic emissions. However, the emissions of PM_{10} and $PM_{2.5}$ would decrease by 6.3% and 6.7%, even with economic development and growth in energy consumption. The main reason is that emissions of cement plants (accounting for 27% of PM_{10} and 25% of $PM_{2.5}$ emissions in 2005) would be reduced by more than 60% during the five years, as a result of the replacement of shaft kilns by precalciner kilns.

The broad reduction of fossil fuel consumption in the carbon tax scenario would decrease the emissions of SO_2, NO_X, PM_{10}, $PM_{2.5}$, BC, and OC by 6–14% compared to the base case, implying widespread potential environmental benefits of the policy. In particular, emissions of SO_2, PM_{10}, and $PM_{2.5}$ would be 5%, 17%, and 16% lower, respectively, than even those in 2005; the lower growth in NO_X emissions, however, would still lead to a rise by 15% compared to 2005.

In the 11th FYP scenario, representing what actually has happened, national emissions of SO_2 are estimated to have declined to 26.0 Tg, approximately 12% lower than that in 2005. This outcome indicates that the compulsory measures in emission control (mainly in the power sector) accomplished the objective of SO_2 abatement successfully and the national target of a 10% emission reduction was likely achieved or exceeded. One should, of course, take into account the sudden global recession in 2008–2009 in comparing plans made in 2006 with actual outcomes.

Regarding the sector contributions, SO_2 emissions from power and cement plants are estimated to have decreased by 46% and 17%, respectively, while iron and steel, other industry (i.e., excluding cement and iron and steel), and domestic uses have increased by 26%, 32%, and 49%, respectively.

In a change similar to that for SO_2, the emissions of PM_{10} and $PM_{2.5}$ under the 11th FYP are estimated to have declined by 10% and 9%, respectively, compared to 2005. For NO_X, BC, and OC, however, the emissions are estimated to have grown over time; the 11th FYP controls generated little difference compared to the base case. The carbon tax scenario, in contrast, has these 2010 emissions noticeably lower than the levels in the base case scenario even if they too are higher than the 2005 levels.

For NMVOC, we estimate that emissions from solvent use and industry have increased fastest among all six source categories, and total emissions in 2010 reach 24.1, 23.2, and 24.1 Tg in the base, carbon tax, and 11th FYP scenarios, respectively, compared to 20.6 Tg in 2005. We assumed no differences in agricultural activities of the three scenarios, and thus NH_3 emissions are estimated at 18.4 Tg for all cases, 11% higher than in 2005.

Regarding CO_2, recall that without need for any spatial distribution, the economic model of chapter 9 uses the 2005 base year emissions to estimate CO_2 emissions in 2010 under the three scenarios.[6] The effects on CO_2 are therefore reported separately in that chapter, including in tables 9.3, 9.8, and 9.10.

6.3.4 Gridded Emissions

All the emissions calculated at the provincial level were allocated into a 0.5° by 0.67° grid, matched to the nested grid of the GEOS-Chem model (see chapter 7). This was done using spatial administrative boundaries matched to the categorization of underlying data, the location of large point sources, and social and economic characteristics to guide the disaggregation of provincial totals.

For large point sources (specifically plants producing coal-fired electric power, cement, and iron and steel), the emissions were directly assigned into the grid cell where the plant was located. (See chapter 4 for an example of gridded emissions from coal-fired power plants.)

Emissions from nonpoint and smaller point sources were allocated by species and sector according to their emission characteristics. Provincial emissions were first divided into county-level emissions by economic or population parameters, as listed in table 6.4. "Primary industry GDP" refers to the GDP (or value added) due to agriculture, while "secondary industry" covers mining, manufacturing, utilities, and construction.

If a grid cell covers more than one county, the emissions of each county were apportioned based on the percentage of the county in the cell. The emissions of each grid cell were then determined by aggregating these apportioned emissions. Since the area of one GEOS-Chem grid cell is close to or larger than most counties in eastern and south-central China, where the intensity of energy consumption and emissions is high, we believe this method of emission allocation is reasonable.

Table 6.4
Parameters for the spatial allocation of emissions by species and sector

Species	Sector	Parameter
NH_3	All	Primary industry GDP
NMVOC	Biofuel combustion	Agricultural population
	Industry (including solvent use)	Secondary industry GDP
	Waste treatment	Primary industry GDP
	Other (fossil fuel combustion, transport, etc.)	Total population
Other species	Industry	Secondary industry GDP
	Transport	Total population
	Domestic	Total population
	Open burning	Primary industry GDP

Gridded emissions of SO_2, NO_X, PM_{10}, $PM_{2.5}$, BC, OC, NMVOC, and NH_3 in 2005 are shown in figures 6.9A–H for five ranges of emission levels, with darker shades representing higher intensities. These gridded emissions were applied in the GEOS-Chem simulations of chapter 7 and can be used in other global or regional air quality modeling.

Large SO_2 emissions were found throughout the industrialized eastern part of China and in southwestern China including the Sichuan Basin. Emissions in the regions targeted for SO_2 control under the "two-control zone" policy (Hao et al. 2000) remained much higher than other areas.

For NO_X and NMVOC (figures 6.9B and 6.9H), emissions were considerable in eastern and part of southern China, mainly Guangdong. This result is partly attributed to the tremendous increase in vehicle numbers there. Besides industrialized areas, relatively high BC and OC emissions were also found in part of southwestern China, where burning of wood and crop wastes for cooking and heating is common in the countryside. Generally, the lowest emissions of all species were found in west, northwest, and northeast China, notably Tibet, Qinghai, Xinjiang, and Inner Mongolia.[7]

Regarding spatial emission distributions for 2010 under the three scenarios, the gridded emissions from the power sector show considerable changes, based on specific construction plans and emission control policies as discussed in chapter 4. The spatial patterns of emissions from other sectors, however, were very similar in 2010 to those of 2005.

6.3.5 Implications for Policy

Table 6.5 summarizes the historical emission trends of different species determined by other studies along with those of this book. The 2010 estimates listed in the table for this work are those of the 11th FYP SO_2 control scenario, since it is our representation of what actually happened. Some modest discrepancies aside, the results for years up to and including 2000 are comparable except perhaps for relatively high estimates of SO_2 by Ohara et al. (2007).

Since there are few studies published to date on Chinese emissions for 2005 aside from our own updated work (Zhao et al. 2011; Lei, Zhang, He, et al. 2011), it is hard to check our estimates against others. One recent study, Klimont et al. (2009), estimated Chinese emissions in 2000 and 2005; our results are lower for SO_2 but higher for other species.

From 1980 to 2000, emissions of SO_2, NO_X, and NMVOC are seen to have increased by approximately 70%, 195%, and 115%, respectively, while energy consumption doubled. Relatively minor increases were also found for BC and OC. Several studies (e.g., Ohara et al. 2007; Streets et al. 2001, 2003; and Zhang et al. 2009) have estimated that emissions of SO_2, BC, and OC actually declined in the late 1990s, a finding which may in part be attributable to problems in the official energy data for that period (Ho and Nielsen 2007, chapter 1).

Between 2000 and 2005, energy consumption rose sharply by 60%, particularly in the electric power and transport sectors; coal consumption by power plants increased by 85%; and oil use by on-road vehicles rose by 70%, as shown in figure 6.10. Even though control policies and advanced technologies were beginning to penetrate these sectors at this time, these could not compensate for the effect of rapid energy growth, and large emission increases were still observed for all species. For example, energy-related emissions (i.e., excluding emissions from noncombustion industrial processes and open burning of biomass) of SO_2 and NO_X increased by 34% and 48%, indicating that control policies at that time were inadequate. If China's economy and energy practices continue on such a path, energy consumption would double again between 2005 and 2020. One study analyzed such a scenario of no additional control, predicting that Chinese national emissions of SO_2, NO_X. and NMVOC would reach 40.8, 25.5, and 38.6 Tg in 2020 (Ohara et al. 2007).

In the face of such challenges, the 11th FYP policies on air pollution control (along with those on energy) were conceived and carried out. These measures appear to have had a major effect: according to official statistics, SO_2 emissions declined continuously for four years (2006–2009), in contrast with substantial increases each

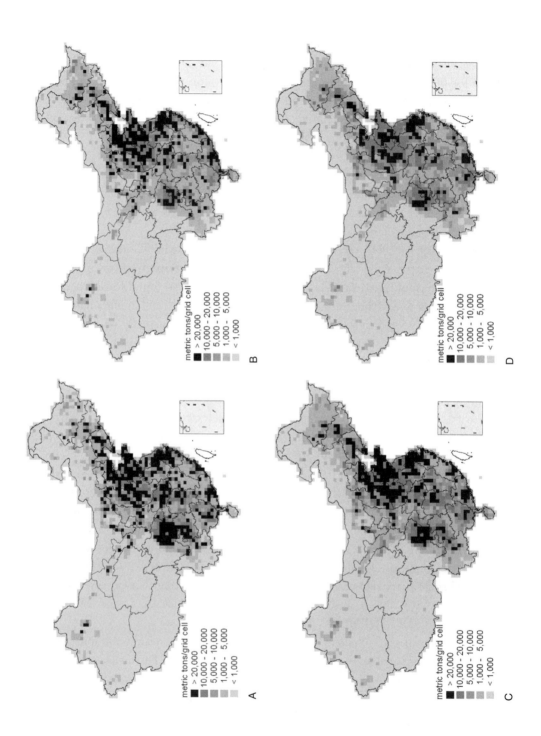

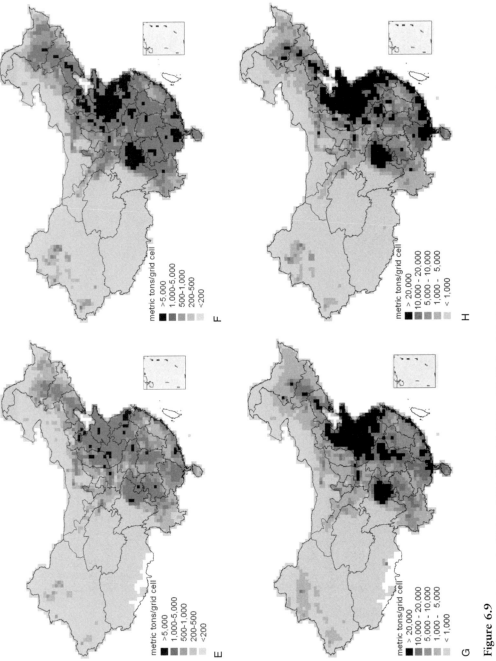

Figure 6.9

Gridded emissions in 2005 of (A) SO_2, (B) NO_x, (C) PM_{10}, (D) $PM_{2.5}$, (E) BC, (F) OC, (G) NH_3, (H) NMVOC.

Table 6.5
Chinese emission trends from anthropogenic sources (in teragrams)

		1980	1985	1990	1995	2000	2005	2006	2010[a]
SO$_2$	MEP (Chinese official)		13.2	14.9	18.9	20.0	25.5	25.9	
	Ohara et al. (2007)	14.9	18.5	21.6	27.1	27.6			30.0
	Streets et al. (2000, 2003); Zhang et al. (2009)		17.9	23.0	25.2	20.4		31.0	
	Klimont et al. (2001, 2009)			20.9	23.9	23.2	34.4	33.2	35.6
	Lu et al. (2010)					21.7	32.3		
	Zhao et al. (2011)[b]						31.1		
	This work (11th FYP case)						29.4		26.0
NO$_X$	Ohara et al. (2007)	3.8	4.6	6.6	9.3	11.2			14.0
	Streets and Waldhoff (2000); Streets et al. (2003); Zhang et al. (2009)			9.5	12.0	11.3		20.8	
	Zhang et al. (2007a)				10.9	12.6			
	Klimont et al. (2001, 2009)			7.2	9.7	11.7	16.9		20.5
	Hao et al. (2002); Tian (2003)			8.4	11.3	11.2			
	van Aardenne et al. (1999)			8.3		13.7			21.9
	Zhao et al. (2011)[b]						19.8		
	This work (11th FYP case)						18.8		23.9
PM$_{10}$	Zhang et al. (2009)							18.2	
	Lei, Zhang, He, et al. (2011)[b]			13.5	18.4	16.1	18.8		
	Zhao et al. (2011)[b]						19.2		
	This work (11th FYP case)						18.1		16.3
PM$_{2.5}$	Zhang et al. (2009)							13.3	
	Lei, Zhang, He, et al. (2011)[b]			9.3	12.1	10.8	12.9		
	Zhao et al. (2011)[b]						13.1		
	This work (11th FYP case)						12.5		11.3

BC							
Ohara et al. (2007)	1.03		1.42	1.39	1.09		1.11
Streets et al. (2001); Streets et al. (2003); Zhang et al. (2009)		1.28		1.34	1.05		1.81
Cao, Zhang, and Zheng (2006)					1.50		
Klimont et al. (2009)					1.34	1.37	1.52
Lei, Zhang, He, et al. (2011)[b]			1.13	1.27	1.18	1.51	
Zhao et al. (2011)[b]						1.70	
This work (11th FYP case)						1.70	1.86
OC							
Ohara et al. (2007)	2.37		3.02	3.19	2.56		2.19
Streets et al. (2003); Zhang et al. (2009)		2.73			3.38		3.22
Cao, Zhang, and Zheng (2006)					4.24		
Klimont et al. (2009)					3.20	2.81	2.82
Lei, Zhang, He, et al. (2011)[b]			2.87	2.86	2.54	3.19	
Zhao et al. (2011)[b]						3.20	
This work (11th FYP case)						3.54	3.76
NMVOC							
Ohara et al. (2007)	6.8	8.2	9.7	12.2	14.7		22.4
Streets et al. (2003); Zhang et al. (2009)				13.1	17.4		23.2
Klimont et al. (2002)			11.1		15.6		17.2
This work (11th FYP case)						20.6	24.1
NH₃							
Streets et al. (2003)			9.7	11.7	13.6		
Klimont et al. (2001)							
This work (11th FYP case)						16.6	18.4

[a] 2010 estimates based on REF scenario for Ohara et al. (2007); GAINS baseline scenario for Klimont et al. (2009); and the 11th FYP scenario for this work.

[b] Zhao et al. (2011) and Lei, Zhang, He, et al. (2011) report research by the current authors that was completed after the work of this chapter.

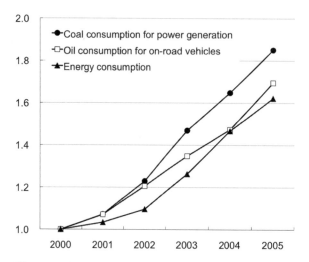

Figure 6.10
Coal consumption by power plants, oil consumption by on-road vehicles, and total energy consumption for 2000–2005 (normalized to year 2000).

year in the prior decade. According to our independent estimates, the national target of a 10% reduction in SO_2 emissions from 2005 to 2010 was not only achieved but exceeded, based on an assumption that all the measures were successfully implemented.

However, emissions of other species such as NO_X and NMVOCs (both of which are important precursors of ozone pollution) increased significantly during the five years. Moreover, although there is considerable uncertainty, NH_3 emissions also likely continued to rise, since no effective control measures were undertaken. These increases in nitrogen emissions could negate the benefits to abatement of soil acidification from SO_2 emission control (Zhao et al. 2009). Therefore, mitigation of nitrogen emissions is an urgent task for Chinese air pollution control in the near future, and NO_X has laudably been targeted for reduction in the 12th FYP of 2011–2015, subject to specific implementing measures. Regarding the different sectors, industrial and domestic sources (particularly small area sources, for which effective control policy is lacking) should be specifically targeted for stringent control efforts.

In addition to the command-and-control measures like those of the 11th FYP, this study suggests that market-based instruments like a carbon tax could also be effective at nitrogen control because they broadly reduce fossil energy consumption. These also merit consideration by the government, although more research should first be conducted on alternative tax levels, the time paths of implementation, and other factors influencing the effectiveness of such policies.

6.4   Conclusions

To serve the needs of both scientific modeling and policy analyses, an emission inventory of primary air pollutants in mainland China was developed based on available information on activity levels and emission factors, using Chinese domestic sources as much as possible. For a base year of 2005, emissions from all anthropogenic sources including power generation, industry, transport, domestic (residential and commercial), and agricultural sectors were evaluated for 31 regions. Total emissions in 2005 were estimated at 29.4 Tg of SO_2, 18.8 Tg of NO_X, 32.3 Tg of TSP, 18.1 Tg of PM_{10}, 12.5 Tg of $PM_{2.5}$, 1.70 Tg of BC, 3.54 Tg of OC, 20.6 Tg of NMVOC, 16.6 Tg of NH_3, and 5516 Tg of fossil fuel CO_2. In addition, NMVOC was speciated into 40 subcategories according to reactivity, toxicity, and importance to chemical processes assessed by our atmospheric model discussed in chapter 7.

We estimated emissions in 2010 for three scenarios: a no-policy base case, a carbon tax case, and a case with the power-sector policies of the 11th FYP. We found that SO_2 and PM emissions likely decreased between 2005 and 2010 with the successful implementation of 11th FYP control policies, while emissions of NO_X and NMVOC likely continued to increase. As an alternative to the command-and-control policy approach used in the 11th FYP, a market-based policy like a carbon tax could also be considered for future emission control; our analysis of such a hypothetical policy during the years of the 11th FYP suggests it could lead to considerable reductions in fossil fuel combustion and pollutant emissions.

Finally, the emissions under all three scenarios were spatially allocated into a 0.5° by 0.67° grid system using information on the location of large point sources and data on GDP and population distribution for area sources. These gridded emissions can be applied in atmospheric transport and chemistry models, notably in the use of GEOS-Chem to estimate pollutant concentrations and associated damages as described in the chapters that follow.

Acknowledgments

This chapter was developed from research originally funded by the Hi-Tech Research and Development Program (No. 2006AA06A305) and the National Basic Research Program of China (No. 2005 CB422206). Further development for the purposes of this book was supported by the China Sustainable Energy Program of the Energy Foundation and the U.S. National Science Foundation (Grant ATM-1019134).

Notes

1. Black carbon is pure carbon formed from incomplete combustion of fossil fuels, biofuels, and biomass. Organic carbon is the carbon bound in an organic compound, in contrast to inorganic carbon in carbonic acid salts. NMVOCs include volatile compounds such as benzene and formaldehyde.

2. We remind readers that all emission data, whether in official statistics or generated by research teams such as ours, are estimates, as discussed in section 1.3. There are no "true" emission data available to research.

3. It appears with benefit of hindsight that our projection for transportation was an underestimate, but a modest one. Note that even if the passenger vehicle stock grew faster than 7%, fuel efficiency of average vehicles also grew, and diverse other categories of transportation are also included.

4. The teragram is a conventional scientific unit for emissions of a chemical species. One Tg is equivalent to one million metric tons (Mt), a unit conventionally used for energy inputs (e.g., coal) and industrial production. In this book, Mt is also used in those contexts.

5. The research aims of this book do not require us to estimate emissions of CO_2 for each policy case using the bottom-up methods of this chapter. Changes in CO_2 due to policy—our interest—are overwhelmingly determined by changes in quantities of fossil energy consumed, not by technology or other factors. Additionally, we are interested in CO_2 for its global (not local) impacts, it plays no role in the atmospheric chemistry driving the damage assessments of chapters 7 and 8, and we therefore have no need to estimate the spatial distribution of CO_2 emissions. We use bottom-up methods only to set our estimate of total CO_2 emissions in the base year of 2005, while the economic model of chapter 9 determines CO_2 emissions for the policy cases, based on its estimates of fuel consumption and scaled to the 2005 value.

After the research of this chapter, the lead author and colleagues completed new bottom-up research on emissions of CO_2, further developing the approach employed here for 2005 using updated inputs and testing its spatial distribution against instrumental observations (Zhao, Nielsen, and McElroy 2012). The newer study yields somewhat higher estimates for the 2005 base year than those generated originally for this chapter. It is simple for us to adopt the updated values as the basis for analysis of the effects of policies on CO_2 in this book. In fact, a difference in the absolute CO_2 emission level in the base year is inconsequential to our final results, which concern percentage changes in emissions due to policies that are nearly unaffected by a modest shift in the underlying baseline. We opt to incorporate the newer CO_2 results for the informational value to the reader of more up-to-date estimates of total emissions.

6. See the preceding endnote.

7. Inclusion of Inner Mongolia may surprise those readers who know it has been one of China's fastest growing provinces. It has a small population and economy relative to other provinces, however, and is geographically enormous. High emissions occur in spots, but surrounded by few emissions at all. Also note that energy- and emission-intensities for resource

extraction, a main source of Inner Mongolia's economic growth, are typically quite low compared to those of many industries.

References

Akimoto, H., T. Ohara, J. Kurokawa, and N. Horii. 2006. Verification of energy consumption in China during 1996–2003 by using satellite observational data. *Atmospheric Environment* 40(40):7663–7667.

Andreae, M. O., and P. Merlet. 2001. Emission of trace gases and aerosols from biomass burning. *Global Biogeochemical Cycles* 15(4):955–966.

Bouwman, A. F., D. S. Lee, W. A. H. Asman, F. J. Dentener, K. W. VanderHoek, and J. G. J. Olivier. 1997. A global high-resolution emission inventory for ammonia. *Global Biogeochemical Cycles* 11(4):561–587.

China Association of Automobile Manufacturers (CAAM). 2006. *China Automobile Industry Yearbook, 2005.* Beijing: China Association of Automobile Manufacturers. In Chinese.

Cao, G. L., X. Y. Zhang, and F. C. Zheng. 2006. Inventory of black carbon and organic carbon emissions from China. *Atmospheric Environment* 40(34):6516–6527.

China Coal Transportation Association (CCTA). 2003. Statistical data on interprovince coal transportation in China. Internal information. In Chinese.

Cheng, P., W. J. Zhan, Y. N. Chu, P. Spanel, and D. Smith. 2003. Analysis of petrol and diesel vapor using selective ion flow tube/mass spectrometry. *Chinese Journal of Analytical Chemistry* 31:548–551. In Chinese.

China National Coating Industry Association (CNCIA). 2006. *China Paint and Coating Industry Annual, 2005.* Internal information. In Chinese.

China National Light Industry Council (CNLIC). 2006. *China Light Industry Yearbook, 2005.* Beijing: China Light Industry Press. In Chinese.

China Petroleum and Chemical Industry Association (CPCIA). 2006. *China Chemical Industry Yearbook, 2005.* Beijing: China National Chemical Information Center. In Chinese.

European Environment Agency (EEA). 2005. EMEP/CORINAIR Atmospheric Emission Inventory Guidebook. Available at http://reports.eea.europa.eu/EMEPCORINAIR4.

Hao, J. M., H. Z. Tian, and Y. Q. Lu. 2002. Emission inventories of NO_X from commercial energy consumption in China, 1995–1998. *Environmental Science & Technology* 36(4): 552–560.

Hao, J. M., S. X. Wang, B. J. Liu, and K. B. He. 2000. Designation of acid rain and SO_2 control zones and control policies in China. *Journal of Environmental Science and Health Part A—Toxic/Hazardous Substances and Environmental Engineering* 35(10):1901–1914.

He, K. B., H. Huo, Q. Zhang, D. Q. He, F. An, M. Wang, and M. P. Walsh. 2005. Oil consumption and CO_2 emissions in China's road transport: Current status, future trends, and policy implications. *Energy Policy* 33(12):1499–1507.

He, Q., X. Wang, L. Zhao, G. Sheng, and J. Fu. 2005. Preliminary study on profiles of VOCs emitted from coking. *Environmental Monitoring in China* 21:61–65. In Chinese.

258   *Chapter 6*

Ho, M. S., and C. P. Nielsen, eds. 2007. *Clearing the Air: The Health and Economic Damages of Air Pollution in China.* Cambridge, MA: MIT Press.

Hwa, M. Y., C. C. Hsieh, T. C. Wu, and L. F. W. Chang. 2002. Real-world vehicle emissions and VOCs profile in the Taipei tunnel located at Taiwan Taipei area. *Atmospheric Environment* 36(12):1993–2002.

Irie, H., K. Sudo, H. Akimoto, A. Richter, J. P. Burrows, T. Wagner, M. Wenig, S. Beirle, Y. Kondo, V. P. Sinyakov, and F. Goutail. 2005. Evaluation of long-term tropospheric NO2 data obtained by GOME over East Asia in 1996–2002. *Geophysical Research Letters* 32(11), Art. No. L11810.

Jiang, J. K. 2004. Preliminary studies on emission and control of atmosphere mercury in China. Master diss., Tsinghua University, Beijing, China. In Chinese.

Kato, N., and H. Akimoto. 1992. Anthropogenic Emissions of SO_2 and NO_X in Asia— Emission Inventories. *Atmospheric Environment Part A—General Topics* 26(16): 2997–3017.

Klimont, Z., J. Cofala, W. Schopp, M. Amann, D. G. Streets, Y. Ichikawa, and S. Fujita. 2001. Projections of SO_2, NO_X, NH_3 and VOC emissions in East Asia up to 2030. *Water Air and Soil Pollution* 130(1–4):193–198.

Klimont, Z., J. Cofala, J. Xing, W. Wei, C. Zhang, S. Wang, K. Jiang, P. Bhandari, R. Mathur, P. Purohit, P. Rafaj, A. Chambers, and M. Amann. 2009. Projections of SO_2, NO_X and carbonaceous aerosols emissions in Asia. *Tellus Series B—Chemical and Physical Meteorology* 61(4):602–617.

Klimont, Z., D. G. Streets, S. Gupta, J. Cofala, L. X. Fu, and Y. Ichikawa. 2002. Anthropogenic emissions of non-methane volatile organic compounds in China. *Atmospheric Environment* 36(8):1309–1322.

Lei, Y. 2008. Research on anthropogenic emissions and control of primary particles and its key chemical components. PhD diss., Tsinghua University, Beijing, China. In Chinese.

Lei, Y., Q. Zhang, K. B. He, and D. G. Streets. 2011. Primary anthropogenic aerosol emission trends for China, 1990–2005. *Atmospheric Chemistry and Physics* 11:931–954.

Lei, Y., Q. Zhang, C. P. Nielsen, and K. B. He. 2011. An inventory of primary air pollutants and CO_2 emissions from cement production in China, 1990–2020. *Atmospheric Environment* 45(1):147–154.

Li, X. H. 2007. Characterization of air pollutants emitted from biomass burning in China. PhD diss., Tsinghua University, Beijing, China. In Chinese.

Li, X. H., L. Duan, S. X. Wang, J. C. Duan, X. M. Guo, H. H. Yi, J. N. Hu, C. Li, and J. M. Hao. 2007. Emission characteristics of particulate matter from rural household biofuel combustion in China. *Energy & Fuels* 21(2):845–851.

Li, X. H., S. X. Wang, L. Duan, J. Hao, C. Li, Y. S. Chen, and L. Yang. 2007. Particulate and trace gas emissions from open burning of wheat straw and corn stover in China. *Environmental Science & Technology* 41(17):6052–6058.

Lu, Z., D. G. Streets, Q. Zhang, S. Wang, G. R. Carmichael, Y. F. Cheng, C. Wei, M. Chin, T. Diehl, and Q. Tan. 2010. Sulfur dioxide emissions in China and sulfur trends in East Asia since 2000. *Atmospheric Chemistry and Physics* 10:6311–6331, doi:10.5194/acp-10-6311-2010.

Ministry of Agriculture (MOA). 2006. *China Animal Industry Yearbook, 2005.* Beijing: China Agriculture Press. In Chinese.

National Bureau of Statistics (NBS). 2001. *China Statistical Yearbook, 2000.* Beijing: China Statistics Press. In Chinese.

————. 2006a. *China Energy Statistical Yearbook, 2005.* Beijing: China Statistics Press. In Chinese.

————. 2006b. *China Industry Economy Statistical Yearbook, 2005.* Beijing: China Statistics Press. In Chinese.

————. 2006c. *China Rural Statistical Yearbook, 2005.* Beijing: China Statistics Press. In Chinese.

————. 2006d. *China Statistical Yearbook for Regional Economy, 2005.* Beijing: China Statistics Press. In Chinese.

Ohara, T., H. Akimoto, J. Kurokawa, N. Horii, K. Yamaji, X. Yan, and T. Hayasaka. 2007. An Asian emission inventory of anthropogenic emission sources for the period 1980–2020. *Atmospheric Chemistry and Physics* 7(16):4419–4444.

Richter, A., J. P. Burrows, H. Nuss, C. Granier, and U. Niemeier. 2005. Increase in tropospheric nitrogen dioxide over China observed from space. *Nature* 437(7055):129–132.

Schauer, J. J., M. J. Kleeman, G. R. Cass, and B. R. T. Simoneit. 2002. Measurement of emissions from air pollution sources. 5. C1–C32 organic compounds from gasoline-powered motor vehicles. *Environmental Science & Technology* 36(6):1169–1180.

State Environmental Protection Administration (SEPA). 1996. *Handbook of Industrial Pollution Emission Rates.* Beijing: China Environmental Science Press. In Chinese.

————. 2001. Chinese environment communique, 2000. Available at http://www.sepa.gov.cn/. In Chinese.

————. 2006. Chinese environment communique, 2005. Available at http://www.sepa.gov.cn/. In Chinese.

Shang, C. P., Y. M. Zhang, and Y. Shen. 2006. Production and consumption status of tetrachloroethylene at home and abroad and its prospective market. *China Chlor-Alkali* 3:4–7. In Chinese.

Streets, D. G., T. C. Bond, G. R. Carmichael, S. D. Fernandes, Q. Fu, D. He, Z. Klimont, S. M. Nelson, N. Y. Tsai, M. Q. Wang, J. H. Woo, and K. F. Yarber. 2003. An inventory of gaseous and primary aerosol emissions in Asia in the year 2000. *Journal of Geophysical Research—Atmospheres* 108, Art. No. D218809.

Streets, D. G., S. Gupta, S. T. Waldhoff, M. Q. Wang, T. C. Bond, and Y. Y. Bo. 2001. Black carbon emissions in China. *Atmospheric Environment* 35(25):4281–4296.

Streets, D. G., N. Y. Tsai, H. Akimoto, and K. Oka. 2000. Sulfur dioxide emissions in Asia in the period 1985–1997. *Atmospheric Environment* 34(26):4413–4424.

Streets, D. G., and S. T. Waldhoff. 2000. Present and future emissions of air pollutants in China: SO_2, NO_X, and CO. *Atmospheric Environment* 34(3):363–374.

Tian, H. Z. 2003. Studies on present and future emissions of nitrogen oxides and its comprehensive control policies in China. PhD diss., Tsinghua University, Beijing, China. In Chinese.

U.S. Environmental Protection Agency (USEPA). 1999. Emissions Factors & AP 42, Compilation of Air Pollutant Emission Factors. Available at http://www.epa.gov/ttnchie1/ap42/, last accessed March 25, 2013.

van Aardenne, J. A., G. R. Carmichael, H. Levy, D. Streets, and L. Hordijk. 1999. Anthropogenic NO_X emissions in Asia in the period 1990–2020. *Atmospheric Environment* 33(4):633–646.

Wang, L. T. 2006. Air quality modeling and control scenario analysis in Beijing area. PhD diss., Tsinghua University, Beijing, China. In Chinese.

Wang, S. X. 2001. Technical options and planning on control of sulfur dioxide emitted from coal-fired power plants in China. PhD diss., Tsinghua University, Beijing, China. In Chinese.

Wang, Y. X., M. B. McElroy, R. V. Martin, D. G. Streets, Q. Zhang, and T. M. Fu. 2007. Seasonal variability of NO_X emissions over east China constrained by satellite observations: Implications for combustion and microbial sources. *Journal of Geophysical Research—Atmospheres* 112, Art. No. D06301.

Wang, Y. X., M. B. McElroy, T. Wang, and P. I. Palmer. 2004. Asian emissions of CO and NO_X: Constraints from aircraft and Chinese station data. *Journal of Geophysical Research-Atmospheres* 109, Art. No. D24304.

Wei, W., S. X. Wang, S. Chatani, Z. Klimont, J. Cofala, and J. M. Hao. 2008. Emission and speciation of non-methane volatile organic compounds from anthropogenic sources in China. *Atmospheric Environment* 42(20):4976–4988.

Xie, Y. M. 2005. Market analysis of trichloroethylene at home and abroad. *Chemical Techno-Economics* 23(4):20–23. In Chinese.

Yi, H. H., J. M. Hao, L. Duan, X. H. Li, and X. M. Guo. 2006. Characteristics of inhalable particulate matter concentration and size distribution from power plants in China. *Journal of the Air & Waste Management Association* 56(9):1243–1251.

Yi, H. H., J. M. Hao, L. Duan, X. L. Tang, P. Ning, and X. H. Li. 2008. Fine particle and trace element emissions from an anthracite coal-fired power plant equipped with a bag-house in China. *Fuel* 87(10–11):2050–2057.

Zhang, J., K. R. Smith, Y. Ma, S. Ye, F. Jiang, W. Qi, P. Liu, M. A. K. Khalil, R. A. Rasmussen, and S. A. Thorneloe. 2000. Greenhouse gases and other airborne pollutants from household stoves in China: A database for emission factors. *Atmospheric Environment* 34(26):4537–4549.

Zhang, Q. 2005. Study on regional fine PM emissions and modeling in China. PhD diss., Tsinghua University, Beijing, China. In Chinese.

Zhang, Q., D. G. Streets, G. R. Carmichael, K. B. He, H. Huo, A. Kannari, Z. Klimont, I. S. Park, S. Reddy, J. S. Fu, D. Chen, L. Duan, Y. Lei, L. T. Wang, and Z. L. Yao. 2009. Asian emissions in 2006 for the NASA INTEX-B mission. *Atmospheric Chemistry and Physics* 9(14):5131–5153.

Zhang, Q., D. G. Streets, K. B. He, Y. X. Wang, A. Richter, J. P. Burrows, I. Uno, C. J. Jang, D. Chen, Z. Yao, and Y. Lei. 2007a. NO_X emission trends for China, 1995–2004: The view from the ground and the view from space. *Journal of Geophysical Research—Atmospheres* 112 (D22), Art. No. D22306.

Zhang, Q., D. G. Streets, K. B. He, and Z. Klimont. 2007b. Major components of China's anthropogenic primary particulate emissions. *Environmental Research Letters* 2(4), Art. No. 045027.

Zhao, Y., L. Duan, J. Xing, T. Larssen, C. P. Nielsen, and J. M. Hao. 2009. Soil acidification in China: Is controlling SO_2 emissions enough? *Environmental Science & Technology* 43(21):8021–8026.

Zhao, Y., C. P. Nielsen, Y. Lei, M. B. McElroy, and J. M. Hao. 2011. Quantifying the uncertainties of a bottom-up emission inventory of anthropogenic atmospheric pollutants in China. *Atmospheric Chemistry and Physics* 11:2295–2308.

Zhao, Y., C. P. Nielsen, and M. B. McElroy. 2012. China's CO_2 emissions estimated from the bottom up: Recent trends, spatial distributions, and quantification of uncertainties. *Atmospheric Environment* 59:214–223.

Zhao, Y., S. X. Wang, C. P. Nielsen, X. H. Li, and J. M. Hao. 2010. Establishment of a database of emission factors for atmospheric pollutants from Chinese coal-fired power plants. *Atmospheric Environment* 44(12):1515–1523.

7

Atmospheric Modeling of Pollutant Concentrations

Yuxuan Wang

7.1 Introduction

Pollutant concentrations near the surface of the earth are of central concern in protecting public health, agricultural productivity, and ecosystems. The mapping of surface concentrations from estimates of emissions like those developed in chapter 6 is seldom straightforward, however, as the atmosphere is a highly complex system in terms of both its physical transport mechanisms (or dynamics) and its chemistry.

Chemical species emitted at or near the surface will mix into the surrounding air and will be carried away by winds and other forms of atmospheric transport such as convection. Species emitted in their final pollutant form—that is, those transported but not chemically transformed—are referred to as *primary* pollutants. Other emitted species undergo chemical transformations in the air, leading to the formation of *secondary* pollutants such as ozone (O_3, a gas) and sulfate, nitrate, and other aerosols (a scientific term for solid particles or liquid droplets suspended in gas). Numerical models that incorporate our current understanding of the physical and chemical processes controlling atmospheric composition are designed to predict the relationships between emissions and the resulting pollution distributions in the atmosphere.

In this chapter we discuss the results of a state-of-the-art numerical model of atmospheric dynamics and chemistry quantifying the effects on air quality over China of the policy scenarios described in previous chapters. The atmospheric modeling approach adopted here is a major advance over the simplified one used in the preceding book of the current research program, *Clearing the Air: The Health and Economic Damages of Air Pollution in China* (Ho and Nielsen 2007).

In section 7.2, we first provide a brief overview of numerical model types. Some general but fundamental differences between the modeling approaches adopted in this chapter and in *Clearing the Air* will be discussed. A deeper description of the model employed in this study is then presented in section 7.3. Section 7.4 presents the model results, focusing on the effects of the different policy scenarios on air quality. A summary and discussion are given in section 7.5.

7.2   Overview of Numerical Atmospheric Models

Four general types of processes determine the variation in space and time of the concentrations of chemicals in the atmosphere: emissions, transport, chemistry, and deposition. The numerical models that predict the relationship from emissions to pollution concentrations need to simulate all these processes. As inputs, they require emission data (including location and amount) and meteorological information (such as wind, pressure, solar radiation, and rainfall). Model outputs are distributions of pollutant concentrations in the atmosphere. The spatial and temporal resolutions of the outputs are determined by the resolutions of the input data and by computational constraints. In this section, we provide a general description intended for nontechnical readers of the two types of models used most commonly in air quality studies: atmospheric dispersion models and Eulerian gridded models.

7.2.1   Atmospheric Dispersion Models

Dispersion models are designed to simulate the transport and evolution of a single pollutant plume emitted from a point source or other sources that can be approximated as a point source. A typical example of a point source is the smokestack of an industrial plant. Given atmospheric wind and stability, the dispersion model calculates the rate at which the plume is transported downwind of the source, considering also the diffusion of the plume that results from atmospheric turbulence. As the mass of the pollutant plume is conserved in the atmosphere, the diffusion process that enlarges the volume of the plume results in decreased concentrations with time in three dimensions of the atmosphere, that is, in downwind, crosswind, and vertical directions. Dispersion models use mathematical algorithms that simplify dispersion and dilution phenomena to predict the evolution of the plume in the atmosphere. Some advanced dispersion models may include pollutant deposition at the surface by assuming a certain loss frequency in response to terrain topography and surface roughness. Simple chemical processes may also be included during the dispersion process through a characteristic chemical decay time scale that is a function of temperature, pressure, humidity, and other atmospheric conditions.

Dispersion models are used mainly to simulate the evolution of a point-based plume. Although some dispersion models can have multiple plumes ("puffs") from a source or multiple source locations, the dispersion processes of these plumes/sources are independently calculated, and the predicted concentration distribution is a linear addition of all the plumes. Dispersion models are commonly used to quantify ground-level concentrations and deposition of single pollutants less than 50 km downwind of point sources, although some dispersion models such as the CALPUFF model endorsed by the U.S. Environmental Protection Agency are designed for long-range assessment. In general, the advantages of dispersion models are simplicity, fast calculation, and strong representation of the plume structure.

The CALPUFF and ISC models employed in *Clearing the Air* are dispersion models. The choice of dispersion modeling in that study was dictated by (1) a lack at the time of sufficiently comprehensive emission data to support a Eulerian gridded model matched to the structure of the team's economic model and (2) adoption of a methodological alternative (termed "intake fraction") that, instead of estimating pollutant exposure risks for each source, employs the average characteristics of sources by sector (Ho and Nielsen 2007, chapters 4–6). The relative computational ease of dispersion modeling made this intake fraction approach feasible, as the calculation of the sector intake fractions required modeling pollutant transport at around 800 representative emission sources across China.

Because of the simplicity, however, dispersion models omit some dynamic and chemical processes affecting pollutant distribution. These include convection, entrainment of pollutants from the background air into the plume, full chemical interaction of different species within a plume, nonlinear chemical and aerosol processing, and detailed representations of deposition processes where nonlinearities in chemical transformation may be important. Therefore, dispersion models cannot sufficiently predict secondary pollutants in the atmosphere, notably ozone and fine aerosols implicated in health and other damages, and particularly not at the regional scale needed for a national study. Although the predicted concentrations can be mapped onto a three-dimensional (3-D) domain, dispersion models are fundamentally different from the gridded numerical models described in the next section because they do not model all atmospheric processes in 3-D space.

7.2.2 Eulerian Gridded Chemical Transport Models
Eulerian numerical models divide the atmosphere into discrete grid boxes through which air flows. Most of the commonly used Eulerian models in air quality research are 3-D, capable of resolving atmospheric processes by latitude, longitude, and

altitude, and are referred to as 3-D atmospheric chemical transport models (CTMs). Because each grid box represents a physical domain of the air, a CTM is capable of simulating many complicated emission and chemical processes within each one. The abundance of a pollutant inside a grid box is affected by emissions, chemical production and loss, deposition, and transport through the boundaries, both in and out. Point and area emissions at the surface and at any elevation within the grid are assumed to be fully mixed in the grid box. A variety of chemical species, both gases and particulates of primary and secondary origins, react with each other in each grid box. Deposition upon contact with the surface (dry deposition) or through rainfall and clouds (wet deposition) is considered. The amount of pollutants transported out of a grid box adds to their abundance in neighboring ones.

Mathematical algorithms are applied to model these processes through the mass balance principle, using a continuity equation. Solving for the concentration of individual chemicals in each grid box is computationally intensive because the continuity equation involves many species and processes and the transport term links the equation of each grid box to those of adjoining ones. The resulting calculated distribution of concentrations must satisfy all the physical, chemical, and mathematical constraints in every grid box. A 3-D CTM is thus not the type of numerical model that can be run practically on a desktop or laptop computer. It needs multiple, high-performance processors with fast data communication speeds, running in parallel for an extended period of computer time (typically for hours).

The advantages of a 3-D CTM are obvious. First, because it aims to model the actual entire lower atmosphere (termed the troposphere) in terms of dynamics and chemistry, calculated pollutant concentrations represent the realistic evolution of emissions being added into the background atmosphere. Importantly, modeled concentrations can be evaluated by comparing to measurements (or observations) taken in the actual atmosphere by field instruments. In contrast, dispersion models can be evaluated with field observations only inside a modeled plume, and not far from the source, to ensure that the background atmosphere does not play an important role.

Second, a 3-D CTM is designed to model accurately secondary pollutants such as ozone and particulates because it includes the full mechanisms of their formation and transformation. Third, a 3-D CTM is capable of simulating the interaction between different chemical species. An example of the chemical interaction is that between cation and anion components in aerosols.

A disadvantage is that a gridded CTM does not resolve the spatial distribution of the concentration of a pollutant within a given grid box, as it assumes each one is uniformly mixed. If its grid cell dimensions exceed the extent of a plume, for example, a gridded CTM will not capture the contrast between the plume and the background atmosphere; instead, it will dilute the plume over the entire grid box and predict an elevated concentration throughout. Therefore, the grid size of a CTM must be reduced to a few kilometers or less if one intends to simulate, for instance, the distribution of pollution inside an individual urban area, where emission can vary substantially over short distances. A grid size of tens of kilometers is generally sufficient for a study at regional scale and can be hundreds of kilometers for a global study. The numerical model used in this study is a regional 3-D CTM embedded (or "nested") in a global CTM, with the advantage of capturing the temporal and spatial variability in the influence of global background conditions to regional air quality that is typically not included in a stand-alone regional CTM. The resolution of the nested-grid regional CTM is appropriate for simulating China's air quality at the national scale that the current analysis requires. Specifics of the model are introduced in the next section.

7.3   Methods

7.3.1   GEOS-Chem Model

In this study we use a regional-scale 3-D CTM named GEOS-Chem (http://www.geos-chem.org/) to simulate pollutant concentrations over China resulting from different emission scenarios. The GEOS-Chem model was originally developed as a global CTM. To model the region of interest, we developed a higher-resolution window over East Asia (centered on China) inside the global GEOS-Chem model. We refer to this as the nested-grid version of GEOS-Chem. The window has a uniform horizontal resolution of 0.5° by 0.67° (approximately 50 km by 67 km at the equator) embedded in the lower-resolution (4° by 5°) global background. The nested domain is set at 70°E–150°E and 11°S–55°N and includes all of China, parts of neighboring countries, and a significant portion of the northwestern Pacific, as shown in figure 7.1 (Y. X. Wang, McElroy, Jacob, et al. 2004; Chen et al. 2009). The high-resolution regional simulation is coupled dynamically to the low-resolution global model through lateral boundary conditions that are updated every three hours of the simulation.

The GEOS-Chem model is driven by meteorological data from the Goddard Earth Observing System (GEOS) at the NASA Global Modeling and Assimilation Office (GMAO). The meteorological data used in this study include 3-D parameters updated every three hours for surface fluxes and mixing depths, and every six hours

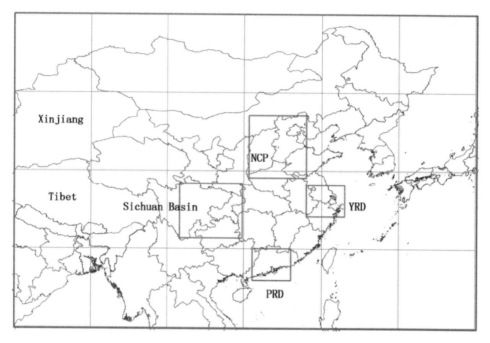

Figure 7.1
A map of China showing provincial boundaries and the regions discussed in the text: the North China Plain (NCP), the Yangtze River Delta (YRD), the Pearl River Delta (PRD), the Sichuan Basin, and Tibet and Xinjiang provincial-level autonomous regions.

for other variables such as winds, pressure, and clouds. The horizontal resolution is 0.5° latitude by 0.67° longitude (dictating the resolution of the CTM), with 72 levels in the vertical extending from the surface to 80 km. The lowest 2 km are resolved into 14 layers with midpoints at altitudes of 70, 200, 330, 460, 600, 740, 875, 1015, 1157, 1301, 1447, 1594, 1770, and 2000 m for a column based at sea level. For inputs to the global GEOS-Chem model, the horizontal resolution of the meteorological fields is degraded to 4° latitude by 5° longitude. The nested-grid GEOS-Chem retains the higher horizontal resolution of the GEOS-5 data for the nested regional domain over China and East Asia.

The GEOS-Chem model includes a detailed tropospheric simulation of the complex chemical interactions between ozone, nitrogen oxides (NO_X), hydrocarbons, and aerosols. The model's chemical mechanism includes about 100 chemical species and more than 200 chemical reactions; in this study we use it to simulate 43 chemical tracers. The aerosol and oxidant chemistry are coupled through the formation of sulfate and nitrate, heterogeneous chemistry, and aerosol effects on photolysis rates. Besides detailed tropospheric chemistry, the model also simulates wet and dry deposition, dynamic interaction between the troposphere and stratosphere (the next layer up in the atmosphere), and a variety of atmospheric dynamics such as advection, convection, boundary layer mixing, and diffusion.

Anthropogenic emissions over the nested East Asia domain outside of China, some of which influence conditions within China, are taken from Zhang et al. (2009) for the year 2006. (The one-year difference from the base year is not critical, for reasons indicated in section 7.4.1.) Emissions over China, drawn from chapter 6, are described in the next section. Natural emissions of ozone precursors include NO_X emitted from bacterial activities in soils and organic wastes, NO_X produced from lightning, and non-methane volatile organic compounds (NMVOCs) from vegetation such as pine trees. Emissions from biomass burning are taken from the GFED-2 inventory constrained by satellite-derived fire spot observations (van der Werf et al. 2006). Biogenic emissions of NMVOCs are adopted from the MEGAN inventory (Guenther et al. 2006).

Application and evaluation of the nested-grid GEOS-Chem model over and around China have been conducted previously for carbon monoxide (CO), NO_X, and ozone. The model has been found to accurately reproduce the temporal and spatial variability (1) of CO and NO_X observed in aircraft campaigns downwind of China (Y. X. Wang, McElroy, Jacob, et al. 2004) and at Chinese surface sites (Y. X. Wang, McElroy, Wang, et al. 2004); (2) of tropospheric column abundance of NO_2 derived from satellite remote sensing data (Y. X. Wang, McElroy, Martin, et al. 2007; Y. X. Wang, McElroy, Boersma, et al. 2007); and (3) of ozone observed at surface sites, mountain sites, and satellite-derived tropospheric columns (Y. X. Wang et al. 2010). Detailed comparison of the model's simulation of ozone and $PM_{2.5}$ (particulate matter under 2.5 microns in diameter) with a few surface measurements will be presented in section 7.4.

7.3.2   Model Setup

Day-to-day meteorological conditions naturally vary from one year to the next. While most atmospheric modeling studies analyze the past and can often use meteorological data from the time period of interest, this is not an option when simulating effects of (sometimes hypothetical) policies in the future, as at the time of research in the current study. The standard practice is to use a representative meteorological data set for a given year in all scenarios, so that comparative effects on air quality result only from the emission changes and not from variation of meteorological inputs to the scenarios.

In this study, the nested-grid GEOS-Chem model uses the meteorological fields from May 2005 to December 2006 and thus simulates the actual 3-D distribution of chemicals in the atmosphere for the same period. The initial conditions (those assumed to describe the atmosphere at the start of the time period being modeled) can affect the results of a simulation, particularly in its initial months. To remove this influence, it is standard practice to simulate a time period that begins well before the period of interest, to allow the model to spin up. In the current study, simulation of the period from May 2005 to December 2005 is used only for this purpose, and results from January 2006 to December 2006 are used for actual analysis. The model is run four times corresponding to four emission scenarios developed for China as described in prior chapters: one for the base year of 2005 and three for analyzing the effects of policy alternatives in 2010. The same meteorology is thus applied in our four scenarios.

We analyze the model results for ozone and $PM_{2.5}$, as they are the focus of the damage assessments of chapters 8 and 9. Note that $PM_{2.5}$ is a category of pollutants defined not by chemical composition but by physical dimensions (i.e., particles less than 2.5 microns in diameter), and it is a pollutant characterization of strongest interest to environmental epidemiologists. Atmospheric chemists normally focus instead on the variety of chemical species that make up $PM_{2.5}$. In our model the components of $PM_{2.5}$ include sulfate, nitrate, ammonium, fine crustal dust, black carbon, organic carbon, and sea salt. The mass of total $PM_{2.5}$ is the summation of all these species simulated separately by the model.

7.4   Results

7.4.1   2005 Base-Year Simulation

As described in chapter 6, 2005 is chosen as the base year of Chinese emissions for the current analyses. Model results driven by the base-year emissions and the meteorology of 2006 are evaluated with available field observations in this section. In fact the 2006 meteorological data set was chosen for model simulation mainly because observations (i.e., field measurements of the atmosphere) are available for 2006; therefore, we have an opportunity to explore how well the model simulates actual conditions. (Given our ultimate interest in the effects on air quality of changes in policy, not the total damages of air pollution levels, we are most concerned with the model's capacities to capture variations, seasonality, and regional gradients. These are primarily determined not by the absolute emission levels that are input into the model—year-to-year incremental changes of which are in fact quite small, typically less than 10%—but by the model's built-in processes of chemistry and transport. Therefore, although we are using emissions in 2005, we can use 2006 meteorology to evaluate the model output because it matches the timing of available observations.)

Since our study focuses on air quality at regional and national scales and given the model's spatial resolution, we are particularly interested in observations that are not sensitive to, nor strongly affected by, local, near-source emissions. For example, observations at urban centers or near major roads are not suitable. Year-round measurements of O_3 and $PM_{2.5}$ at only a few nonurban sites in China are available from the literature. We selected the following three representative sites for model comparison:

1. Hok Tsui, a remote coastal site in Hong Kong (22°13′N, 114°15′E, 60 meters above sea level [a.s.l.]) (T. Wang et al. 2009)

2. Linan, a rural site in the Yangtze River Delta region (30°25′N, 119°44′E, 132 m a.s.l.) (T. Wang et al. 2001, 2002)

3. Miyun, a rural site in the North China Plain roughly 100 km northeast of Beijing (40°29′N, 116° 46.45′E, 152 m a.s.l.) (Y. X. Wang et al. 2008, 2010)

The Miyun observatory is our own station, established in 2004 through the joint cooperation of the Harvard China Project and Tsinghua University; observations from the Hok Tsui and Lin An stations are available through collaboration with other research scientists. The three sites, which are roughly 10° latitude apart and spread from south to north China, cover three rapidly developing regions in China, namely the Pearl River Delta, the Yangtze River Delta, and the Beijing-Tianjin-Tangshan city cluster of the North China Plain, respectively. Figure 7.1 shows the locations of these regions in China.

Seasonality of surface O_3 at the three sites is summarized in figure 7.2, with observations shown in figure 7.2A and model results in figure 7.2B. Surface ozone exhibits distinctively different seasonality at the three sites (figure 7.2A). It peaks in autumn at Hok Tsui, in May at Lin An, and in June at Miyun. At the peak month, the mean mixing ratio (another metric of concentration that defines the mole fraction of a species in the air) of O_3 increases from south to north, from 50 parts per billion by volume (ppbv) at Hok Tsui, to 60 ppbv at Lin An, and to 70 ppbv at Miyun. Annual mean O_3, on the other hand, decreases from south to north.

In contrast to the pattern observed typically in North America and Europe, where there are O_3 maxima in summer, a common feature of the three surface sites is the relatively low O_3 levels in mid- to late summer (July–August). It has been suggested

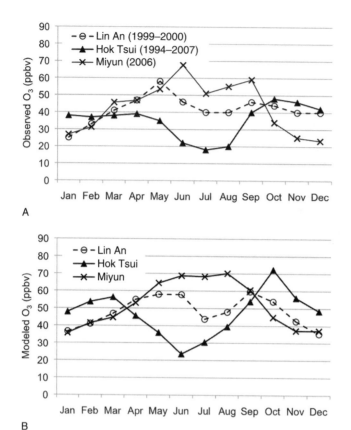

Figure 7.2
Monthly mean O_3 at three representative surface sites in China: (A) observations, (B) base-year model results.

that the monsoonal circulation in East Asia has significant influence on seasonal variability of ozone in China, but research on ozone in China is at a comparatively early stage, and the mechanism controlling its levels in summer is not well under-stood. The inflow of clean maritime air in summer due to monsoonal circulation is one possible explanation, as primary pollutants such as CO exhibit minima in summer at these sites (not shown). Also, production of ozone is a photochemical process, which means the chemical reactions require ultraviolet solar radiation as an energy input; Y. X. Wang et al. (2008) attributed the rapid decrease in O_3 from June to July at Miyun to seasonal increases in monsoonal rainfall and optically thick cloudiness in July, which would limit incidence of solar radiation at the surface and suppress photochemical production.

As shown by comparison of figure 7.2A and figure 7.2B, the model successfully reproduces not only the general features in seasonality of O_3 at the three surface sites, but also the gradients in both CO and O_3 between the sites. The troughs of both species in the middle summer months are captured by the model. One major deficiency in the O_3 simulations is that the model overestimates the minimum concentration levels in summer and underestimates the duration of the summer trough. The discrepancy is due to the model's overestimate of ozone in marine boundary layers (Liu et al. 2006). In addition, for the Lin An and Hok Tsui cases in particular, the period of the model simulation (2006) did not match exactly the years of the measurement (1999–2000 for Lin An and 1994–2007 for Hok Tsui); these differences would likely heighten the discrepancies between model results and observations.

Observed seasonality of $PM_{2.5}$ and its chemical composition is available for comparison only at the Miyun site. The observation method is weekly sampling by a filter system. Figures 7.3A and 7.3B summarize the comparisons between observed and simulated monthly mean mass concentrations at Miyun of $PM_{2.5}$ and sulfate, respectively. Observations of $PM_{2.5}$ mass concentrations show peaks in April–May, August, and October. The model successfully reproduces the timing of the three peaks, although it overestimates the concentrations in spring and fall by a factor of about two.

For several reasons, this overestimation is less of a concern than it might first seem. Examining the composition of $PM_{2.5}$ simulated by the model, we find that natural dust—unrelated to the policies we evaluate—contributes 60–90% of $PM_{2.5}$ mass in the model except in the summer. The model may overestimate dust loading at the site, leading to overestimation of $PM_{2.5}$. It suggests that the peak of $PM_{2.5}$ in the spring and fall results from elevated levels of dust whereas the peak in the summer is attributed to increases in the abundance of inorganic aerosols. The latter are associated with increases in atmospheric oxidation and relative humidity, which lead both to their enhanced production and to hydrophilic growth (i.e., expansion by absorption of water).

Figure 7.3B displays the comparison of model results and observations for sulfate, the form of $PM_{2.5}$ that is most critical to our policy assessments. The model successfully captures the observed seasonality, particularly the major peak in August, the enhancement in spring and early fall, and the sharp decrease in March. The model underestimates sulfate observations in most months by less than 30%, an effect which may be attributed to the model's uncertainties in meteorological fields (e.g., precipitation), aerosol chemistry, and SO_2 emissions. Most important for the

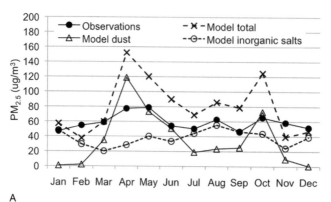

A

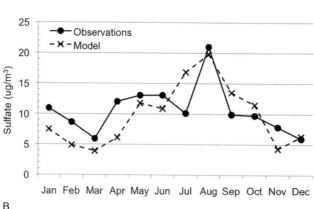

B

Figure 7.3
Comparison of Miyun observations and simulated monthly mean mass concentrations of (A) PM$_{2.5}$ and (B) sulfate.

purposes of the current research, however, is adequately capturing the timing and scale of variations rather than the absolute scale of concentrations, particularly of sulfate aerosols, as we are not seeking to quantify total damages but rather the changes to air quality that result from different policies.

In summary, we conclude that the model does a satisfactory job in simulating the seasonality of PM$_{2.5}$ and sulfate observations. Any deficiency of the model in natural dust simulation will not affect the analysis in the study, because dust emissions are held constant between different policy scenarios.

7.4.2   2010 Base Case
The 2010 base case is the most pessimistic scenario in our study, in which the actual energy saving and emission control policies during 2006–2010 are assumed not to

have been enacted. The base case is used as a reference against which we will measure the impact on air quality due to emission changes of selected policy packages. We first examine the spatial and seasonal distribution of major air pollutants over China for the base case.

The five sets of maps in figure 7.4 summarize monthly mean surface concentrations of (A) NO_X, (B) SO_2, (C) sulfate, (D) O_3, and (E) $PM_{2.5}$, all simulated by GEOS-Chem for four months in 2010. By convention, the months of January, April, July, and October are chosen to represent individual seasons.

For relatively short-lived primary gaseous pollutants such as NO_X (figure 7.4A) and SO_2 (figure 7.4B), the spatial distributions of their concentrations are similar to those of their emissions, with high concentrations in eastern China, particularly over the North China Plain, Yangtze River Delta, and Sichuan Basin. The distribution of SO_2 is more heterogeneous than NO_X and concentrated in pockets, reflecting the dominance of point sources (especially power plants) in total SO_2 emissions and the near-source concentrations that result. The seasonal changes in NO_X and SO_2 are similar. Concentrations are highest in January (winter) and lowest in July (summer). The oxidation rate of NO_X to nitric acids and other reactive nitrogen species, which is the dominating loss mechanism of NO_X, is highest in summer and lowest in winter. The same seasonal pattern holds for oxidation of SO_2 to sulfate by both gaseous and aqueous pathways. More active convection and an unstable atmosphere in the summer also contribute to lower concentrations at the surface.

It is important to note that the temporal and spatial distribution of secondary sulfate aerosols (figure 7.4C) is different from that of primary SO_2. This result occurs partly because sulfate has a longer atmospheric lifetime than gaseous SO_2 and can be transported farther downwind from emission locations of SO_2. As a result, sulfate is less likely to form localized "hot spots" like those of SO_2 shown in figure 7.4B, but rather spreads out over eastern China. The main removal mechanism of sulfate from the atmosphere is through rainfall washout (called wet deposition) and uptake by cloud droplets. Rainfall in China is heavier in the south and during the summer. Consistent with this seasonal and spatial distribution of rain, sulfate levels are higher over the areas north of the Yangtze River than south of it, and higher in the autumn than in the summer.

As noted earlier, ozone is a secondary pollutant produced by a photochemical process: oxidation of volatile organic compounds (VOCs) and CO in the presence of NO_X. The spatial and seasonal distribution of ozone over China (figure 7.4D) is different from that of its precursors such as NO_X. The relatively high concentrations of surface ozone in western and northwestern China (especially the Tibetan Plateau

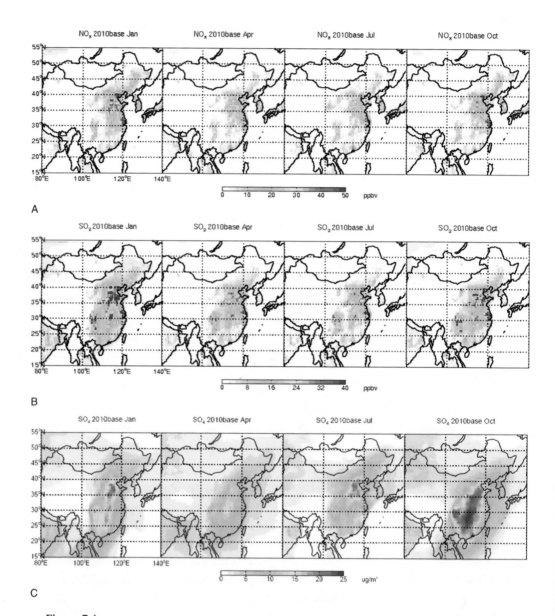

Figure 7.4
Simulated surface concentrations of (A) NO$_X$, (B) SO$_2$, (C) sulfate, (D) O$_3$, and (E) PM$_{2.5}$ for the 2010 base case, by representative months.

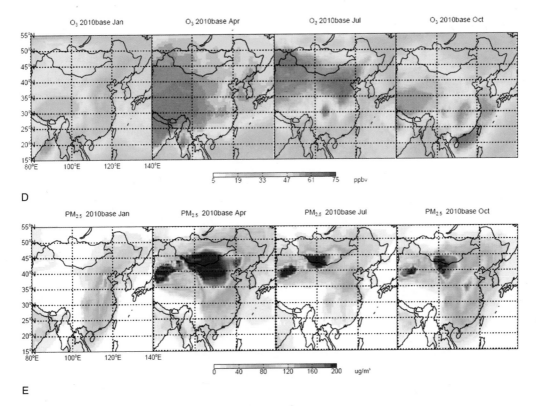

Figure 7.4 (continued)

and Xinjiang) are less relevant to our simulations than they may first appear, because they are chiefly natural in origin and will be the same across emission scenarios. The relatively high elevation of this region leads to significant natural background concentrations resulting from the influence of the stratosphere, which usually peaks in spring (the April panel of figure 7.4D). Natural sources of NO_X from lightning and bacterial activities in soils also contribute to elevated ozone levels over western China in summer.

Surface ozone over eastern China (east of 110°E), however, has a much larger anthropogenic influence, and will be more strongly affected by the policy options. Surface ozone peaks in summer over the North China Plain, with a mean level of 70 ppbv, and in spring over south China, with a mean level of 50 ppbv. In autumn, high ozone is concentrated over central south China and the southeast coastal regions.

The distribution of PM$_{2.5}$ mass concentrations shown in figure 7.4E differs significantly from SO$_2$, NO$_X$, and O$_3$. High PM$_{2.5}$ centers are located over northwest China, particularly the Gobi Desert and desert regions in Xinjiang. This pattern reflects the dominance of fine particles from dust storms in total PM$_{2.5}$ mass. The dust mobilization scheme in the model depends on surface characteristics and meteorological parameters such as wind speed and precipitation. The model predicts that dust levels peak in spring, reaching up to 200 µg/m^3 over northwest China, consistent with observations (the dark brown areas in the map for April). The model also predicts elevated dust loading in summer and autumn, but much less than in spring. In this study, we regard dust as a natural component of PM$_{2.5}$, and thus it will be consistent across our 2010 emission scenarios. (We recognize that human influences on land cover may have led to increases in frequency and severity of dust storms; these complex phenomena, however, will be ignored in our policy analysis.)

Contributions from anthropogenic combustion emissions (of both primary pollutants and secondary precursors) are more significant in total PM$_{2.5}$ mass over east China, although fugitive dust from bare soils, roads, and construction sites is also an important component of PM$_{2.5}$. Although the formation of secondary aerosols is more rapid in summer, the anthropogenic PM$_{2.5}$ mass peaks in autumn reflecting less efficient removal of secondary aerosols in autumn as a result of reduced rainfall. Note some spatial similarity of figures 7.4C and 7.4E if concentrations of natural dust are ignored, and recognizing the different scales of the two figures; these similarities reflect a major contribution of sulfate to total anthropogenic PM$_{2.5}$ mass.

7.4.3   2010 11th FYP Case

The 11th Five-Year Plan (FYP) case is the 2010 scenario closest to actual outcomes, in which we assume full implementation of official measures to save energy and abate emissions in the power sector. As discussed in chapter 6, the differences in emissions between the 2010 base case and the 11th FYP case are most significant for SO$_2$, with lesser effects on NO$_X$, PM$_{2.5}$, or NMVOCs. Therefore, we discuss only the differences between the base and 11th FYP cases in surface concentrations of SO$_2$ and its secondary oxidation product, sulfate aerosols (again, a key form of PM$_{2.5}$), as simulated by the model.

Figure 7.5 presents the changes in (A) SO$_2$, and (B) sulfate concentrations, at the surface by month. The differences are expressed as the 11th FYP case minus the base case levels, denoted as $\Delta SO_{2,FYP} = SO_{2,FYP} - SO_{2,base}$ and $\Delta SO_{4,FYP} = SO_{4,FYP} - SO_{4,base}$, respectively. Negative changes indicate that the 11th FYP scenario reduces

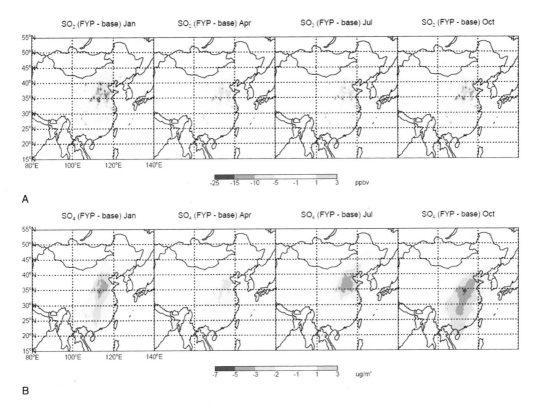

Figure 7.5
Simulated changes in surface concentrations of (A) SO_2 and (B) sulfate resulting from the 11th FYP policies, by representative months. Negative values are reductions due to the policies.

pollutant levels at the surface as compared with the base case. We find that both SO_2 and sulfate concentrations are significantly reduced in the 11th FYP case over China and in all seasons, indicating the average air quality benefit in the two species for the four months (and thereby four seasons) shown.

The seasonal and spatial heterogeneities in the extent of the reductions of both species are noteworthy. The largest reduction of SO_2 in terms of absolute concentrations occurs at locations where emissions and consequently SO_2 levels are high, such as the Beijing-Tianjin-Tangshan urban cluster, the Yangtze River Delta, and the Sichuan Basin. This spatial pattern does not change significantly from month to month. The spatial correlation coefficients between $\Delta SO_{2,FYP}$ (figure 7.5A) and $SO_{2,base}$ (figure 7.3B) are 0.88, 0.85, 0.83, and 0.87 in January, April, July, and October, respectively. Seasonal variations in $\Delta SO_{2,FYP}$ follow the same pattern as in

$SO_{2,base}$, with the largest reductions in January followed in order by October, April, and July.

Turning our attention to effects of the 11th FYP on sulfate levels, we find that $\Delta SO_{4,FYP}$ is most negative over the North China Plain where the reduction in sulfur dioxide concentrations, $\Delta SO_{2,FYP}$, is large. However, $\Delta SO_{4,FYP}$ (figure 7.5B) is comparably much weaker over the Yangtze River Delta and the Sichuan Basin despite the large $\Delta SO_{2,FYP}$ (figure 7.5A) over these regions. The seasonal variability of $\Delta SO_{4,FYP}$ is closer to $SO_{4,base}$ (figure 7.4C) than that of $\Delta SO_{2,FYP}$, with largest reductions occurring in October instead of in January. The figure for ΔSO_4 in January is comparable with that in July even though $\Delta SO_{2,FYP}$ is more negative in January than in July, because the oxidation rate of SO_2 to sulfate is higher in July than in January, likely caused by the higher humidity in summer noted at the end of section 7.4.2.

7.4.4   2010 Carbon Tax Case

The carbon tax case is a hypothetical scenario in which a tax of 100 yuan per ton of carbon is applied to fossil fuels starting in 2006, encouraging increased energy efficiency, shifts to fuels with lower or no carbon content, and structural economic change away from industries that emit a lot of carbon. As discussed in chapter 9, the changing consumption pattern of fossil fuels and the fuel mix in major sectors that results from higher fuel prices affects the prices of goods and services produced by energy inputs, with impacts propagating throughout the entire economy. Therefore, significant differences in emissions between the 2010 base case and the tax case are found for all the atmospheric species considered in this study. For this reason we discuss the differences in surface concentrations of all these primary species and their secondary products simulated by the model (O_3, sulfate, and total secondary $PM_{2.5}$) under the base case and the tax cases.

Figure 7.6 summarizes monthly mean surface concentrations of (A) NO_X, (B) SO_2, (C) O_3, and (D) $PM_{2.5}$ for the four months. The differences are expressed as the level of the carbon tax case minus the level of the base case, denoted as $\Delta NO_{X,tax}$ = $NO_{X,tax} - NO_{X,base}$ and analogously for $\Delta SO_{2,tax}$, $\Delta O_{3,tax}$, and $\Delta PM_{2.5,tax}$. Negative values indicate that the carbon tax reduces pollutant levels at the surface compared to the base case.

Figure 7.6
Simulated changes in surface concentrations of (A) NO_X, (B) SO_2, (C) O_3, and (D) $PM_{2.5}$ resulting from the carbon tax, by representative months. Negative values are reductions due to the carbon tax.

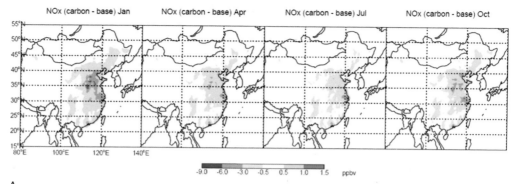

A

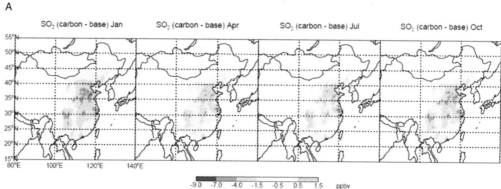

B

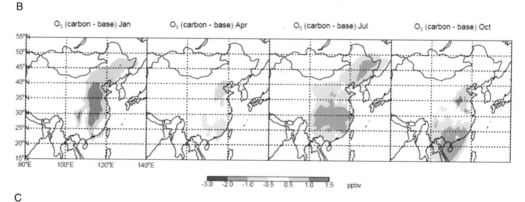

C

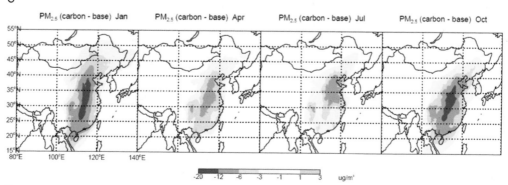

D

Model results show that both $\Delta NO_{X,tax}$ and $\Delta SO_{2,tax}$ are negative over all regions of China and in all seasons, indicating the direct effect of the carbon tax in decreasing the atmospheric concentrations of these primary species. In the carbon tax scenario, total national emissions of NO_X and SO_2 are estimated to decrease by 11% and 14%, respectively, compared to the base case. The $\Delta NO_{X,tax}$ of figure 7.6A and the $\Delta SO_{2,tax}$ of figure 7.6B are high in absolute concentrations over the North China Plain, Yangtze River Delta, Pearl River Delta, and Sichuan Basin, as regions with more developed economies and denser populations are affected more by imposition of a carbon tax. The annual mean decrease over these regions is about 1.7 ppbv for NO_X and 3.5 ppbv for SO_2. $\Delta NO_{X,tax}$ and $\Delta SO_{2,tax}$ show the highest reduction in the winter and lowest in the summer. $\Delta SO_{2,tax}$ is 42% smaller in absolute concentrations than $\Delta SO_{2,FYP}$ over all seasons, reflecting the specific targeting of SO_2 emissions in the 11th FYP case. Because the emission data used in the model are annual averages distributed uniformly across months, the seasonal variability of $\Delta NO_{X,tax}$ and $\Delta SO_{2,tax}$ is driven entirely by the seasonal variability of the atmospheric processes (chemistry and dynamics) that affect the lifetimes of these two gases. (The laborious process of estimating annual emissions by sector and at the 0.5° by 0.67° grid scale is described in chapters 4–6. Disaggregating the emissions by season would be even more difficult to do given the lack of quarterly output data, but it would be a valuable contribution should such data become available.)

Emissions of primary $PM_{2.5}$ in the carbon tax case are lower than in the base case by about 10%. Reductions in NO_X and SO_2 in the carbon tax case also result in decreases in secondary sulfate and nitrate aerosols, leading to a further reduction of $PM_{2.5}$ mass concentrations. As a result of these reductions of both primary and secondary $PM_{2.5}$ levels, $\Delta PM_{2.5,tax}$ is again negative throughout China and in all seasons. Relatively large reductions are simulated for all of east China extending from the North China Plain to the Pearl River Delta, where annual mean reduction of $PM_{2.5}$ can reach $10\,\mu g/m^3$. The seasonal changes in $\Delta PM_{2.5,tax}$ are again driven by atmospheric dynamics and chemistry affecting secondary $PM_{2.5}$ formation. The maximum reduction in absolute levels of $PM_{2.5}$ occurs in October instead of January, because of larger reductions in secondary $PM_{2.5}$ (sulfate and nitrate in particular), which is more sensitive to precursor emission changes as the oxidation rate is higher in October than in January.

In contrast to $\Delta SO_{2,tax}$, $\Delta NO_{X,tax}$, and $\Delta PM_{2.5,tax}$, the change in ozone concentration, $\Delta O_{3,tax}$, is not always negative (figure 7.6C). Instead, it shows large spatial and seasonal variations. Over east China, $\Delta O_{3,tax}$ is generally positive in January and negative in July. The changes reach +3 ppbv in January (i.e., an increase in ozone concentration due to the policy) and −2.5 ppbv in July (a reduction), about 6.1% and 4.9% of the base case concentration level, respectively. The largest decrease in monthly mean ozone occurs around 30°N in east China in July, with 1–2 ppbv reduction or about 5% of the base case level. In both April and October, $\Delta O_{3,\text{tax}}$ shows a transition from positive values (increased ozone) over the North China Plain to negative values south of the Yangtze River. However, more reduction in ozone is found over south China in October than in April. The largest reduction in ozone over the Pearl River Delta is found in October, reaching 2 ppbv.

The patterns of change in ozone are notably more complicated than those of the other pollutants and deserve some detailed discussion. Most nonscientist readers will find it counterintuitive that under some conditions, a *decline* in emissions of the ozone precursor NO_X can sometimes lead to an *increase* in ozone levels, as projected in some regions of China in January, April, and October of figure 7.6C. This is a critically important characteristic of the photochemical smog challenge, not just in China, but also throughout the world. (It is no accident that scientists, policy makers, and citizens in southern California have invested more than 50 years of effort to understand and incrementally address the emissions and mechanisms raising ozone levels in the Los Angeles air basin as its population and economy have grown.) As China successfully mitigates less complex primary air pollution forms, the nature of its emissions and air quality problems are now evolving toward the more complex secondary ones observed in highly industrialized and urbanized nations. In these countries, ozone is often considered the greatest current air pollution risk and the leading focus of control policies, a foreshadowing of what China likely faces. Yet research on ozone in China is still at a very early stage, and systematic monitoring and reporting of ozone levels are not yet available at the country scale. The simulations of the current study provide an opportunity to introduce some of the complexities of ozone pollution to readers unaware of its central importance to China's environmental future.

Ozone is produced by nonlinear photochemical reactions involving both NO_X and VOCs (or CO, which has a much simpler and less efficient oxidation pathway to produce ozone). The phenomenon particularly needing explanation is the overall increase in ozone in some regions, especially in winter, despite reductions in emissions of a key precursor, NO_X. (We should note at the outset that an increase of this magnitude, 1–2 ppbv, will not raise the risks to public health or ecosystems close to those already observed in early summer, as the base ozone levels in winter are generally low, 30–40 ppbv.) To explain the apparent discrepancy, we need to introduce the ozone chemistry (reactions 2–5 in box 7.1) and the hydroxyl radical (OH),

Box 7.1
Chemical reactions affecting ozone formation and removal

Photochemical OH production:

$$H_2O + hv \rightarrow OH + H \qquad (1)$$

Photochemical ozone production:

$$RH + OH + O_2 \rightarrow RO_2 + H_2O \qquad (2)$$

$$RO_2 + NO \rightarrow RO + NO_2 \qquad (3)$$

$$NO_2 + hv \rightarrow NO + O \qquad (4)$$

$$O + O_2 + M \rightarrow O_3 \qquad (5)$$

Regeneration of OH:

$$RO + O_2 \rightarrow R'CHO + HO_2 \qquad (6)$$

$$HO_2 + NO \rightarrow OH + NO_2 \qquad (7)$$

Competing NO_X reactions:

$$NO_2 + OH \rightarrow HNO_3 \qquad (8)$$

$$NO + O_3 \rightarrow NO_2 + O_2 \qquad (9)$$

In these reactions, hv denotes photon, RH denotes volatile organic carbons (VOCs), RO_2 denotes organic peroxy radicals, RO denotes organic oxy radicals, R'CHO denotes carbonyl compounds, and M denotes any inert molecule in the atmosphere (generally O_2 and N_2) absorbing excess energy. The other species are in standard molecular notation.

generated by photochemical dissociation of water vapor (reaction 1). OH is often called the primary "cleansing agent" in the troposphere because it is highly reactive and easily reacts with a number of pollutant species to initiate their chemical removal from the atmosphere. But OH is also the key oxidant in initiating ozone production (reaction 2) and can be regenerated in the process of ozone production (reaction 7). Given low temperatures and weaker solar radiation in the winter, the photochemical reactivity of the atmosphere is not large enough to generate an abundant enough supply of the OH radicals needed to produce a lot of ozone at the surface. This fact explains why ozone concentrations are generally much smaller in the winter than in the summer, as shown in figure 7.4D. The photochemical production rate of ozone at this time of year is controlled not by precursor emissions but by the availability of the OH radical.

When NO_X, consisting of NO_2 and NO, is emitted at the surface, it typically undergoes three major reaction pathways (reactions 4, 8, and 9). As the photolysis pathway leading to ozone formation (reaction 4 followed by 5) is not favored in wintertime, addition of NO_X emissions will either consume OH radicals (reaction 8) or provide fresh NO to react with ozone itself in urban centers (reaction 9), neither of which is favorable for ozone production. Therefore, decreasing emissions of NO_X will reduce the ozone-limiting influence of these two competing reactions, leading instead to enhanced ozone levels in the winter.

This mechanism also explains the small $\Delta O_{3,tax}$ over the North China Plain in particular. Although the reduction in NO_X concentration, $\Delta NO_{X,tax}$, is largest over the North China Plain in all seasons, $\Delta O_{3,tax}$ over this region is only negative in July and becomes positive in other seasons. Even in July, $\Delta O_{3,tax}$ over the North China Plain is smaller in absolute concentrations (0.5–1 ppbv) compared with those over the Yangtze and Pearl River Deltas. Because of its higher latitude (thus less solar radiation for reaction 1) and drier climate (thus less water vapor for reaction 1), the North China Plain is expected to have a lower abundance of OH radicals than either the Yangtze or Pearl River Delta, especially in spring and autumn. Because of a lower OH concentration and higher NO_X emissions, NO_X concentrations are typically higher over the North China Plain than over the Yangtze and Pearl River Deltas (cf. figure 7.4A). This suggests that the North China Plain is in a NO_X-saturated regime. Reducing NO_X emissions by about 10%, as indicated in the tax case versus the base case, is insufficient to change the NO_X-saturated state and decrease the extent to which reaction 8 suppresses OH and to which reaction 9 consumes ozone, leading to only slight increases in ozone. (Keep in mind that this chemistry also depends on our projections of VOCs, as described in section 6.3.3 and figure 6.8 of chapter 6.)

In the summertime, however, OH is in abundant supply over the North China Plain when the solar zenith angle is relatively high. In this case the abundance of OH radicals is not critically dependent on reaction 8 and therefore NO_x levels. Reducing NO_X emissions will decrease the photolysis rate of NO_2 (reaction 4) and consequent production of ozone (reaction 5). Hence, the change in ozone concentration, $\Delta O_{3,tax}$, is negative over the North China Plain only in the summer while negative over the Yangtze and Pearl River deltas in all seasons except winter.

7.4.5 Daily Maximum Ozone

It is worth mentioning that this chapter has mainly focused on the monthly-mean and regional-mean changes in air quality resulting from the policy scenarios, instead of the maximum daily effects that are expected to be larger. To illustrate this point, figure 7.7 compares the simulated daily maximum ozone concentrations over China in July 2006 between the 2010 base case and carbon tax case. The difference on

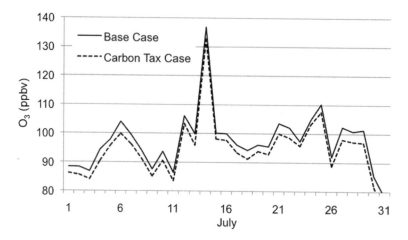

Figure 7.7
Simulated daily peak ozone level over China in July 2006. The solid line is the 2010 base case, and the dashed line is the 2010 tax case.

most of the days is about 2–5 ppbv, larger by a factor of two than the monthly-mean difference of 1–2 ppbv. The difference matters most on the days (e.g., 25–29 July) when ozone levels are predicted to be on the verge of exceeding the national ambient air quality standard (102 ppbv in China) in the base case, whereas those in the tax case are well below the standard. The larger effect of the 2010 tax case on daily maximum ozone levels should have important implications for acute exposure studies.

7.5   Summary and Conclusions

This chapter discusses the effects on air quality over China of the policy scenarios described in previous chapters using a state-of-the-art, regional, 3-D chemical transport model (CTM) embedded in a global CTM. The atmospheric modeling approach is a substantial improvement over the more simplified one used in the preceding book of the current research program, *Clearing the Air* (Ho and Nielsen 2007). Through detailed comparison of the model's base-year simulation of ozone and $PM_{2.5}$ with a few surface measurements, we show that the model does a satisfactory job in capturing the timing and scale of the observed variations, lending confidence to the model's ability to simulate the changes to air quality that result from different policies.

We first examine the spatial and seasonal distribution of major air pollutants (NO$_x$, SO$_2$, O$_3$, sulfate aerosol, PM$_{2.5}$) over China for the 2010 base case, which is a reference against which we will quantify the impact on air quality of emission changes under selected policy packages. For NO$_X$ and SO$_2$, their spatial distributions are similar to those of their emissions. The distribution of secondary pollutants, such as sulfate and ozone, however, exhibits less localized "hot spots" but rather spreads out in elevated concentrations over eastern China. High PM$_{2.5}$ centers located over northwest China are attributed to high dust loadings over this region, whereas contributions from anthropogenic combustion emissions are more significant in total PM$_{2.5}$ mass over eastern China.

The different effects on air quality between the 2010 base case and the 11th FYP case mainly concern SO$_2$ and sulfate aerosols. We find that both SO$_2$ and sulfate concentrations are significantly reduced in the 11th FYP case over China and in all seasons, indicating the average air quality benefit in the two species. The largest reduction of SO$_2$ in terms of absolute concentrations occurs at locations where emissions and consequently SO$_2$ levels are high. Seasonal variations in ΔSO$_{2,FYP}$ (ΔSO$_{2,FYP}$ = SO$_{2,FYP}$ − SO$_{2,base}$ and analogously for ΔSO$_{4,FYP}$) follow the same pattern as in SO$_{2,base}$, with the largest reductions in January followed in order by October, April, and July. The reduction in sulfate concentration, ΔSO$_{4,FYP}$, is largest over the North China Plain where the reduction in sulfur dioxide concentration, ΔSO$_{2,FYP}$, is large. The seasonal variability of ΔSO$_{4,FYP}$ is closer to SO$_{4,base}$ than that of ΔSO$_{2,FYP}$, with the largest reductions occurring in October instead of in January.

We discuss the differences in surface concentrations of all the primary species and their secondary products simulated by the model under the base case and the carbon tax case. Model results show that both ΔNO$_{X,tax}$ (ΔNO$_{X,tax}$ = NO$_{X,tax}$ − NO$_{X,base}$ and analogously for ΔSO$_{2,tax}$, ΔO$_{3,tax}$, and ΔPM$_{2.5,tax}$) and ΔSO$_{2,tax}$ are negative over all regions of China and in all seasons, indicating the direct effect of the carbon tax in decreasing the atmospheric concentrations of these primary species. ΔNO$_{X,tax}$ and ΔSO$_{2,tax}$ are highest in absolute levels in the winter and lowest in the summer. As a result of the reductions of both primary and secondary PM$_{2.5}$ levels, ΔPM$_{2.5,tax}$ is again negative throughout China and in all seasons. For all of east China extending from the North China Plain to the Pearl River Delta, annual mean reduction of PM$_{2.5}$ can reach $10 \mu g/m^3$.

In contrast, the change in ozone concentration, ΔO$_{3,tax}$, is not always negative. Over east China, ΔO$_{3,tax}$ is generally positive in January and negative in July. The largest decrease in monthly mean ozone occurs around 30°N in east China in July, with 1–2 ppbv reduction or about 5% of the base case level. As projected in some regions of China in January, April, and October, a critically important characteristic of the photochemical smog challenge is that a *decline* in emissions of ozone precursors can sometimes lead to an *increase* in ozone levels, which is a common problem throughout the world because of the nonlinear nature of ozone formation chemistry. Through the explanation of ozone chemistry and the role of the hydroxyl radical (OH), we show that as the photolysis pathway leading to ozone formation is not favored in wintertime, addition of NO_x emissions will either consume OH radicals or provide fresh NO to react with ozone itself in urban centers, neither of which is favorable for ozone production. Therefore, decreasing emissions of NO_x will reduce the ozone-limiting influence of these two competing reactions, leading instead to enhanced ozone levels in the winter.

References

Chen, D., Y. X. Wang, M. B. McElroy, K. B. He, R. M. Yantosca, and P. Le Sager. 2009. Regional CO pollution and export in China simulated by the high-resolution nested-grid GEOS-Chem model. *Atmospheric Chemistry and Physics* 9(11):3825–3839.

Guenther, A., T. Karl, P. Harley, C. Wiedinmyer, P. I. Palmer, and C. Geron. 2006. Estimates of global terrestrial isoprene emissions using MEGAN (Model of Emissions of Gases and Aerosols from Nature). *Atmospheric Chemistry and Physics* 6:3181–3210.

Ho, M. S., and C. P. Nielsen, eds. 2007. *Clearing the Air: The Health and Economic Damages of Air Pollution in China.* Cambridge, MA: MIT Press.

Liu, X., K. Chance, C. E. Sioris, T. P. Kurosu, R. J. D. Spurr, R. V. Martin, T. M. Fu, J. A. Logan, D. J. Jacob, P. I. Palmer, M. J. Newchurch, I. A. Megretskaia, and R. B. Chatfield. 2006. First directly retrieved global distribution of tropospheric column ozone from GOME: Comparison with the GEOS-CHEM model. *Journal of Geophysical Research* 11, Art. No. D02308, doi:10.1029/2005JD006564.

van der Werf, G. R., J. T. Randerson, L. Giglio, G. J. Collatz, P. S. Kasibhatla, and A. F. Arellano. 2006. Interannual variability in global biomass burning emissions from 1997 to 2004. *Atmospheric Chemistry and Physics* 6:3423–3441.

Wang, T., T. F. Cheung, Y. S. Li, X. M. Xu, and D. R. Blake. 2002. Emission characteristics of CO, NO_x, SO_2 and indications of biomass burning observed at a rural site in eastern China. *Journal of Geophysical Research* 107, Art. No. D12, 4157, doi:10.1029/2001JD000724.

Wang, T., V. Cheung, C. M. Anson, and Y. S. Li. 2001. Ozone and related gaseous pollutants in the boundary layer of eastern China: Overview of the recent measurements at a rural site. *Geophysical Research Letters* 28(12):2373–2376.

Wang, T., X. L. Wei, A. J. Ding, C. N. Poon, K. S. Lam, Y. S. Li, L. Y. Chan, and M. Anson. 2009. Increasing surface ozone concentrations in the background atmosphere of Southern China, 1994–2007. *Atmospheric Chemistry and Physics* 9:6217–6227.

Wang, Y. X., M. B. McElroy, K. F. Boersma, H. J. Eskes, and J. P. Veefkind. 2007. Traffic restrictions associated with the Sino-African summit: Reductions of NO_X detected from space. *Geophysical Research Letters* 34, Art. No. L08814, doi:10.1029/2007GL029326.

Wang, Y. X., M. B. McElroy, D. J. Jacob, and R. M. Yantosca. 2004. A nested grid formulation for chemical transport over Asia: Applications to CO. *Journal of Geophysical Research* 109, Art. No. D22307, doi:10.1029/2004JD005237.

Wang, Y. X., M. B. McElroy, R. V. Martin, D. G. Streets, Q. Zhang, and T. M. Fu. 2007. Seasonal variability of NO_X emissions over east China constrained by satellite observations: Implications for combustion and microbial sources. *Journal of Geophysical Research* 112, Art. No. D06301, doi:10.1029/2006JD007538.

Wang, Y. X., M. B. McElroy, J. W. Munger, J. M. Hao, H. Ma, C. P. Nielsen, and Y. S. Chen. 2008. Variations of O_3 and CO in summertime at a rural site near Beijing. *Atmospheric Chemistry and Physics* 8:6355–6363.

Wang, Y. X., M. B. McElroy, T. Wang, and P. I. Palmer. 2004. Asian emissions of CO and NO_X: Constraints from aircraft and Chinese station data. *Journal of Geophysical Research* 109, Art. No. D24304, doi:10.1029/2004JD005250.

Wang, Y. X., J. W. Munger, S. Xu, M. B. McElroy, J. M. Hao, C. P. Nielsen, and H. Ma. 2010. CO_2 and its correlation with CO at a rural site near Beijing: Implications for combustion efficiency in China. *Atmospheric Chemistry and Physics* 10:8881–8897, doi:10.5194/acp-10-8881-2010.

Zhang, Q., D. G. Streets, G. R. Carmichael, K. B. He, H. Huo, A. Kannari, Z. Klimont, I. S. Park, S. Reddy, J. S. Fu, D. Chen, L. Duan, Y. Lei, L. T. Wang, and Z. L. Yao. 2009. Asian emissions in 2006 for the NASA INTEX-B mission. *Atmospheric Chemistry and Physics* 9:5131–5153, doi:10.5194/acp-9-5131-2009.

8

Benefits to Human Health and Agricultural Productivity of Reduced Air Pollution

Yu Lei

We have seen in prior chapters how the SO_2 controls of the 11th Five-Year Plan (11th FYP) or a tax of 100 yuan per ton of carbon (27 yuan per ton of CO_2) did reduce or would reduce emissions and concentrations of a variety of both primary and secondary air pollutants. Among the most powerful implications of either of these policies would be their very large effects on public health. To the extent that they affect concentrations of ground-level ozone, they would also impact the productivity of major grain crops. In this chapter, we estimate the health and agricultural benefits that would result from the improved air quality under the two policy options. We consider these topics in sequence, first addressing the centrally important health benefits and then turning to the crop benefits.

8.1 Health Benefits

8.1.1 Methodology and Data
Framework of BenMAP We analyze health benefits under the base case and policy scenarios described in the previous chapters using the Benefit Mapping and Analysis Program (BenMAP). This is a tool developed by the U.S. Environmental Protection Agency (U.S. EPA) for estimating the health and associated economic impacts of changes in ambient air pollution levels (Hubbell, McCubbin, and Hallberg 2003). BenMAP was originally adapted for China by Xiaochuan Pan of Peking University, as part of the Integrated Environmental Strategies program sponsored by the U.S. EPA. Figure 8.1 illustrates the flow of calculations in BenMAP.

   The first step in evaluating the health impacts is estimating changes in *population exposure* to air pollution levels due to a given policy scenario compared to the base case. This is accomplished by combining a spatial projection of the population distribution with pollutant concentrations simulated by the air quality model. In

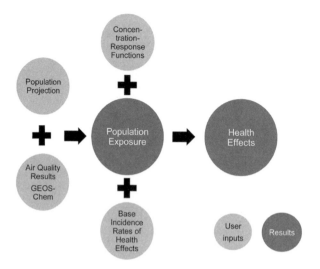

Figure 8.1
BenMAP flow diagram. *Note:* BenMAP has an additional component for monetizing health effects, a topic discussed in chapter 9.

the next step, BenMAP calculates the *health effects*, meaning the changes in various health endpoints such as the number of cases of premature mortality or outpatient hospital visits per year. These are estimated by applying concentration-response (C-R) functions from the epidemiological literature to the population exposure, along with the base incidences in the population of those health endpoints. (A C-R function, sometimes alternatively called an exposure-response function, relates a change in exposure to a pollutant with a change in the incidence of a health endpoint; see U.S. EPA [2013] regarding the terminology. Base incidence refers to the baseline number of cases of the health endpoint in the population due to all causes, expressed in terms of cases per person per unit time—for example, the total deaths due to all causes per million people per year. The incidence of all-cause mortality, for instance, follows directly from total deaths due to all causes per million people per year.)

The first two steps are represented by the following formula for pollutant *x*:

$$\Delta HE_{hx} = f_{hx}(\Delta C_x) \times Pop \times BI_h \tag{8.1}$$

where ΔHE_h denotes the change in the number of cases of health endpoint *h*; *f* is the C-R function and ΔC_x denotes the change of concentration of air pollutant *x*; *Pop* represents the population exposed to the pollutant; and BI_h represents the baseline incidence of the health endpoint h. We thus require the four elements on the right-hand side of equation (8.1) to estimate the health effects of a policy.

In our study, the changes in concentrations of air pollutants are provided by the output of the GEOS-Chem modeling described in chapter 7. Data on the population distribution and baseline incidence are drawn from official statistical sources, which are described later in this section. The C-R functions, critical factors in any estimation of health effects of pollution, will be discussed in section 8.1.2.

Air Pollutants of Concern   A large number of epidemiological studies have shown that particulate matter under 10 microns in aerodynamic diameter (PM_{10}) leads to adverse health effects because it contains toxic and hazardous matter. (See Levy and Greco [2007] for a full summary of key findings in environmental epidemiology briefly introduced here.) $PM_{2.5}$, the subset of PM_{10} made up of finer particles, is believed to pose higher health risks because the smaller size allows deeper penetration into the human respiratory system and the larger specific surface area on a mass basis permits greater exposure to hazardous compounds. The United States and Europe have shifted from PM_{10} to $PM_{2.5}$ as the primary form of particulate matter to target as a criterion air pollutant, and attention to $PM_{2.5}$ has escalated swiftly since 2011 in China (see the subsection on $PM_{2.5}$ in section 8.1.2 and section 1.4.3). For these reasons, and taking advantage of the ability of the GEOS-Chem atmospheric model to simulate concentrations of the major species that comprise $PM_{2.5}$, we assess it rather than PM_{10} in the current study.

Ozone (O_3), a gas formed in the air from precursors through complex secondary chemistry discussed in chapter 7, has been recognized as another critical air pollutant leading to adverse effects in the human respiratory system (Bell, Dominici, and Samet 2005; Ito, De Leon, and Lippmann 2005; Jerrett et al. 2009; Levy, Chemerynski, and Sarnat 2005; Ostro, Tran, and Levy 2006). Although Chinese national protocols have only recently added monitoring of ground-level ozone and thus it is only beginning to be measured nationwide, in recent years several megacities such as Beijing and Shanghai have focused on it as a major new pollutant of concern. Given the capabilities of GEOS-Chem to simulate the unusually complex photochemistry generating this important but poorly understood pollutant form in China, we include ozone in our health impact assessment (and also in the crop productivity assessment that follows in section 8.2).

Although China has historically recorded high concentrations of sulfur dioxide (SO_2) and the 11th FYP focused on emission control of this gas, we do not estimate *direct* health effects of gaseous SO₂ in the current study, although its *indirect* effects, through secondary formation of sulfate particles, are important in our PM$_{2.5}$ calculations. As described in Levy and Greco (2007), some studies in China suggest evidence of a direct acute mortality effect of SO₂ on respiratory health that is independent of the larger effect as a precursor to sulfate PM$_{2.5}$. Other studies, including more recent ones such as Qian et al. (2007), draw an opposite conclusion. Because of the uncertainties reflected in this ongoing epidemiological debate, we therefore choose to exclude the direct SO₂ effect from our health assessment.

Similarly, there is some evidence of a link between nitrogen oxide exposure and disease, as reviewed in U.S. EPA (2008). However, because of the close correlations between NO$_X$ and PM, it is again difficult to form a firm conclusion (e.g., see COMEAP 2009). We thus also ignore any modest direct effect of NO$_X$ and consider only the impacts of NO$_X$ emissions on secondary nitrate particles and ozone.

Population   For our analysis, the exposed population includes all residents in mainland China, both urban and rural. The total population in the target year of 2010, 1.358 billion, is the same as that used in the economic model described in chapter 9.

The spatial distribution of the population is critical to the quantification of health effects. Chapters 4–6 of Ho and Nielsen (2007) describe how the estimated health impact of air pollution changes with the location of population relative to the source of emissions. China's fast economic development has been accompanied by swift urbanization, resulting in large changes in the population distribution. This process is difficult to track or predict, and an updated representation of the population distribution will not be available until the 2010 census is made available (usually two or three years after the census is conducted). Thus our current assessment is based on the most recent high-resolution population distribution we could acquire at the time of research. This GIS-based data set was developed by the Institute of Geographical Sciences and Natural Resources Research of the Chinese Academy of Sciences in 2005, and represents the population in 2003 at a very fine 1 km by 1 km grid scale. It was developed using methods described in Jiang, Yang, and Liu (2002) to integrate land-use information from remote sensing with census-based population data at the county level.

We scaled up the population in each grid cell based on the ratio of estimated total populations for 2003 and 2010, and then aggregated the cells to match the $0.5° \times 0.67°$ resolution of the GEOS-Chem atmospheric output. The resulting population distribution is shown in figure 8.2.

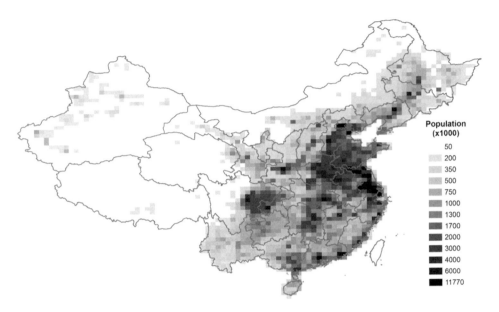

Figure 8.2
Distribution of the Chinese population in 2010.

Baseline Incidence Exposure to outdoor air pollution is associated with a broad spectrum of acute and chronic health effects ranging from bodily irritations to death. The health effects are categorized according to codes of the International Classification of Diseases (ICD) compiled by the World Health Organization (WHO). The endpoints chosen for inclusion in health effects analysis ultimately depend on analytical objectives and the availability of appropriate concentration-response functions and incidence data. As 2005 serves as the base year of the current study, we use baseline incidence rates from 2005 health statistics.

The Chinese Ministry of Health (MOH) conducted a sample-based investigation as part of the "Third National Review on Cause of Death" in 2006, the latest thorough investigation of mortality in China that is available (Chen 2008). The incidence of all-cause, nonaccident mortality is reported from this investigation, corresponding to the disease codes A00-R99 from ICD version 10. To associate the mortality estimation with the incidence, all concentration-response relationships used to analyze mortality in this study are likewise based on nonaccident mortality. The morbidity incidence, including hospital admissions due to cardiovascular disease (disease codes I00–I99) and respiratory disease (J00–J99), and outpatient visits for all reasons (which do not have an ICD code), are drawn from China's health

Table 8.1
Baseline incidence for 2005 (cases per person per year)

Health Endpoints	Baseline Incidence
Mortality, all-cause nonaccident	0.0055
Hospital admission, cardiovascular	0.0062
Hospital admission, respiratory	0.0042
Outpatient visits	3.4788

statistical yearbook for 2006 (MOH 2006). The incidence rates used in this study are listed in table 8.1.

The differences in incidence may be large over different regions of China, such as north versus south or urban versus rural areas, especially for morbidity endpoints. Lacking any convenient geographical differentiation of the incidence rates, however, we assume that they are the same nationwide. We also assume that the baseline incidence in 2010 is the same as in 2005, lacking any official projections of the rates over time. Based on these assumptions, we estimate the health benefits in 2010 by comparing incidences of premature mortality and morbidities in the two policy scenarios to those of the base case.

8.1.2   Concentration-Response Relationships

This section offers a brief introduction to key characteristics of air pollution epidemiology to help explain the research challenges of choosing appropriate C-R functions and applying them in studies of Chinese air quality and health risk. For a more expansive introduction, see Levy and Greco (2007) in the earlier book by the current research program (Ho and Nielsen 2007). Other overviews of relevant issues, with reviews of the literature, include Aunan and Pan (2004) and HEI (2004).

The impacts of air pollution on human health are usually characterized as either chronic effects of long-term exposures or acute effects of short-term exposures. Many studies around the world have derived C-R relationships between exposures to air pollutants and human health. Most of these are time-series studies estimating acute, short-term effects. Far fewer analyses of chronic, long-term effects have been conducted because such cohort studies are much costlier and can take decades to complete. Studies of long-term effects have been conducted mainly in the United States.

Note that there is uncertainty as to the extent to which long-term health effects derived from cohort studies for a given pollutant coincide with short-term effects detected by time-series studies (Eftim and Dominici 2005; Thomas 2005). To some degree, cohort studies will capture the effects not only of chronic exposures but also of acute ones, and this possibility discourages treating these two pathways as independent and thus additive. We therefore will calculate the impacts on premature mortality in two ways, applying either acute or chronic effect epidemiology, and report the results as two separate estimates.

Since disease prevalence, age structures, exposure patterns, health care systems, average ambient concentrations, and the chemistry of pollutants (in the case of PM) are different in various countries, one should ideally use C-R functions from epidemiological studies done in the country or region concerned (Levy and Greco 2007). Only where there is no reliable set of local studies should one turn to estimates from other countries.

Fortunately, a literature of time-series air pollution epidemiology in Chinese cities has been growing in the last decade. We have reviewed most, if not all, of the published studies conducted in mainland China, Hong Kong, and Taiwan, many of which are cited below. Although none of these has a national scope, the city-level research is sufficient to give a defensible representation of acute health effects of short-term exposures to the pollutants of concern across the nation as a whole. That no cohort studies of long-term effects exist for any developing country, let alone China, however, poses a significant analytical challenge that we will consider carefully.

PM$_{2.5}$ As is the case in many developing countries, the National Ambient Air Quality Standards (NAAQS) of China had long identified PM$_{10}$, not PM$_{2.5}$, as a criterion air pollutant. For this reason, routine monitoring by local authorities of ambient PM$_{10}$ concentrations has been required and standard in Chinese cities, while monitoring of PM$_{2.5}$ has occurred in comparatively few cities. The central government, however, has recently revised the NAAQS and formulated a plan to improve local air-quality-monitoring capacities and to implement the new standards. Based on this plan, monitoring of PM$_{2.5}$ was initiated in 74 cities beginning in 2013 and is to be extended to all prefectural cities by the end of 2015. Given the relative availability of PM$_{10}$ data to date, most studies of particulate C-R relationships in China also analyze concentrations of PM$_{10}$, not PM$_{2.5}$.

As discussed in the subsection "Air pollutants of concern" in section 8.1.1, however, we want to analyze the impacts of PM$_{2.5}$ because of its higher, and more precisely estimated, risks, and to capitalize on our atmospheric research capabilities. To do so, we will need to adjust the C-R functions derived in China for PM$_{10}$

concentrations into C-R functions for $PM_{2.5}$. We recalculate them by assuming a ratio of $PM_{2.5}$ to PM_{10} of 0.6, which is considered the default ratio in U.S. studies (Dockery and Pope 1994; Levy and Greco 2007).

$PM_{2.5}$ Acute Effects on Premature Mortality   A growing number of time-series studies of short-term, acute health effects from PM exposures in China have been published in peer-reviewed journals. The first papers, from the 1990s, used observations of concentrations not of PM_{10} but of total suspended particulates (TSP), the form of PM mandated at that time to be monitored by municipal governments. For the current review we ignore those results and instead focus on studies published after 2000, because (1) sufficient literature on PM_{10} and in some cases $PM_{2.5}$ is now available, allowing us to avoid large uncertainties introduced by conversions of TSP-based C-R functions, and (2) pollution characteristics, exposure patterns, and health care provision have changed so fast in China that even the 1990s may be unrepresentative of current conditions.

Applying this criterion, we identify eight time-series studies to inform our development of an appropriate C-R function for acute mortality from $PM_{2.5}$ exposures. All eight were conducted in large cities with relatively accessible and reliable air quality monitoring and mortality data, including

- Beijing (Chang et al. 2003)
- Shanghai (Kan and Chen 2003; Dai et al. 2004; Kan et al. 2007; C. M. Wong, Vichit-Vadakan, et al. 2008)
- Hong Kong (T. W. Wong, Tam, et al. 2002; C. M. Wong, Vichit-Vadakan, et al. 2008)
- Wuhan (Qian et al. 2007; C. M. Wong, Vichit-Vadakan, et al. 2008)
- Taiyuan (Y. P. Zhang et al. 2007)

These studies are summarized in figure 8.3, which plots relative risk (RR) of all-cause mortality per $10\,\mu g/m^3$ increase in concentration versus the mean concentration of $PM_{2.5}$. The RR is the ratio of the probability that the event will occur in the group exposed to the higher concentration to that of a baseline group. One can see that there is a wide range of estimated effects from these studies. After fitting a logarithmic relation for the all-cause mortality, we find a weakly negative relationship between RR and mean concentration (i.e., one in which the relative risk declines as the concentration rises), as shown in figure 8.3.

In a comparison of time-series mortality studies between China and the United States, Levy and Greco (2007) found that the dose-response of all-cause mortality

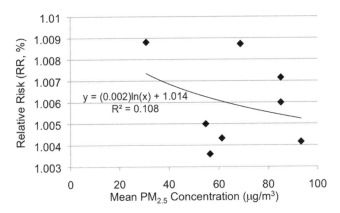

Figure 8.3
The relative risks of all-cause mortality versus mean $PM_{2.5}$ concentration. The *RR*s are based on $10\,\mu g/m^3$ increases of $PM_{2.5}$ concentration. The mean concentrations refer to when and where the corresponding studies (cited in the text) were conducted.

from PM_{10} in China (a 0.3–0.4% increase in risk per $10\,\mu g/m^3$ increase in concentration) is slightly lower than that estimated for the United States (0.3–0.6%). If we convert these figures into *RR* for $PM_{2.5}$ by assuming the aforementioned $PM_{2.5}{:}PM_{10}$ ratio of 0.6, we get 1.005–1.007 for China and 1.005–1.01 for the United States.

The average ambient $PM_{2.5}$ concentration in the United States decreased from $13.5\,\mu g/m^3$ in 2000 to $9.9\,\mu g/m^3$ in 2009 (U.S. EPA 2011). In China, the State Environmental Protection Administration (now Ministry of Environmental Protection) reported that in 2005, 59.5% of cities in China met the national PM_{10} standard, which is $100\,\mu g/m^3$ (SEPA 2006). Among these cities, larger ones with higher populations almost always have higher PM concentrations, so we assume that the population-weighted PM_{10} concentration in Chinese cities in 2005 is $100\,\mu g/m^3$. This assumption is also consistent with the average PM_{10} concentrations reported for 31 mid- to large-sized cities presented in figure 1.4. Applying the default $PM_{2.5}{:}PM_{10}$ ratio, $100\,\mu g/m^3$ of PM_{10} converts to $60\,\mu g/m^3$ of $PM_{2.5}$. The urban estimates for both the United States and China are consistent with the fitted curve in figure 8.3.

Given this reasonable fit of the logarithmic curve, we use it to choose a value for our national (i.e., not just urban) study as opposed to, for instance, applying an average of the *RR*s from the eight studies of urban areas. To apply this fitted curve, we must first estimate the average national $PM_{2.5}$ concentration in China. Although there is very little measurement of ambient $PM_{2.5}$ concentrations in rural areas, we can safely suggest that rural concentrations are generally somewhat lower than the

$60 \mu g/m^3$ noted previously for China's cities, which is similarly true in the United States ($\sim 10 \mu g/m^3$ versus $\sim 15 \mu g/m^3$, as given in U.S. EPA [2011]). With this assumption in mind, $50 \mu g/m^3$ is a reasonable value to represent the national average PM$_{2.5}$ concentration. The corresponding *RR* value from the fitted curve of figure 8.3 is 1.0065. This is the value we adopt for estimating acute premature mortality in our estimates of health effects.

While our choice of $50 \mu g/m^3$ to be the national average concentration may seem rough, we note first that it is being used to locate a point on figure 8.3 to use as the relative risk. This curve is defined by points estimated with a large range of uncertainty, at a level that swamps that for the national average concentration. Second, an alternative procedure such as averaging of a small sample of cities is not obviously more precise for this purpose. Similarly, while the atmospheric model in chapter 7 does calculate concentrations, it is developed less to project absolute values of those concentrations than to capture their gradients and variations (including in response to perturbations, such as new emission policies). It would not generate an obviously superior estimate of the national average concentration faced by the population.

PM$_{2.5}$ Chronic Effects on Premature Mortality   Cohort studies in the United States and Europe have revealed strong evidence of effects of long-term exposure to PM on the risk of premature mortality (Pope et al. 2002; Schwartz et al. 2008). So far there is no similar cohort study conducted in China.

Comparing cohort and time-series studies in the United States, the estimated C-R coefficients of chronic mortality from PM$_{2.5}$ are around an order of magnitude greater than those for acute mortality from PM$_{2.5}$. Although PM-related acute effects in China and the United States are similar, it is inappropriate to simply extrapolate the rate of chronic effects in China from U.S. data. As already noted, the concentration levels of PM$_{2.5}$ in China are several times higher than those in the United States. Applying some chronic C-R values from American studies (for example, the 1.10 *RR* of Levy et al. [2009] adopted by Zhou et al. [2010] in a study of the Yangtze River Delta) to Chinese national PM$_{2.5}$ concentration data would imply that fully 40% of all deaths in China are due to chronic exposure to PM pollution and that the death rates due to PM are several times higher than those in the United States. Such national death rates are unrealistic, as the all-cause annual death rate of China, roughly 0.6%, is not much different from the U.S. rate. This simple analysis indicates that the C-R values for chronic mortality at the high concentrations in China should be lower than the values estimated at low U.S. concentrations. Pope et al. (2009) investigate this general issue by integrating data sets on exposures both to ambient $PM_{2.5}$ pollution and to cigarette smoke, strongly indicating a nonlinear C-R function with declining marginal mortality effect at higher $PM_{2.5}$ concentrations. Pope et al. (2009) do not provide a specific function to apply, however, and making use of this result for our purposes must await such a function. In the meantime we choose from among existing American cohort studies of $PM_{2.5}$ exposures the one with the lowest *RR*, the 61-cities study of Pope et al. (2002) with an *RR* of 1.04, and make sure to include caveats about assumptions when we report the results that follow. Such a chronic rate is six times higher than that implied in the 1.0065 *RR* chosen by us for the acute effect.

$PM_{2.5}$ Effects on Morbidity   In general, the air pollution epidemiological literature characterizes morbidity (incidence of disease) in two very different categories of health endpoints. One is by type of disease, such as chronic bronchitis and asthma, as assessed in an early study of air pollution damages in China by the World Bank (1997). The other is by more general health-care-related activities of patients, such as hospital admissions and outpatient visits to hospitals and clinics (e.g., T. W. Wong et al. 1999; C. M. Wong et al. 2002). Because many of those suffering from these diseases will at the same time receive health care as outpatients, or after being admitted to hospitals, we should not regard the two endpoint categories as additive or we risk double-counting the morbidity impacts.

Estimating population risk of morbidity from air pollution would be more precise if we could rely on disease endpoints alone, as these are the ultimate morbidity effects that we seek to understand. To date, however, there has been little research in China estimating disease-based morbidity C-R relationships. The reason for this lack is presumably that such studies are relatively difficult to design and to conduct because of challenges of sampling, especially for rural residents. We know of only two available research results estimating disease-based C-R functions in Chinese populations, from Hong Kong (Ko et al. 2007) and Taipei (Bell, Levy, and Lin 2008). These, however, considered chronic obstructive pulmonary disease (COPD) and asthma only on the basis of hospital admissions, ignoring patients diagnosed with these conditions who were treated as outpatients.

Given both the scarcity of disease-based morbidity C-R functions in China and the risk of double counting, we therefore estimate the morbidity effects of $PM_{2.5}$ exposures based on the less precise but larger literature of activity-based C-R functions. As noted in the subsection "Baseline incidence" in section 8.1.1 and in table 8.1, hospital admissions and outpatient visits are the major activities for which underlying incidence statistics are available.

In the 1990s, Xu, Li, and Huang (1995) reported C-R relationships between PM and overall outpatient visits based on research in Beijing, and this report provides one applicable source for the current study in the absence of newer research on a Chinese population.

Because air pollution is recognized as highly associated with both cardiovascular and respiratory diseases, morbidity health endpoints such as hospital admissions and outpatient visits are sometimes subcategorized according to disease-specific causes when data are available. The body of research on disease-specified morbidity C-R relationships for China is smaller than that for acute mortality described in the previous section and is again somewhat dated and focused on Hong Kong and Taiwan. Most of these studies estimated C-R relationships for activities caused by respiratory diseases, both hospital admissions and outpatient visits (T. W. Wong et al. 1999, 2006; T. W. Wong, Wun, et al. 2002; C. M. Wong et al. 2002; Hwang and Chan 2002). Studies estimating C-R relationships for activities prompted by cardiovascular-related disease are sparser. There is one report on hospital admissions for cardiovascular disease (T. W. Wong et al. 1999) and one for cardiac disease (C. M. Wong et al. 2002), both from Hong Kong.

Based on the results of these studies, we choose three activity-based morbidity endpoints due to $PM_{2.5}$ exposures as listed in table 8.2: hospital admissions (cardiovascular), hospital admissions (respiratory), and outpatient visits (all-cause). Results from other studies are not chosen because matching baseline incidence data are unavailable: some of the studies focused on outpatient visits caused by specific diseases (as opposed to all causes) while others focused on patients of specific ages.

Ozone Effects on Acute Premature Mortality   While the international epidemiological literature on ozone is less developed than that for $PM_{2.5}$, ozone has been firmly associated with respiratory health outcomes. A number of time-series studies have shown that short-term exposure to ozone can lead to higher premature mortality (e.g., Ostro, Tran, and Levy 2006). Three independent meta-analyses (studies that combine results from prior studies) in the United States have yielded very

Table 8.2
Relative risk for morbidity health endpoints if $PM_{2.5}$ concentration increased by $10\,\mu g/m^3$

Endpoint	RR	Reference
Hospital admissions, cardiovascular	1.0100	T. W. Wong et al. (1999)
Hospital admissions, respiratory	1.0268	T. W. Wong et al. (1999)
Outpatient visits, all causes	1.0039	Xu, Li, and Huang (1995)

similar findings on the acute mortality C-R relationship, with central estimates of 0.41%, 0.39%, and 0.34% increases in all-cause mortality per 10 parts per billion (ppb) increase in daily 1-hour maximum O_3 concentrations (respectively, Bell, Dominici, and Samet 2005; Ito, De Leon, and Lippmann 2005; Levy, Chemerynski, and Sarnat 2005). One cohort study of long-term effects (Jerrett et al. 2009) found that even after taking PM into account, exposure to ozone has a significant positive effect on the risk of mortality from respiratory causes.

Although the effect of ozone on mortality is a very active area of current air pollution epidemiological research in Western countries, in China there are fewer time-series studies of C-R relationships for exposures to ozone than to PM because ozone is not routinely monitored in most Chinese cities, as described under "Air pollutants" in section 8.1.1. Independent studies have been conducted in Hong Kong (T. W. Wong, Tam, et al. 2002) and Shanghai (Y. H. Zhang et al. 2006), and a series of coordinated studies in Hong Kong, Shanghai, and Wuhan have been completed, as part of the Public Health and Air Pollution in Asia (PAPA) program of the U.S.-based Health Effects Institute (HEI) (C. M. Wong, Ou, et al. 2008; Kan et al. 2008; Qian et al. 2008). The conclusion drawn by the PAPA project comparing its three studies indicates an *RR* of 1.0031 for all-cause acute mortality if daily exposure to ozone increases by $10 \,\mu g/m^3$ (C. M. Wong, Vichit-Vadakan, et al. 2008). Given the relative comprehensiveness of the PAPA program and its review of the prior literature, we adopt this value for our estimation of the acute mortality effects of ozone exposure.

Ozone Effects on Morbidity   As with the PM literature in China, studies of the C-R relationship for morbidity effects of ozone exposures are again fewer than those for mortality. Our literature review indicates that all known studies on Chinese populations have been conducted in Hong Kong or Taipei. T. W. Wong et al. (1999) reported the health impacts of ozone on hospital admissions in Hong Kong, both for cardiovascular and respiratory diseases. Wong also led studies estimating the C-R relationship of ozone exposures and outpatient visits for respiratory symptoms (T. W. Wong, Wun, et al. 2002; T. W. Wong et al. 2006). Besides general respiratory-related hospital admissions, studies of hospital admissions for specific diseases including COPD (Ko et al. 2007) and asthma (Bell, Levy, and Lin 2008) have been conducted in Hong Kong and Taipei, respectively. Considering the limited availability of baseline incidence data (which is necessary to apply the *RRs* in our calculations and must match the definition used for the relevant C-R function), we limit our ozone morbidity assessment to its effect on hospital admissions, using results from T. W. Wong et al. (1999). The corresponding *RR* values are 1.0496 and 1.0850 for cardiovascular and respiratory hospital admissions, respectively, due to an increase in daily exposure to ozone of $10\,\mu g/m^3$.

Sources of Uncertainty   Some $PM_{2.5}$ *RR* values are derived from PM_{10} *RR* values based on the aforementioned default assumption that $PM_{2.5}$ concentrations are simply 60% of the observed PM_{10} concentrations. Although this is considered an appropriate average ratio of ambient $PM_{2.5}$ mass concentration to PM_{10} mass concentration, the actual ratio for a given location is determined by the characteristics of local air pollution and could change considerably from day to day (Jia et al. 2008). Therefore, the daily variation of $PM_{2.5}$ concentrations could show quite different trends from those of PM_{10}, implying different C-R relationships. This is a limitation of available data for which we currently have no solution; the uncertainties introduced by this assumption will be reduced in future research of this sort as more systematic monitoring of $PM_{2.5}$ concentrations takes place and as time-series (and ultimately cohort) studies are conducted of the health effects of $PM_{2.5}$ exposures in China.

Another limitation is that the time-series mortality and morbidity C-R functions used in our analysis are all from field research in large or very large cities in China. We believe these are more representative of conditions in China than if we relied instead on time-series studies from Western countries. But we must note that uncertainty is nevertheless introduced because the sample populations yielding the C-R functions are disproportionately urban. Since the health care patterns in rural areas of China are dramatically different from those of its cities, especially large and wealthy ones such as Beijing, Shanghai, and Hong Kong, the rural C-R relationships could differ from urban ones. For instance, people in rural areas with less health care availability are less likely to visit a hospital when experiencing some respiratory symptoms. Using coefficients based on urban samples would overestimate *hospital visits*, though it may in fact better capture pollution-induced *illness* in rural residents. Our application of the C-R functions is national in scope.

All the *RR* values used in this study are derived from single-pollutant models, which treat health effects from one pollutant as independent of those from others. In the real atmospheric environment, however, some types of pollution episodes will yield similar trends in different pollutants. Evaluating such an episode with a single-pollutant model could overestimate the *RR* of an endpoint that is due to exposure to a given pollutant; the actual observed health effect could result from exposures to several air pollutants at the same time. Although this approach could result in some overestimation, this error should not be large because we only consider here the effects of two key pollutants ($PM_{2.5}$ and O_3), and we will soon see that the overall concentrations and health effects of $PM_{2.5}$ in China are much higher than those of ozone.

Construction of Applicable C-R Functions   C-R functions represent the relationship between the concentration of a pollutant (x) and the population health response (y). The epidemiological studies cited use different functional forms to represent this relationship, including linear, log-linear, logarithmic, and Cox proportional risk models. For the purposes of our estimation of the benefits of small changes in concentration, we are not directly interested in the nature of the C-R function itself. Our main focus lies in the relationship between the *change* in concentration of the pollutant, Δx, and the corresponding change in the population health response, Δy, at current levels of pollution. However, many epidemiological studies do not estimate the C-R function over an entire range, but instead report the *RR* associated with a given change in the pollutant concentration of the region studied.

Most of the C-R relationships cited in this study are derived using a log-linear model in which the natural logarithm of the health response is a linear function of the pollutant concentration. We convert the relative risk into a coefficient applicable in BenMAP in the following way.

The log-linear model defines the population health response (y) as

$$y = B \times e^{\beta x} \quad \text{or} \quad \ln(y) = \alpha + \beta \times x \tag{8.2}$$

where the parameter B represents the incidence rate y when the concentration of the air pollutant is zero, the parameter β is the coefficient of the air pollutant concentration that BenMAP requires, $\ln(y)$ is the natural logarithm of y, and $\alpha = \ln(B)$.

Thus the population health responses under a base concentration (x_0) and relative concentration (x_r) are presented as $y_0 = B \times e^{\beta x_0}$ and $y_r = B \times e^{\beta x_r}$. According to the definition of *RR*,

$$RR = \frac{y_r}{y_0} = \frac{B \times e^{\beta x_r}}{B \times e^{\beta x_0}} = e^{\beta(x_r - x_0)} \tag{8.3}$$

and the coefficient in the C-R function underlying the relative risk is derived as

$$\beta = \frac{\ln(RR)}{x_r - x_0} = \frac{\ln(RR)}{\Delta x} \tag{8.4}$$

The β coefficients that are derived from the *RR*s discussed earlier and used in this study are listed in table 8.3.

Table 8.3

Coefficients β used in BenMAP for this study

Health Endpoint	$PM_{2.5}$	Ozone
Acute mortality	0.000646	0.001161
Chronic mortality	0.003922	—
Hospital admissions, cardiovascular	0.000997	0.004844
Hospital admissions, respiratory	0.002646	0.008161
Outpatient visits, all causes	0.000391	—

Note: The β for acute mortality from ozone exposure requires an additional conversion step from the $RR = 1.0031$ noted in the text because the ozone concentrations in C. M. Wong, Vichit-Vadakan, et al. (2008) are expressed in 8-hour rather than 24-hour averages.

8.1.3   Health Benefit Results

We now report the estimates of health effects of the policies using the population data and C-R coefficients just discussed. The health benefits for each endpoint are summarized in table 8.4. In addition to the mean estimates, lower and upper bounds representing the 95% confidence interval (CI), calculated based on the published CIs for RR from the corresponding literature, are listed in table 8.4. Compared with the base case scenario, improvement of air quality under the carbon tax policy would have avoided a mean estimate of 18,600 premature deaths from acute effects of $PM_{2.5}$ and O_3 exposures in 2010, or 103,000 premature deaths from chronic effects from $PM_{2.5}$. These would account for 0.22% or 1.25%, respectively, of all-cause deaths. Our assessment of the 11th FYP SO_2 control policies yields a mean estimate of 12,400 avoided premature deaths from acute effects of $PM_{2.5}$ and O_3 exposure, or 73,900 premature deaths from chronic effects from $PM_{2.5}$, accounting for 0.15% or 0.89% of all-cause deaths, respectively. Keep in mind that we report both estimates because the acute mortality estimate is based on epidemiology conducted in China, as is preferred, while the chronic mortality estimate captures the dominant mortality pathway, but on the basis of epidemiology from the very different conditions of the United States.

Benefits in the morbidity endpoints are also significant although not as striking: the carbon tax would have avoided mean estimates of 97,700 cases of hospital admissions and 6,620,000 cases of outpatient visits due to pollution exposures in 2010, and the 11th FYP SO_2 control policies may have avoided 61,100 cases of hospital admissions and 4,730,000 cases of outpatient visits.

BenMAP also provides the geographical distribution of the health benefits, which we show for avoided acute mortality due to changes in both pollutants for the two

Table 8.4
Health benefits of the carbon tax and the 11th FYP SO₂ controls (number of cases avoided compared to the base case in 2010)

Pollutant	Health Endpoints	Carbon Tax			11th FYP SO$_2$ Controls		
		Mean	Lower	Upper	Mean	Lower	Upper
PM$_{2.5}$	Acute mortality	17,200	10,600	23,800	12,300	7,550	17,000
	Chronic mortality	103,000	29,500	176,000	73,900	21,100	126,000
	Hospital admissions (cardiovascular)	30,000	7,390	52,500	21,400	5,280	37,500
	Hospital admissions (respiratory)	53,700	37,000	70,400	38,400	26,500	50,300
	Outpatient visits, all causes	6,620,000	2,640,000	10,600,000	4,730,000	1,880,000	7,570,000
Ozone	Acute mortality	1,380	374	2,390	123	33	213
	Hospital admissions (cardiovascular)	6,530	4,630	8,420	582	413	751
	Hospital admissions (respiratory)	7,440	6,030	8,850	664	538	790

policy cases in figure 8.4. The geographical distribution of the benefits of other health outcomes would have roughly similar patterns as the figure (albeit with different absolute values) since the incidence rates and C-R functions for any of our given endpoints are assumed constant throughout the whole domain.

The benefits represented by avoided acute mortality shown in figure 8.4 are determined largely by the very large reductions of $PM_{2.5}$ under both policies; ozone health effects are much smaller in comparison. The indicated total benefits occur throughout China, with particular concentrations around the North China Plain, the Yangtze River Delta, Hunan and Hubei provinces in the interior south, and the Sichuan Basin, all of which not only emit a lot of primary PM and PM precursors but are densely populated over large areas.

Compared with the carbon tax, the smaller total health benefits under the 11th FYP SO_2 control policy would be somewhat more concentrated in northern China than in the south. This difference between the north and south is mainly due to the existence of more flue gas desulfurization (FGD) installations under the policy in the north. Comparatively more power plants in southern Chinese provinces had already installed FGD because acid rain control zones had been designated across that region previously. The lower preexisting FGD penetration rate in northern provinces such as Shandong, Shanxi, Hebei, and Henan allowed more installations in the 2005–2010 period, and thus comparatively more health benefit from the 11th FYP policy. The carbon tax, by contrast, would be less influenced by previous pollution control efforts and would reduce emissions and health damages in a more distributed manner across the entire economy, as well as more broadly in geographical terms.

The geographical distributions of the much smaller effects on acute mortality from O_3 exposures, illustrated in figure 8.5, are entirely different from the $PM_{2.5}$-dominated patterns of figure 8.4. Because of increased ozone concentrations in Beijing, Tianjin, Hebei, Henan, and Shandong resulting from the carbon tax policy, the health benefits due to O_3 changes are negative in these areas. The positive health benefits from O_3 reduction in the rest of the country under the carbon tax, however, are much larger in aggregate.

The health benefits from reducing O_3 exposure in the 11th FYP SO_2 control policy case are similar in that they are mainly positive across the country, but negative in some of the same areas of the North China Plain and the Yangtze River Delta, and also in a few areas of Guangxi and the coastal southeast. Importantly, though, the total O_3 effect on premature mortality under the 11th FYP case is roughly an order of magnitude less than that under the carbon tax because emissions of NO_x, which

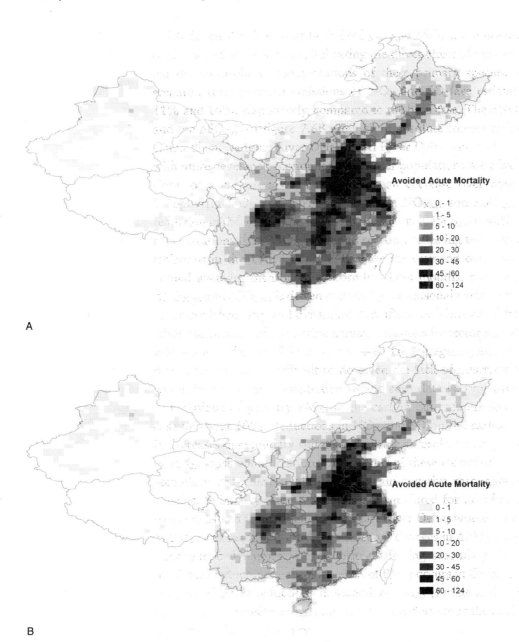

Figure 8.4
Cases of acute mortality avoided under the two policy scenarios: (A) carbon tax policy, (B) 11th FYP SO_2 control policy.

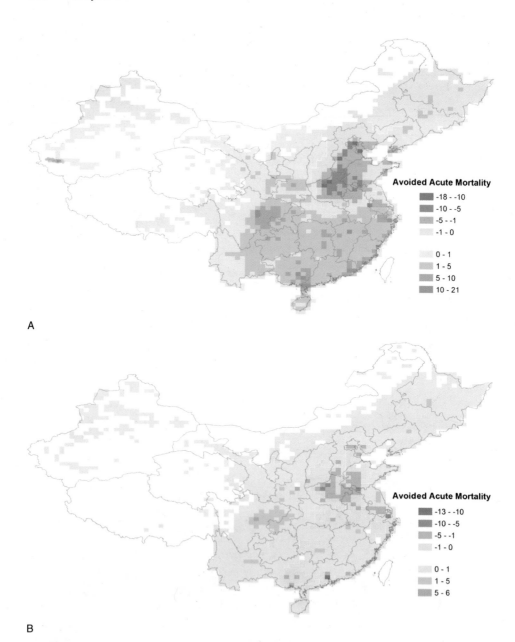

Figure 8.5
Cases of acute mortality avoided from changes in O₃ exposure under the two policy scenarios:
(A) carbon tax policy, (B) 11th FYP SO₂ control policy.

play an important role in the formation of O_3, are only minimally affected. Indeed the net health benefit across the country of O_3 changes under the 11th FYP case is negligible, as also reflected in table 8.4.

8.2   Agricultural Benefits

Ozone is recognized as the air pollutant that causes the greatest damage to agricultural productivity in most parts of the world. A recent global meta-analysis indicates that observed O_3 concentrations of 31–50 ppb, compared to base levels of less than 26 ppb, reduced the yields of potatoes, barley, wheat, rice, beans, and soybeans by 5.3%, 8.9%, 9.7%, 17.5%, 19.0%, and 7.7%, respectively (Z. Z. Feng and Kobayashi 2009). As regional O_3 concentrations in China increase with industrialization and urbanization, their adverse effects on crop yields could be increasingly damaging to China's food security in the future. In the following sections, the impacts of ozone on grain production are analyzed for the base case and carbon tax scenarios. By assigning market values to the crop production that would have been saved from O_3 damage had the carbon tax been in effect, the agricultural benefits of the policy can be quantified in monetary terms.

We do not conduct all the same calculations to quantify the impacts on crops of the 11th FYP SO_2 control scenario, because, as indicated in the preceding health benefit assessment, the net effect of that policy on ozone levels is very small and effects on crop productivity will be similarly negligible.

8.2.1   Main Grain Crops in China

It is a considerable national accomplishment that China, responsible for feeding 20% of the world's population, has increased its grain production in recent years. According to the Food and Agriculture Organization of the United Nations (FAO 2009), and as shown in figure 8.6, the annual yields of the four largest grain crops in China have collectively been over 400 million metric tons for most of the last decade, an average production of over 300 kg of all grains per capita. Rice is the leading cereal grain in China, with annual yields of around 180 million tons, accounting for some 30% of global production. Wheat is an important grain in northern China, and its annual production of around 100 million tons accounts for 15%–18% of global output. China's annual maize yields have increased substantially from 106 million tons in 2000 to 166 million in 2008, accounting for 18%–20% of global production. Compared with these three grain crops, soybeans are not as important in China in terms of both annual yields (around 15 million

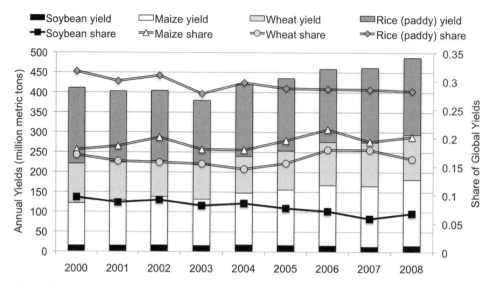

Figure 8.6
Annual yields of major grain crops in China and their share in global yields, 2000–2008. The annual yields are indicated by the vertical bars (left scale), and the shares are indicated by the lines (right scale). *Source:* FAO 2009.

tons) and share of global production (8%). With the production of rice, wheat, and maize an order of magnitude higher than that of soybeans, we limit our current analyses to the impacts of O_3 on the first three crops.

The spatial distributions of these crops are critical inputs in assessing the impact of O_3 on their yields, as is the timing of crops in relation to the strong seasonal variations of O_3 formation in the atmosphere. To fit the resolution of O_3 concentrations generated by the atmospheric modeling in chapter 7, county-level yields of rice, wheat, and maize are compiled from provincial statistical yearbooks (Provincial Statistical Bureaus 2008) and mapped to match the $0.5° \times 0.67°$ grid scale of the model. The outputs of rice and wheat have been relatively stable in recent years, but as noted before, maize output has been rising at a rate of 7.3% per year since 2002.

In the other chapters we use 2005 as the base year for calculating emissions and output. Here we use 2007 as the base year for grain yields because it is the latest year in which county-level crop production data are available and it provides a reasonable 2006–2010 average for the swiftly rising maize production, while not appreciably affecting estimates of rice and wheat. The geographical distributions of yields of rice, wheat, and maize at the county level are shown in figure 8.7.

Figure 8.7A indicates that rice is mostly produced in south China, the Sichuan Basin in the southwest, and in northeastern China. Double-cropped rice—that which can be planted and harvested twice in one year—has been grown in south China for several decades, but early season rice accounted for only 17% of the total national rice yield of 2007 (Ministry of Agriculture 2008). Late season rice and single-cropped rice dominate China's rice production, generally ripening at the end of September and beginning of October, and we use it to represent total rice production.

Wheat is mainly planted in the North China Plain, shown in figure 8.7B. There are two types, spring wheat and winter wheat. Spring wheat is planted in cold regions, such as northeastern China, where the planted grain cannot survive winter. Winter wheat is widely planted in the warmer areas south of the Great Wall. Generally speaking, spring wheat ripens in late August whereas winter wheat ripens in late May.

Maize is planted widely in north and northeastern China, as indicated in figure 8.7C. Although maize ripens in the fall in most cases, the precise time of year varies according to differences in both the type of maize and the condition of planting. Lacking specific information to make such distinctions, however, we assume in this study that all maize in China ripens in late September.

8.2.2   Exposure Methodology, Data, and Assessment

Ozone Exposure Indices   As explained in chapter 7, the formation of O_3 is highly dependent on meteorological conditions such as temperature, humidity, and solar radiation. The diurnal variation of O_3 concentrations (i.e., the variation that occurs over the course of the day) is generally very large, with high levels in the afternoon and low levels at night. The studies that have attributed reduced crop productivity to exposure to O_3 have indicated that this damage occurs during times of the day when both O_3 concentrations and rates of photosynthetic growth are highest.

A variety of indices have been developed to differentiate O_3 exposures that are most damaging to crops from those believed to be less damaging or harmless. Most field experiments evaluate O_3 exposures by seasonal mean concentrations for only specified hours of the day. For instance, the National Crop Loss Assessment Network (NCLAN) studies supported by the U.S. EPA used mean O_3 concentrations during the seven hours from 9 A.M. to 4 P.M. during the three-month peak ozone season, an index labeled M7. Analogous twelve-hour seasonal means (8 A.M. to 8 P.M., termed M12) were employed in later NCLAN studies (Hogsett, Tingey, and Lee 1988). A number of Asian studies have considered O_3 exposures limited to other

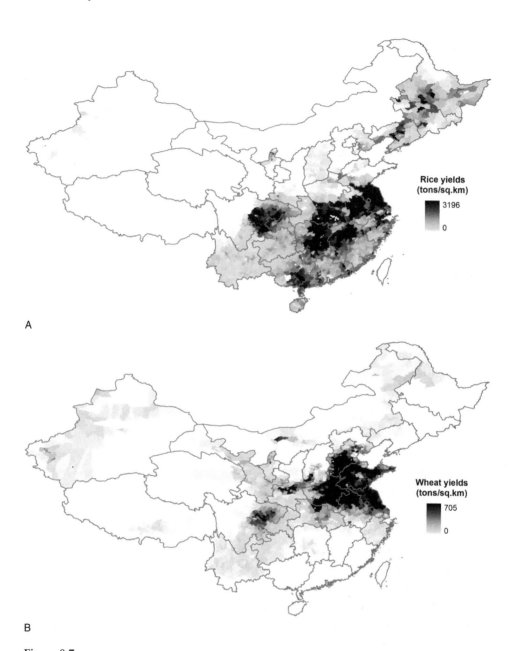

A

B

Figure 8.7
County-level distributions of 2007 yields of major grains in China: (A) rice, (B) wheat,
(C) maize. *Source:* Data are compiled from 2008 statistical yearbooks for all provinces and
national municipalities (Provincial Statistical Bureaus 2008).

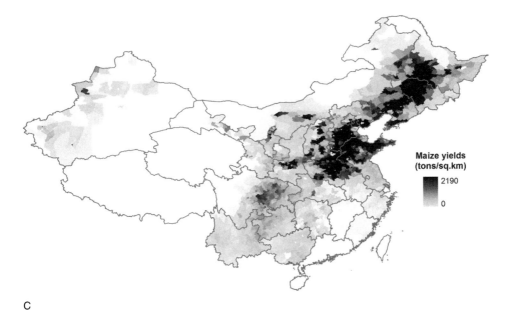

C

Figure 8.7 (continued)

stretches of the day, including four-hour (Agrawal 2005), six-hour (Wahid et al. 2001), and eight-hour periods (Tiwari, Agrawal, and Manning 2005).

Shortcomings in indices based on mean concentrations have been recognized, notably that they insufficiently account for the harmful effects of cumulative exposures (Finnan, Burke, and Jones 1997). Alternative, cumulative indices were introduced in the postexperimental, data-analysis stage of the NCLAN studies. The SUM06 index stresses both exposure to high O_3 concentrations and the duration of exposures, and is defined as the sum of the hourly ozone values when it exceeds 0.06 parts per million (ppm):

$$\text{SUM06} = \sum_{i=1}^{n}(C_{O_3})_i \quad \text{for} \quad C_{O_3} \geq 0.06 \text{ ppm} \tag{8.5}$$

where $(C_{O_3})_i$ is the mean concentration of ozone in parts per million during hour i, n is the number of hours in a three-month peak ozone season, and SUM06 is in parts per million-hours (ppm·h).

In European studies of O_3 exposures, a similar index, AOT40 (accumulated exposure to ozone above a threshold of 40 parts per *billion* [ppb]) has been used, expressed in ppb·h:

$$\text{AOT40} = \sum_{i=1}^{n} (C_{O_3} - 40)_i \quad \text{for} \quad C_{O_3} \geq 40 \text{ ppb} \tag{8.6}$$

These types of cumulative indices have also been criticized for treating an increment of O_3 at all concentrations above the threshold as equally effective at eliciting a biological response (Finnan, Burke, and Jones 1997). The W126 index is similar to SUM06 except that the hours above a threshold concentration, in this case 40 ppb, are multiplied by a weighting factor:

$$\text{W126} = \sum_{i=1}^{n} w_i \times (C_{O_3})_i \tag{8.7}$$

expressed in ppb·h, with the weighting factor defined as

$$w_i = \frac{1}{1 + 4403 \exp[-0.126(C_{O_3})_i]} \quad \text{for} \quad C_{O_3} \geq 40 \text{ ppb} \tag{8.8}$$

The summation is again over the hourly concentration during a three-month peak ozone season.

As introduced by the U.S. EPA and the European Environment Agency (EEA), these three cumulative indices (SUM06, AOT40, and W126) consider the 12-hour (8 A.M. to 8 P.M.) hourly average ozone concentrations during the three-month peak ozone season. However, when the SUM06 and AOT40 indices were recalculated in China as the hourly average ozone concentration over a 24-hour period, they were found to be 3–13% and 5–23% higher, respectively, than the 12-hour values (H. X. Wang et al. 2005). This finding illustrates that high nighttime O_3 concentrations and exposures can occur in some Chinese regions, although the effect of exposures when photosynthesis is weak is unclear.

Despite the considerable shortcomings of the M7 and M12 indices, we include them here because they are used in some of the concentration-response studies that we use to estimate crop damages. How the indices are combined with the C-R coefficients and applied to estimate improved crop productivity is described in sections 8.2.3 and 8.2.4.

Ozone Exposure Changes Due to the Carbon Tax   As discussed in chapter 7, reductions in emissions of NO_X and VOCs that result from the carbon tax policy lead to changes in O_3 exposures throughout China. Based on hourly concentrations simulated by GEOS-Chem in chapter 7, the exposure to surface O_3 according to different indices is calculated, and differences of O_3 exposure between the base and carbon tax policy cases are quantified.

According to atmospheric observations in some regions (Y. X. Wang et al. 2010), surface O_3 concentrations usually peak around 4 P.M. in spring and summer in China, and the daytime period of high concentrations is longer than that in the United States. To ensure that we consider ozone exposures throughout the highest daily concentrations, we choose the 12-hour mean M12 as a better seasonal mean concentration index for China than the M7 index that stops at 4 P.M.

Figure 8.8 contrasts the differences of M12 exposure risks, independent of the crop distributions, between the base and carbon tax cases in two seasons: spring and summer of 2010. The areas colored green are those where the average concentration under the carbon tax case is reduced, while the red ones have increased average concentrations. Generally speaking, O_3 exposure risks would be reduced in most parts of China under a carbon tax, but are increased in springtime in a region extending from Shanghai to north of Beijing and from Shandong to Shanxi provinces, and in summertime in the metropolitan areas around Shanghai and Guangzhou.

8.2.3 Concentration-Response Relationships

Quantifying the risks of crop exposures to O_3 also requires C-R relationships or, more precisely, exposure–relative yield relationships. These are determined by experimental field investigations of crop responses to controlled fumigation with a range of pollutant concentrations.

The first effort investigating such C-R relationships was initiated in the United States, where the aforementioned NCLAN, a coordinated experimental program using standardized protocols at six study sites across the country, was carried out in the early 1980s (Heck et al. 1988). This program was followed by the European Open Top Chamber (EOTC) Project in the late 1980s and early 1990s (Jäger et al. 1992).

Researchers in the United States and Europe have developed a series of C-R relationships for a number of key agricultural crops. These relationships from Western countries have been used to estimate yield losses of crops from exposure to ozone in China (Aunan, Berntsen, and Seip 2000) and East Asia (X. P. Wang and Mauzerall 2004). However, as it is recognized that factors such as climate, agricultural management practices, and pollutant exposure patterns will alter plant response to O_3 (Fuhrer and Booker 2003), field studies conducted in China or Asia to derive local C-R relationships should improve yield loss analyses.

Dozens of experimental studies have been conducted in recent years in Asian countries to determine C-R relationships for local crop species and agricultural

A

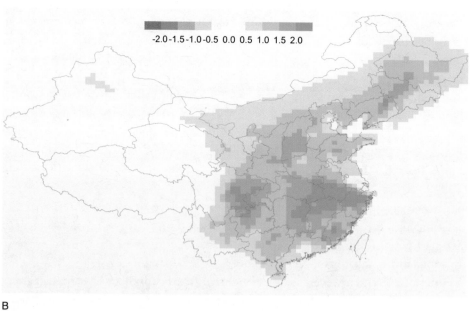

B

Figure 8.8
Reduction of M12 O_3 exposure (ppb) due to implementation of the carbon tax policy in 2010: (A) spring (March, April, and May), (B) summer (June, July, and August). The values represent the difference of M12 O_3 exposure between the base case and carbon tax case.

Table 8.5
The exposure–relative yield functions applied in this study

Crop	Exposure–Relative Yield Relationship	Reference	Country of Field Research
Rice	$RY = \exp[-(M7/202)^{2.47}]$	Adams et al. (1989)	U.S.
	$RY = 1 - AOT40 \times 0.526/100$	Z. W. Feng et al. (2003)	China
Wheat	$RY = \exp[-(M7/136)^{2.56}]$ (winter wheat)	Lesser et al. (1990)	U.S.
	$RY = \exp[-(M7/186)^{3.2}]$ (spring wheat)	Adams et al. (1989)	U.S.
	$RY = \exp[-(SUM06/52.32)^{2.176}]$	X. P. Wang and Mauzerall (2004)	U.S.
	$RY = \exp[-(W126/51.2)^{1.747}]$	X. P. Wang and Mauzerall (2004)	U.S.
	$RY = 1 - AOT40 \times 1.296/100$	Z. W. Feng et al. (2003)	China
Maize	$RY = \exp[-(M12/124)^{2.83}]$	Lesser et al. (1990)	U.S.
	$RY = \exp[-(SUM06/93.485)^{3.5695}]$	X. P. Wang and Mauzerall (2004)	U.S.
	$RY = \exp[-(W126/93.7)^{3.392}]$	X. P. Wang and Mauzerall (2004)	U.S.

practices, mainly in India and Pakistan but to some degree in China. Emberson et al. (2009) compared North American estimates with pooled Asian data and suggested that Asian rice and wheat cultivars may be more vulnerable to O_3 exposures than American ones, although they noted that the C-R varies from place to place. They estimated that China's rice yield loss from O_3 exposures is at a similar level to that of the United States, while its wheat yields are more sensitive to O_3 exposure.

In the current study, we employ C-R functions from a variety of studies to estimate the impact of O_3 on grain production. For each type of grain cultivar, we must apply functions based on the different O_3 exposure indices used in the C-R studies. We prefer the two available functions derived in studies conducted in China, but must apply those from U.S. research when appropriate Chinese functions are unavailable. In order to consider all of the grain types and to favor C-R functions from China, we use a range of exposure indices including the noncumulative ones, M7 and M12. The C-R functions used in this study are listed in table 8.5.

8.2.4 Agricultural Benefit Results
We now report the estimated benefits to rice, wheat, and maize production that are due to the carbon tax policy. Estimates of grain production benefits are calculated

Table 8.6

Grain production benefits of carbon tax policy (increase of crop yields in thousand tons compared to the base case in 2010)

	M12	SUM06	W126	AOT40	Average
Wheat	−59.6	208.0	596.3	−47.4	174.3
Rice	495.2	—	—	1,491.7	993.5
Maize	1,035.3	2,072.4	1,255.7	—	1,454.5

for each O_3 exposure index and corresponding C-R function from the literature. This procedure results in more than one estimate for each crop type, as summarized in table 8.6. Each of these estimates has strengths and weaknesses. Because their relative merits are difficult to judge, we use the average of the several estimates for each grain as the best estimate of the overall benefits of the policy to the production of that grain.

The estimates of benefits to wheat production of the carbon tax vary from −59,600 tons (a loss, or disbenefit) to 596,000 tons. Figures 8.9A–D show the regional distributions of the estimated effect on output of wheat from the carbon tax for the four indices (M12, AOT40, SUM06, and W126) and their associated C-R functions. The dark blue areas are where output rises the most as a result of the tax, whereas the red areas are where it falls the most. One can see that the four indices are relatively consistent: in all cases the red-yellow zones are in the Shanghai to Beijing region where the O_3 concentrations are elevated by the policy. The magnitude of changes in output differs by location, and overall, the effect estimated by the M12 index is the smallest.

When the benefits are summed nationally, the total changes in output differ substantially across the indices. These differences are due not only to the indices but also to the paired exposure–relative yield functions, which are based on experiments in different O_3 exposure conditions. The four indices stress differing features of O_3 exposure. For example, the M12 index treats an incremental ppm at high O_3 concentrations as equal to one at low concentrations, while the SUM06 index gives no weight to an incremental ppm at low concentrations. A decrease in the O_3 concentration from 61 ppb to 59 ppb barely affects O_3 exposure under the M12 measure, while it substantially changes the result when assessed with SUM06, which ignores O_3 concentrations under 60 ppb entirely.

The average estimate of benefits to wheat yields from the carbon tax policy is 174,000 tons. The average benefit to rice yields is much higher at 993,000 tons, with a lowest estimate of 495,000 tons and a highest estimate of 1,490,000 tons.

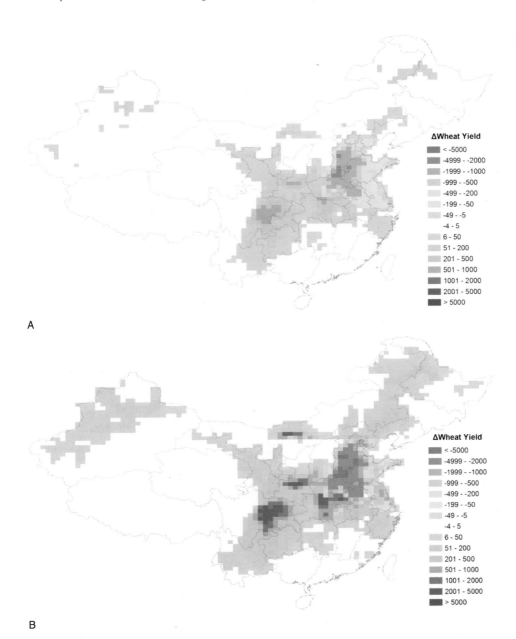

A

B

Figure 8.9
Change of wheat yields in metric tons under the carbon tax policy scenario compared to the base case: (A) M12, (B) AOT40, (C) SUM06, (D) W126. Estimates are based on each of four O$_3$ exposure indices indicated, and their corresponding exposure–relative yield functions. Blue and green represent increases in wheat yields under the carbon tax, and red and yellow represent decreases. The linear feature apparent in north China results from the latitude (40°N) assumed to differentiate regions of spring wheat and winter wheat cultivation.

(Figure continued)

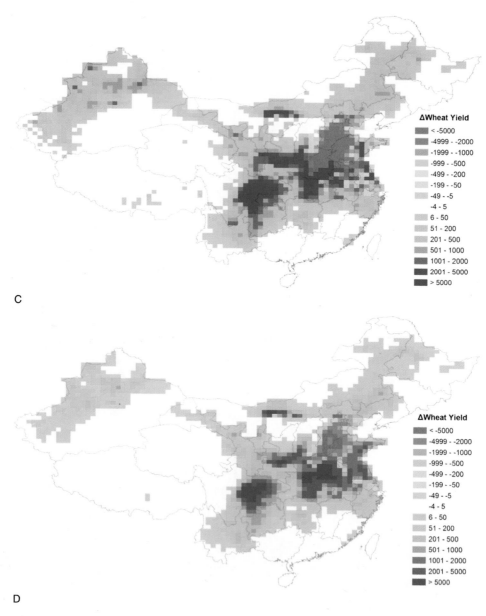

C

D

Figure 8.9 (continued)

Table 8.7
Monetary benefits of increase in grain yields from carbon tax policies

Crop	Increased Yields (tons)	Price (yuan/ton)	Monetary Benefits (billion yuan)
Wheat	174,300	1,403	0.25
Rice	993,500	2,630	2.615
Maize	1,454,500	1,550	2.255

The average benefit to maize yields (1,450,000 tons) is the highest of the three grains in terms of total tonnage, with a lowest estimate of 1,040,000 tons and a highest estimate of 2,070,000 tons.

We compare our estimates to China's total grain production in 2007. We find that the carbon tax would increase wheat, rice, and maize production in China by around 0.16%, 0.53%, and 0.95%, respectively. While this increase is not large compared to total national production, in some regions the effect could be as high as 3.75%, which would be significant locally.

Finally, we also calculate the monetary value of the grain yield benefits from the carbon tax by applying China's average national price of each grain in 2005 (FAO 2010). The total value of the grain yield benefits (table 8.7) is 5.11 billion yuan, which we will see in chapter 9 and appendix B is the same order of magnitude as the monetary value of the benefits from reducing acute mortality from ozone exposures.

While these estimated benefits are modest, it is important to stress that they are based on a relatively small tax of 100 yuan per ton of carbon (or 27 yuan per ton of CO_2). Higher taxes or other aggressive multipollutant control policies (at least those reducing emissions of NO_X and VOCs) could, accordingly, produce substantially larger benefits to grain production. The results indicate that China's future energy and environmental policy making should recognize a significant impact on agriculture, as well as on human health, when the policies affect O_3 concentrations.

8.3 Conclusions

We have seen that the two policies we considered, a hypothetical tax on carbon and the SO_2 controls actually implemented in the 11th FYP, would have resulted or did result in large total benefits from reduced air pollution.

The benefits under both policies are dominated by the avoided health impacts of reduced ambient $PM_{2.5}$. The full scale of this effect, however, depends on which epidemiological evidence concerning premature mortality is applied: that concerning acute $PM_{2.5}$ exposures derived in China, or that concerning chronic $PM_{2.5}$ exposures estimated in the United States. This is a longstanding quandary for analysts of health impacts of air pollution in developing countries, and we believe it is important to report both results so readers are aware of such analytical uncertainties and trade-offs.

In comparison, the benefits to health and grain productivity from the effects of the two policies on ozone levels are smaller. They are in fact negligible under the 11th FYP SO_2 controls, as the policy has little net effect on mean concentrations of ozone nationwide. Because a carbon tax reduces a wider range of pollutants including precursors to ozone, however, it leads to larger changes in ozone concentrations across China, both positive and negative. The net national effect is significant and positive, and thus even a small carbon tax would offer additional benefits to public health and also improve the productivity of major grain crops in China.

Having estimated such large benefits of the two policy options, the logical next question concerns how much it costs to achieve them. Addressing this important topic is the subject of the next chapter.

References

Adams, R. M., J. D. Glyer, S. L. Johnson, and B. A. McCarl. 1989. A reassessment of the economic effects of ozone on US agriculture. *Journal of the Air Pollution Control Association* 39:960–968.

Agrawal, M. 2005. Effects of air pollution on agriculture: An issue of national concern. *National Academy of Science Letters—India* 28(3–4):93–106.

Aunan, K., T. Berntsen, and H. M. Seip. 2000. Surface ozone in China and its possible impact on agricultural crop yields. *Ambio* 29(6):294–301.

Aunan, K., and X.-C. Pan. 2004. Exposure-response functions for health effects of ambient air pollution applicable for China: A meta-analysis. *Science of the Total Environment* 329:3–16.

Bell, M. L., F. Dominici and J. M. Samet. 2005. A meta-analysis of time-series studies of ozone and mortality with comparison to the national morbidity, mortality, and air pollution study. *Epidemiology* 16(4):436–445.

Bell, M. L., J. K. Levy, and Z. Lin. 2008. The effect of sandstorms and air pollution on cause-specific hospital admissions in Taipei, Taiwan. *Occupational and Environmental Medicine* 65:104–111.

Chang, G. Q., X. C. Pan, X. Q. Xie, and Y. L. Gao. 2003. Time-series analysis on the relationship between air pollution and daily mortality in Beijing. *Journal of Hygiene Research* 32:565–568. In Chinese.

Chen, Z., ed. 2008. *Report of the Third National Mortality Retrospective Sampling Survey.* Beijing: Peking Union Medical College Press. In Chinese.

Committee on the Medical Effects of Air Pollutants (COMEAP). 2009. Long-term exposure to nitrogen dioxide: Epidemiological evidence of effects on respiratory morbidity in children. United Kingdon Department of Health. Available at http://comeap.org.uk/, last accessed October 12, 2012.

Dai, H. X., W. M. Song, X. Gao, L. M. Chen, and M. Hu. 2004. Study on relationship between ambient PM_{10}, $PM_{2.5}$ pollution and daily mortality in a district in Shanghai. *Journal of Hygiene Research* 33:293–297. In Chinese.

Dockery, D. W., and C. A. Pope. 1994. Acute respiratory effects of particulate air pollution. *Annual Review of Public Health* 15:107–132.

Eftim, S., and F. Dominici. 2005. Multisite time-series studies versus cohort studies: Methods, findings, and policy implications. *Journal of Toxicology and Environmental Health, Part A* 68(13–14):1191–1205.

Emberson, L. D., P. Buker, M. R. Ashmore, G. Mills, L. S. Jackson, M. Agrawal, M. D. Atikuzzaman, S. Cinderby, M. Engardt, C. Jamir, K. Kobayashi, N. T. K. Oanh, Q. F. Quadir, and A. Wahid. 2009. A comparison of North American and Asian exposure–response data for ozone effects on crop yields. *Atmospheric Environment* 43(12):1945–1953.

Feng, Z. W., M. H. Jin, F. Z. Zhang, and Y. Z. Huang. 2003. Effects of ground level ozone (O_3) pollution on the yields of rice and winter wheat in the Yangtze River Delta. *Journal of Environmental Sciences* 15(3):360–362.

Feng, Z. Z., and K. Kobayashi. 2009. Assessing the impacts of current and future concentrations of surface ozone on crop yield with meta-analysis. *Atmospheric Environment* 43(8):1510–1519.

Finnan, J. M., J. I. Burke, and M. B. Jones. 1997. An evaluation of indices that describe the impact of ozone on the yield of spring wheat (*Triticum aestivum* L.). *Atmospheric Environment* 31(17):2685–2693.

Food and Agriculture Organization of the United Nations (FAO). 2009. *FAO Statistical Yearbook 2009.* Available at http://faostat.fao.org/site/567/default.aspx#ancor, last accessed on July 10, 2011.

———. 2010. FAO PriceSTAT database. Available at http://faostat.fao.org/site/570/default.aspx#ancor, last accessed on July 10, 2011.

Fuhrer, J., and F. Booker. 2003. Ecological issues related to ozone: Agricultural issues. *Environment International* 29:141–154.

Health Effects Institute (HEI). 2004. Health effects of outdoor air pollution in developing countries of Asia: A literature review. Special report no. 15, International Scientific Oversight Committee, HEI. Boston.

Heck, W. W., O. C. Taylor, and D. T. Tingey, eds. 1988. *Assessment of Crop Loss from Air Pollutants.* Proceedings of an international conference, Raleigh, North Carolina, 25–29 October, 1987. Barking, UK: Elsevier Applied Science.

Ho, M. S., and C. P. Nielsen, eds. 2007. *Clearing the Air: The Health and Economic Damages of Air Pollution in China.* Cambridge, MA: MIT Press.

Hogsett, W. E., D. T. Tingey, and E. H. Lee. 1988. Ozone exposure indices: Concepts for development and evaluation of their use. In *Assessment of Crop Loss from Air Pollutants*, ed. W. W. Heck, O. C. Taylor, and D. T. Tingey. Barking, UK: Elsevier Applied Science.

Hubbell, B., D. McCubbin, and A. Hallberg. 2003. Assessing the health impacts of air pollution regulations using BenMAP, the Environmental Benefits Mapping and Analysis Program. Annual meeting of the American Geophysical Union, San Francisco, 8–12 December.

Hwang, J. S., and C. C. Chan. 2002. Effects of air pollution on daily clinic visits for lower respiratory tract illness. *American Journal of Epidemiology* 155(1):1–10.

Ito, K., S. F. De Leon, and M. Lippmann. 2005. Associations between ozone and daily mortality: Analysis and meta-analysis. *Epidemiology* 16(4):446–457.

Jäger, H. J., M. Unsworth, L. De Temmermann, and P. Mathy, eds. 1992. Effects of air pollution on agricultural crops in Europe: Results of the European Open-Top Chamber Project. Air Pollution Research Report #46. Commission of the European Communities. Brussels.

Jerrett, M., R. T. Burnett, C. A. Pope, K. Ito, G. Thurston, D. Krewski, Y. L. Shi, E. Calle, and M. Thun. 2009. Long-term ozone exposure and mortality. *New England Journal of Medicine* 360(11):1085–1095.

Jia, Y. T., K. A. Rahn, K. B. He, T. X. Wen, and Y. S. Wang. 2008. A novel technique for quantifying the regional component of urban aerosol solely from its sawtooth cycles. *Journal of Geophysical Research—Atmospheres* 113, Art. No. D21309.

Jiang, D., X. H. Yang and H. H. Liu. 2002. Study on spatial distribution of population based on remote sensing and GIS. *Advance in Earth Science* 17:734–738. In Chinese.

Kan, H. D., and B. H. Chen. 2003. Air pollution and daily mortality in Shanghai: A time-series study. *Archives of Environmental Health* 58(6):360–367.

Kan, H. D., S. J. London, G. H. Chen, Y. H. Zhang, G. X. Song, N. Q. Zhao, L. L. Jiang, and B. H. Chen. 2007. Differentiating the effects of fine and coarse particles on daily mortality in Shanghai, China. *Environment International* 33(3):376–384.

Kan, H. D., S. J. London, G. H. Chen, Y. X. Zhang, G. X. Song, N. Q. Zhao, L. L. Jiang, and B. H. Chen. 2008. Season, sex, age, and education as modifiers of the effects of outdoor air pollution on daily mortality in Shanghai, China: The Public Health and Air Pollution in Asia (PAPA) study. *Environmental Health Perspectives* 116(9):1183–1188.

Ko, F. W. S., W. Tam, T. W. Wong, D. P. S. Chan, A. H. Tung, C. K. W. Lai, and D. S. C. Hui. 2007. Temporal relationship between air pollutants and hospital admissions for chronic obstructive pulmonary disease in Hong Kong. *Thorax* 62:780–785.

Lesser, V. M., J. O. Rawlings, S. E. Spruill, and M. C. Somerville. 1990. Ozone effects on agricultural crops: Statistical methodologies and estimated dose-response relationships. *Crop Science* 30(1):148–155.

Levy, J. I., S. M. Chemerynski, and J. A. Sarnat. 2005. Ozone exposure and mortality: An empiric Bayes metaregression analysis. *Epidemiology* 16(4):458–468.

Levy, J. I., and S. L. Greco. 2007. Estimating health effects of air pollution in China: An introduction to intake fraction and the epidemiology. In *Clearing the Air: The Health and Economic Damages of Air Pollution in China*, ed. M. S. Ho and C. P. Nielsen. Cambridge, MA: MIT Press.

Levy, J. I., S. L. Greco, S. J. Melly, and N. Mukhi. 2009. Evaluating efficiency-equality tradeoffs for mobile source control strategies in an urban area. *Risk Analysis* 29(1):34–47.

Ministry of Agriculture. 2008. *China Agriculture Statistical Report 2007*. Beijing: China Agriculture Press.

Ministry of Health of China (MOH). 2006. *China's Health Statistics Yearbook 2006*. Beijing: Peking Union Medical College Press. In Chinese.

Ostro, B. D., H. Tran, and J. I. Levy. 2006. The health benefits of reduced tropospheric ozone in California. *Journal of the Air and Waste Management Association* 56(7):1007–1021.

Pope, C. A., R. T. Burnett, D. Krewski, M. Jarrett, Y. L. Shi, E. E. Calle, and M. J. Thun. 2009. Cardiovascular mortality and exposure to airborne fine particulate matter and cigarette smoke. *Circulation* 120:941–948.

Pope, C. A., R. T. Burnett, M. J. Thun, E. E. Calle, D. Krewski, K. Ito, and G. D. Thurston. 2002. Lung cancer, cardiopulmonary mortality, and long-term exposure to fine particulate air pollution. *Journal of the American Medical Association* 287:1132–1141.

Provincial Statistical Bureaus. 2008. *Statistical Yearbook of [each of the 31 provinces and national municipalities]*. Beijing: China Statistics Press.

Qian, Z. M., Q. C. He, H. M. Lin, L. L. Kong, C. M. Bentley, W. S. Liu, and D. J. Zhou. 2008. High temperatures enhanced acute mortality effects of ambient particle pollution in the "oven" city of Wuhan, China. *Environmental Health Perspectives* 116(9):1172–1178.

Qian, Z. M., Q. C. He, H. M. Lin, L. L. Kong, D. P. Liao, J. J. Dan, C. M. Bentley, and B. W. Wang. 2007. Association of daily cause-specific mortality with ambient particle air pollution in Wuhan, China. *Environmental Research* 105(3):380–389.

Schwartz, J., B. Coull, F. Laden, and L. Ryan. 2008. The effect of dose and timing of dose on the association between airborne particles and survival. *Environmental Health Perspectives* 116(1):64–69.

State Evironmental Protection Administration (SEPA). 2006. *Report On the State of the Environment In China 2005*. Available at http://english.mep.gov.cn/standards_reports/soe/soe2005/200706/t20070622_105622.htm, last accessed July 6, 2011.

Thomas, D. 2005. Why do estimates of the acute and chronic effects of air pollution on mortality differ? *Journal of Toxicology and Environmental Health, Part A* 68(13 & 14): 1167–1174.

Tiwari, S., M. Agrawal, and W. J. Manning. 2005. Assessing the impact of ambient ozone on growth and productivity of two cultivars of wheat in India using three rates of application of ethylenediurea (EDU). *Environmental Pollution* 138(1):153–160.

U.S. Environmental Protection Agency (U.S. EPA). 2008. Integrated science assessment for oxides of nitrogen—Health criteria (final report). Report no. EPA/600/R-08/071. Available at http://cfpub.epa.gov/ncea/cfm/recordisplay.cfm?deid=194645, last accessed October 12, 2012.

———. 2011. Air trends—National trends in particulate matter. Available at http://www.epa.gov/airtrends/pm.html, last accessed on July 10, 2011.

———. 2013. Human health risk assessment—Step 2, dose-response assessment. Available at http://www.epa.gov/risk/dose-response.htm, last accessed January 2, 2013.

Wahid, A., E. Milne, S. R. A. Shamsi, M. R. Ashmore, and F. M. Marshall. 2001. Effects of oxidants on soybean growth and yield in the Pakistan Punjab. *Environmental Pollution* 113(3):271–280.

Wang, H. X., C. S. Kiang, X. Y. Tang, X. J. Zhou, and W. L. Chameides. 2005. Surface ozone: A likely threat to crops in Yangtze delta of China. *Atmospheric Environment* 39(21):3843–3850.

Wang, X. P., and D. L. Mauzerall. 2004. Characterizing distributions of surface ozone and its impact on grain production in China, Japan and South Korea, 1990 and 2020. *Atmospheric Environment* 38(26):4383–4402.

Wang, Y. X., M. B. McElroy, J. W. Munger, J. M. Hao, H. Ma, and C. P. Nielsen. 2010. Year-round measurements of O_3 and CO at a rural site near Beijing: Variations in their correlations. *Tellus B: Chemical and Physical Meteorology* 62(4):228–241.

Wong, C. M., R. W. Atkinson, H. R. Anderson, A. J. Hedley, S. Ma, P. Y. K. Chau, and T. H. Lam. 2002. A tale of two cities: Effects of air pollution on hospital admissions in Hong Kong and London compared. *Environmental Health Perspectives* 110(1):67–77.

Wong, C. M., C. Q. Ou, K. P. Chan, Y. K. Chau, T. Q. Thach, L. Yang, R. Y. N. Chung, G. N. Thomas, J. S. M. Peiris, T. W. Wong, A. J. Hedley, and T. H. Lam. 2008. The effects of air pollution on mortality in socially deprived urban areas in Hong Kong, China. *Environmental Health Perspectives* 116(9):1189–1194.

Wong, C. M., N. Vichit-Vadakan, H. D. Kan, and Z. M. Qian. 2008. Public health and air pollution in Asia (PAPA): A multicity study of short-term effects of air pollution on mortality. *Environmental Health Perspectives* 116(9):1195–1202.

Wong, T. W., T. S. Lau, T. S. Yu, A. Neller, S. L. Wong, W. Tam, and S. W. Pang. 1999. Air pollution and hospital admissions for respiratory and cardiovascular diseases in Hong Kong. *Occupational and Environmental Medicine* 56:679–683.

Wong, T. W., W. S. Tam, T. S. Yu, and A. H. S. Wong. 2002. Associations between daily mortalities from respiratory and cardiovascular diseases and air pollution in Hong Kong, China. *Occupational and Environmental Medicine* 59:30–35.

Wong, T. W., W. Tam, I. T. S. Yu, Y. T. Wun, A. H. S. Wong, and C. M. Wong. 2006. Association between air pollution and general practitioner visits for respiratory diseases in Hong Kong. *Thorax* 61(7):585–591.

Wong, T. W., Y. T. Wun, T. S. Yu, W. Tam, C. M. Wong, and A. H. S. Wong. 2002. Air pollution and general practice consultations for respiratory illnesses. *Journal of Epidemiology and Community Health* 56:949–950.

World Bank. 1997. *Clear Water, Blue Skies: China's Environment in the New Century*. Washinton DC: World Bank.

Xu, X. P., B. L. Li, and H. Y. Huang. 1995. Air pollution and unscheduled hospital outpatient and emergency room visits. *Environmental Health Perspectives* 103(3):286–289.

Zhang, Y. H., W. Huang, S. J. London, G. X. Song, G. H. Chen, L. L. Jiang, N. Q. Zhao, B. H. Chen, and H. D. Kan. 2006. Ozone and daily mortality in Shanghai, China. *Environmental Health Perspectives* 114(8):1227–1232.

Zhang, Y. P., Z. Q. Zhang, X. H. Liu, X. P. Zhang, B. Q. Feng, and H. P. Li. 2007. Concentration-response relationship between particulate air pollution and daily mortality in Taiyuan. *Journal of Peking University (Health Sciences)* 39:153–157. In Chinese.

Zhou Y., J. S. Fu, G. S. Zhuang, and J. I. Levy. 2010. Risk-based prioritization among air pollution control strategies in the Yangzi River Delta, China. *Environmental Health Perspectives* 118:1204–1210, doi:10.1289/ehp.1001991.

9

The Economics of Environmental Policies in China

Jing Cao, Mun S. Ho, and Dale W. Jorgenson

9.1 Introduction

As described in chapter 1, our overall objective is to develop a methodology for analyzing environmental policies that recognizes the main elements of the complex web of interactions between economic activity, energy use, emissions, air quality, and damages to public health and agriculture. The goal of the analysis is to estimate the economic costs and the health and environmental benefits of a pollution control policy in an integrated framework. The economic effects of interest include the impact on aggregate growth and changes in industry composition.

Our specific aims within this overall goal are to analyze two major air pollution control policies: the sulfur control policies in the 11th Five-Year Plan (11th FYP, 2006–2010) and a carbon tax. We bring together the results of the previous chapters and describe the whole chain of analysis of these policies to show how our methodology is applied. As we have discussed earlier, the 11th FYP policies had the most ambitious targets so far for sulfur dioxide (SO_2) control and were the result of much debate within the Chinese government. The scale and cost of this program make it a natural choice to analyze. Carbon pricing policies are under active consideration under the 12th Five-Year Plan (12th FYP), including pilot implementation of carbon trading in some localities, and are obvious alternative policy options to examine.

Chapters 4 through 6 discuss how the base year emission inventories are constructed from detailed industry data and then how these are used to project future emissions using the industry output and energy use projected by the economic model presented in this chapter. These projections give emissions of a broad range of pollutants for each grid cell in our map covering the whole country. Then chapter 7 discusses how the atmospheric model relates emissions to air quality in each of those grid cells, and how the concentrations of fine particulate matter ($PM_{2.5}$) and ozone are changed by the projected changes in emissions in all the grid cells. Chapter 8 describes how air quality affects public health and agricultural output, including how the changes in mortality, morbidity, and crop productivity due to changes in air pollution levels are estimated. It also translates the estimated agricultural benefits of the policies into monetary values. The valuation of health benefits is more complex and is discussed in appendix B.

In this chapter we show how the components in each of these previous chapters are combined with an economic analysis to provide an integrated framework to study the effects of emission control policy. We first describe how we estimate the impact of the 11th FYP SO_2 control policy on economic growth and energy use using a multisector model of the Chinese economy. To do so, we simulate the model twice for the output and energy use in each of 33 industries for each year during the 11th FYP. The first simulation is for a base case without the SO_2 policy; the second simulation includes the policy requirements for the electric power sector.[1] Comparing these two cases gives us the impact of the SO_2 policy on GDP, aggregate consumption, fuel use, and carbon emissions. Then we compare these estimates of economic costs and other effects with the health and agricultural benefits that were calculated in the previous chapters.

The SO_2 control policy consists of two major components, one requiring the installing of flue gas desulfurization (FGD) equipment and one shutting down small, inefficient power plants. In this chapter we first analyze these two components separately and then report the combined effects. We find that the FGD equipment requirement alone substantially reduces health damages at a small cost in reduced GDP. When both components are combined, we estimate a net positive effect on GDP that results from the fuel savings. The combined policy reduces the incidence of premature mortality due to acute pollution exposures by a conservatively estimated 12,400 cases a year. If we use a less certain estimate of mortality due to chronic exposures to fine particles, the reduction is 73,900 cases a year.

We contrast these effects with the effects of a tax of 100 yuan per ton of carbon, about US$4 per ton of CO_2 at 2010 exchange rates. The carbon tax would encourage higher energy efficiency and a reduction in coal use by 14%. Moreover, the reduction in the use of all fossil fuels leads to a substantial improvement in air quality, reducing the incidence of mortality from chronic exposures to fine particles by 103,000 cases a year, even higher than the 11th FYP SO_2 policies. A sizable net national reduction of ozone from the carbon tax provides modest additional benefits to public health, and also to the productivity of major grain crops. Such a tax on carbon reduces annual emissions of CO_2 by 12% compared to the base case, a substantial global benefit in reducing the risks of severe climate change from what otherwise would be incurred.

In the next section we describe the economic model. In section 9.3 we discuss how we represented the SO_2 policy in our simulation by incorporating the costs of using the required desulfurization equipment and costs of replacing small, inefficient power plants. In section 9.4 we present the simulation results of implementing the 11th FYP SO_2 policies, while the carbon tax effects are discussed in section 9.5. Section 9.6 provides a comparison of the two policies: the carbon tax and the 11th FYP SO_2 controls. Finally, section 9.7 summarizes our conclusions.

9.2   The Economic-Environmental Model

In order to comprehensively analyze the costs of environmental policies, one needs a model of the economy that can trace the interactions among the various sectors. For example, not only does a policy requiring FGD equipment impose higher costs on electricity generation, but also the higher price of electricity brought on by this requirement will have further ramifications throughout the economy. The industries that use electricity intensively will face higher costs and need to charge higher prices to cover them. Such widespread, but uneven, changes in prices will affect consumption and investment.

In this section we describe our model of economic growth that is designed to capture these interindustry interactions. We provide enough detail here to give a good sense of the key drivers of growth and energy use in the model. Readers who are not interested in the technical details may skip this section and go to the results starting in section 9.3. The detailed descriptions of the equations of the model and the data construction are given in appendix A.

Our economic model is a standard, multisector model for one country that is modified to recognize the two-tier plan-market nature of the Chinese economy. That is, while the economy is much less regulated than 30 years ago, the government still directly controls the prices and allocation of inputs and output in some sectors, and this planned component is explicitly recognized in the model.

This model of economic growth is "dynamic recursive," or of the Solow type, which means that certain simple rules determine savings and investment.[2] The key feature is that the "general equilibrium" effects of policies are taken into account in this model. This is unlike "partial equilibrium" analysis, which does not consider feedback effects. In the desulfurization example just given, not only does more expensive electricity result in more expensive inputs for steelmaking, but also higher prices for steel dampen its demand and lower its output, in turn reducing demand for steelmaking inputs such as coal and capital. The resulting changes in coal and capital prices in turn affect the costs of generating electricity. Our general equilibrium model takes into account all these effects.

This version of our model is based on the Chinese input-output table for 2005, the appropriate base year for analyzing the 11th FYP, where 33 production sectors and commodities are identified, including six energy industries. Sector output, value added, and energy use are given in tables 9.1 and 9.2. The largest sector in terms of employment, and until very recently also in terms of gross output, is agriculture, but the largest output now comes from construction, at 4.26 trillion yuan in 2005. The largest user of coal is electricity, consuming close to 50% of the total. The largest emitter of total suspended particulates (TSP) from both combustion and noncombustion sources, is by far the nonmetal mineral products sector (which includes production of cement, glass, and other materials). The estimates of TSP and SO_2 are from chapter 6, and we note again here that our estimates are bigger than the official estimates (MEP 2009) that are shown in figure 1.3. While the finer particulates are the ones most important for health, we report TSP here to allow comparison with the emission data regularly published by the Ministry of Environmental Protection (MEP); PM_{10} by industrial sectors is not estimated officially.

The household sector maximizes a utility function that has all 33 commodities as arguments. The demand for consumption goods is allowed to change over time to represent the "income effect": the share of total expenditures allocated to income-inelastic goods, such as food, falls as income rises, while the share allocated to income-elastic goods, notably services, rises. Income is derived from supplying labor, private capital, and land, and is supplemented by transfers from the government. Unlike most models of the developed market economies where households are assumed to control all the capital and take all the capital income, our model allows for a distinct role for enterprises to reflect the large role of the state in China. Households receive a portion of the capital income, and the remainder is kept by the enterprises as retained earnings.

As in the original Solow model, the private savings rate is set exogenously.[3] Total national savings is composed of private household savings, retained earnings of enterprises, and allocations from the central plan. These savings are used to finance domestic investment, the government deficit, and the current account surplus. The investment in period t increases the stock of capital that is used for production in future periods.

Table 9.1

Output, employment, and fuel use, 2005

Sector	Gross Output (billion yuan)	Employment (million)	Coal Use (million tons)	Oil Use (million tons)	Gas Use (million cubic meters)
1 Agriculture	3936	440.5	38.5	7.2	0.0
2 Coal mining and processing	792	4.6	66.7	2.8	0.0
3 Crude petroleum mining	567	0.6	11.1	5.5	2.0
4 Natural gas mining	36	0.0	0.1	0.3	247.0
5 Nonenergy mining	551	3.3	8.2	8.0	16.3
6 Food products, tobacco	2588	10.7	23.4	1.4	87.6
7 Textile goods	1586	10.9	23.3	1.8	211.6
8 Apparel, leather	1222	14.1	3.5	0.7	0.0
9 Sawmills and furniture	602	6.3	14.5	1.2	0.0
10 Paper products, printing	1085	4.3	20.6	1.8	30.3
11 Petroleum refining and coking	1262	0.8	26.9	17.3	0.0
12 Chemical	2872	7.0	116.9	52.3	12,900.6
13 Nonmetal mineral products	2667	9.2	178.6	12.6	2891.1
14 Metals smelting and pressing	3143	3.9	219.9	21.1	1803.7
15 Metal products	1063	6.0	9.9	2.6	361.9
16 Machinery and equipment	2510	9.6	42.6	4.9	1189.1
17 Transport equipment	1757	6.9	18.7	2.0	1160.6
18 Electrical machinery	1657	9.8	7.1	2.3	214.2
19 Electronic and telecom equipment	2805	3.4	4.2	1.8	212.2
20 Instruments	360	1.4	0.7	0.2	1.2
21 Other manufacturing	497	8.2	12.9	0.9	0.1
22 Electricity, steam, hot water	1845	5.2	1052.0	19.5	14,337.8
23 Gas production and supply	74	0.4	10.5	3.2	97.9
24 Construction	4256	25.6	11.1	13.2	0.0
25 Transportation	2446	20.5	23.0	74.7	58.7
26 Communications	1060	2.3	5.7	0.4	0.0
27 Trade	2909	48.5	9.5	7.6	0.0
28 Accommodation and food	1028	12.5	11.8	1.0	1333.9
29 Finance and insurance	1026	4.2	1.4	0.9	0.0
30 Real estate	1025	3.3	9.5	0.5	6.8
31 Business services	1820	10.3	22.4	5.2	417.2
32 Other services	2873	44.3	74.5	3.7	185.6
33 Public administration	1281	19.7	10.8	2.6	28.6
Households			47.6	4.2	3284.1
Total		758.3	2137.8	285.3	41,079.8

Note: Fuel use is combustion, excluding the transformation to secondary fuels and products.

Sources: 2005 input-output table (NBS 2008), authors' calculations.

Table 9.2
Emissions in 2005

Sector		Gross Output (billion yuan)	TSP Emissions (kilotons)	SO₂ Emissions (kilotons)
1	Agriculture	3936	1014	73
2	Coal mining and processing	792	256	296
3	Crude petroleum mining	567	11	44
4	Natural gas mining	36	0	1
5	Nonenergy mining	551	195	238
6	Food products, tobacco	2588	274	519
7	Textile goods	1586	94	416
8	Apparel, leather	1222	23	51
9	Sawmills and furniture	602	54	72
10	Paper products, printing	1085	190	613
11	Petroleum refining and coking	1262	1584	996
12	Chemical	2872	595	1982
13	Nonmetal mineral products	2667	17,498	1948
14	Metals smelting and pressing	3143	1572	1694
15	Metal products	1063	20	36
16	Machinery and equipment	2510	72	124
17	Transport equipment	1757	44	58
18	Electrical machinery	1657	13	38
19	Electronic and telecom equipment	2805	8	24
20	Instruments	360	3	18
21	Other manufacturing	497	133	378
22	Electricity, steam, hot water	1845	2792	16,241
23	Gas production and supply	74	11	26
24	Construction	4256	478	308
25	Transportation	2446	461	545
26	Communications	1060	133	70
27	Trade	2909	225	154
28	Accommodation and food	1028	276	146
29	Finance and insurance	1026	33	21
30	Real estate	1025	224	116
31	Business services	1820	529	296
32	Other services	2873	1765	918
33	Public administration	1281	254	143
	Households		1509	835
	Total	55,204	32,342	29,439

Work hours are supplied inelastically by households, and labor is assumed to be mobile across sectors.[4] Effective labor is the work hours multiplied by a quality, or composition, index (see appendix A, equation A.7). The capital stock, as noted, is partly owned by households and partly owned by the government. The plan part of the stock is immobile in any given period, while the market part responds to relative returns (further details are in appendix A.1.4). We assume that the capital stock in the first period corresponds to the plan component and any further investment accrues to the market component. Over time, the plan capital is depreciated and the total stock becomes mobile across sectors. An exception to this capital mobility assumption is the electric utilities sector, which will be described further under the heading "Special treatment of the electric power industry."

The government imposes taxes on value added, capital income, sales, and imports, and also derives revenue from a number of miscellaneous fees. On the expenditure side, it buys commodities, makes transfers to households, pays for plan investment, makes interest payments on the public debt, and provides various subsidies. The government deficit is set exogenously and projected for the duration of the simulation period. This exogenous target is met by making government spending on goods endogenous.

Finally, the rest of the world supplies imports and demands exports. World relative prices are set by the data in the last year of the sample period for all commodities except for oil, where we use the forecast of the U.S. Energy Information Agency (U.S. EIA 2009, table 112). The current account balance consists of exports minus imports plus net factor receipts and transfers, and is set exogenously in this one-country model. An endogenous variable for terms of trade clears the current account equation (A.30).

On the production side, each of the 33 producers uses capital, labor, and intermediate goods to produce output, and a cost function with constant returns to scale is used to determine the choice of inputs. The production technology for industry output is allowed to change over time; that is, productivity growth is allowed, and there is "biased technical change." The rate of productivity growth is projected exogenously, initially at a rate equal to the historical average and then declining according to the $g(t)$ term in equation (A.3) in appendix A. Biased technical change refers to changes in input requirements per unit output that are not explained by changes in price. For example, in the United States we see that the use of workers with information technology skills rises even as their wages rise.[5] In particular, we allow energy requirements per unit input to decline over time. This is often called the autonomous energy efficiency improvement (AEEI).

In modeling the energy sector, there are separate industries for coal mining, crude petroleum, natural gas mining, petroleum refining, electric power, and gas supply (including coal gas). Energy from nonfossil fuel sources, including hydropower and nuclear power, are included as part of the electric power sector.

There are 33 markets for the commodities; that is, there are 33 endogenously determined prices that equate supply with demand for the domestic commodities identified in the model. For the few sectors such as coal mining and agriculture that have a plan component (in which the price and quantity of a portion of output is regulated by the government), we allow for two types of prices: a plan price and a market price. We assume that the outputs and inputs corresponding to the plan allocation are only a portion of each firm's total output and input that is exchanged at the plan price; that is, they are inframarginal. With this assumption, the profit-maximizing conditions work in the usual way; inputs are bought until the marginal product equals the market price for the input. This topic is dealt with in more detail in equation (A.2).

The total supply of each commodity consists of domestically produced goods and imported varieties, and the endogenous terms of trade (*e* in equation A.28) clears the international market. There are three markets for the factors of production—land, capital, and labor—and three distinct prices to clear them. Finally, the government budget constraint is met by the endogenous level of government purchases.

The main exogenous variables driving growth in this model are total population, working age population, labor force participation, total factor productivity growth, growth in capital and labor quality, saving rates, dividend payout rates, government deficits, and current account balances. The construction of these exogenous projections and other parameters of the model is described in section A.2. The main source of information for the model is a Social Accounting Matrix (SAM) based on the input-output table for 2005.[6] The summary SAM is given in figure A.1.

The columns of the summary SAM are the expenditures of each agent, and the rows are the receipts (income). For example, the row for households shows that they received 6986 billion yuan in labor income, 787 billion in land rental income, 3058 billion in dividends, 418 billion worth of net transfers and interest from the government, and 21 billion as transfers from the rest of the world. The column for households records 7122 billion of consumption expenditures, 209 billion of fees and taxes to the government, and 3939 billion of personal savings.

These flows of payments combined with the data for number of workers and capital stocks give the wage rate and rental cost of capital. Combined with the industry output and value added, they also give the various tax rates. The tax rates are summarized in table A.2.

Special Treatment of the Electric Power Industry   Given the major contribution of the electric power industry to emissions and the fact that it is directly controlled by government agencies where the prices are not set by ordinary market forces, we need to represent it in a distinct way that recognizes this unusual feature. In particular we need to represent the policy of shutting down small units. This is a nonmarket intervention made by the central government and is thus not as straightforwardly simulated as most other policies using general equilibrium models, including carbon taxes.

Furthermore, our representation of the national power sector is a highly abstract simplification of the actual economy of many different types of power-generating technology, including small (higher cost) thermal plants, large (lower cost) thermal plants, hydroelectric facilities, and nuclear power plants. In addition, we combine the generation and transmission activities into one sector. In a textbook abstraction of a single market, only the lowest-cost technology would exist in the market. The real-world coexistence of different technologies and segmented markets are in part due to the uneven demand for power over the course of the day and over the year, proximity to coal or hydro resources, and the long-distance transmission costs. Certain technologies only exist because of implicit or explicit government subsidies, and some of the small inefficient plants would not be viable without such subsidies. (They can survive in isolated markets where alternative sources face high transmission costs.)

We therefore represent the power sector differently from other sectors in the model for the technology-forcing policy case in which retirement of small plants is mandated. The demand for mobile capital in the other sectors is endogenous, depending on the market price of aggregate capital; for the power sector the capital stock is immobile and set exogenously. Because of this different treatment of capital, we derive an endogenous industry-specific rate of return to capital in the power sector, which differs from the economy-wide rate of return.

It should also be noted that since transmission is included in this sector, the energy cost share of the delivered price is much lower than the cost share of the generators. In 2005, the total energy cost share is 37%, with a coal cost share of 23%, both measured at producer prices in the conventions of the input-output tables.

9.2.1   Base Case Projection

The base case projection of the economy consists of the simulation from year 1 to year T given the initial stocks of debt, capital, and the labor force. The time path of the economy (i.e., the evolution of all the endogenous variables) is determined essentially by the assumed path of the exogenous variables noted earlier. The main aim of the model is to study the impact of policies or, more specifically, to estimate the percentage changes in variables of interest between a policy simulation and the base case. The values of the variables in the base case itself—for example, the growth rate of GDP or emissions—are not our primary interest here; most of the percentage changes resulting from a policy are affected in only a minor fashion by the underlying levels in the base case, such as the population. Here we briefly document the main outcomes of the base case projection for reference, to show how our assumptions of the exogenous variables compare to other projections.

The exogenous variables such as tax rates, energy coefficients, and emission coefficients (as determined in chapters 4–6) are set for each projection year given our assessment of preexisting policies under the base case. These variables are not held fixed at the base year values. For example, we assume an existing policy of requiring FGD equipment in new power plants in the base case, and thus we have a rising desulfurization rate over time in the base case. Then in the 11th FYP policy case we add the retrofit requirement, giving an even higher rate of desulfurization. There is also a continual reduction in particulate emission per unit of output in the base case as envisioned by existing policies.

Given the initial stock of capital and the labor force, we solve for the three factor prices and the 33 commodity prices that clear the markets in the first period. This gives us all the necessary quantities for the first period, including investment, which augments the stock of capital for use in the next period. The solution process is repeated for each period in the simulation horizon. The results for the main variables of interest are given in table 9.3. The path of GDP and total energy from fossil fuel use is also plotted in figure 9.1.

While the model is used here to simulate policies out to 2010 only, we describe the behavior of the model over a longer horizon to allow readers who are familiar with such modeling exercises to compare our projections with those from other models. We project GDP growth at 7.6% per year over the 25 years beginning in 2005, and at 9.6% annually from 2005 to 2015. This projection is in the middle of the range estimated by Holz (2006, table 5) and similar to the projections by the Development Research Center of the State Council (Li 2009). It is somewhat higher than the 6.1% for 2005–2030 projected in IEA (2008) but similar to its 10-year horizon. During this 2005–2030 period, we project effective labor supply (recall that this is the hours worked adjusted for composition of the work force) to grow at 0.84% per year, about twice the growth rate of population.

Coal use is projected to grow at only 3.4% per year, less than half the growth rate of GDP, while natural gas use grows at 7.1%, close to GDP growth. Electricity

Table 9.3

Base case projection

Variable	2005	2006	2010	2030	25-Year Growth Rate
Population (million)	1312	1320	1354	1462	0.43%
Effective labor supply (billion yuan)	6986	7168	7806	8757	0.84%
GDP (billion 2005 yuan)	19,311	21,189	30,933	121,938	7.6%
Consumption/GDP	0.38	0.36	0.41	0.64	2.4%
Fossil energy use (million tons of standard coal equivalent)	2087	2158	2764	5188	3.7%
Coal use (million tons)	2159	2267	2799	4952	3.4%
Oil use (million tons)	329	317	449	863	3.9%
Gas use (billion cubic meters)	56	63	89	312	7.2%
Electricity use (TWh)	2200	2372	3217	9973	6.2%
CO_2 emissions (fossil fuel combustion, million tons)	5516	5718	7255	13,373	3.6%
PM_{10} (million tons)	18.05		17.67		
SO_2 emissions (million tons)	29.43		32.47		
NO_X emissions (million tons)	18.76		24.45		

Note: There is 0.714 ton of standard coal equivalent (sce) in 1 ton of coal.

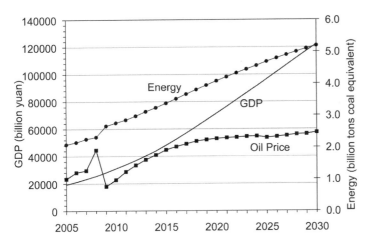

Figure 9.1

Base case projection of GDP and energy consumption. Oil price is world price projected by IEA; it is normalized to 1.0 in 2005 (right scale).

growth is also very rapid at 6.2% per year. Total fossil energy consumption (i.e., the sum of the standard coal equivalents of the three fossil fuels) grows at 3.7% per year. As a result, energy intensity is projected to continue to decline at a rapid rate exceeding 3% per year.

The growth of energy consumption is close to projections by the International Energy Agency over the same 25-year horizon (IEA 2007). However, given our higher GDP projection, this means that our rate of energy conservation is somewhat more optimistic. Given the rate of coal use, carbon emissions are projected at 3.6% over 2006–2030 and at 4.9% over the first 10 years.

The composition of GDP is projected to change substantially with the fall in savings rate. The consumption share of GDP is projected to rise from the current 38% to 64% by 2030. The Chinese export share of GDP, at about 40% today, is unusually high for a large country, and that is projected to fall substantially to 25% by 2030.

9.3   The Economics of the 11th Five-Year Plan Energy and SO$_2$ Policies

In this section we describe how we parameterize and simulate the 11th FYP target to shut down the small power plants and to require the use of FGD equipment.[7]

9.3.1   Small Power Plant Shutdown Policy

At the end of 2005, almost one third of China's thermal power capacity was in small-scale generation units, where small scale is defined as capacities less than 100 megawatts (MW).[8] Most of these small-scale units were coal-fired, but some were oil and diesel units serving localities that had in the past experienced severe electricity shortages. These small units were generally inefficient in their use of energy and were also highly polluting. Table 9.4 gives the cost estimates made by the Energy Research Institute of the National Development and Reform Commission, as reported in JES (2007). The average total cost (operating cost, fuel cost, and capital cost) per kilowatt-hour (kWh) for small plants was more than twice the cost for the average of large and medium plants (and almost three times the cost for large plants), mostly because of the higher fuel requirements per kilowatt-hour of electricity of smaller plants, with diesel-fired plants being particularly inefficient and costly.

Under the plan, 50 gigawatts (GW) of small-scale power plant capacity were targeted for closure by the end of 2010, a target that was ultimately exceeded by an additional 9 GW, as discussed in chapter 4. Implementing this shutdown policy also required that replacement capacity be built. Since this policy was implemented

Table 9.4
Cost structure for thermal power plants, 2005 (yuan per kWh)

Costs	Large and Medium Plants	Small Plants		
		Averaged	Coal Fired	Diesel Fired
Average total costs (including capital cost)	0.286	0.704		
Operating and maintenance costs	0.057	0.068		
Fuel costs	0.189	0.596	0.23	2.52

Source: JES (2007).

gradually over five years, since the retired individual units were proportionately small and widespread geographically, and since electricity connected to the grid is fungible, the actual cost of this replacement capacity can be assumed to be an average for all new capacity installed over the plan period. Thus the direct cost of the shutdown policy would be equal to the cost of producing the replacement electricity less the fuel, operating, and maintenance costs that would have been incurred by operating the small units, plus decommissioning costs.[9] The decommissioning costs could include the shutdown of the small plants themselves and perhaps the retraining and relocating of displaced workers. The value of any scrap materials and the land on which plants were located should be accounted for as negative costs. The estimation of such costs for the heterogeneous units is difficult, and JES (2007) reports some limited effort to make such estimates, which are included here.

According to the plan for the 11th FYP SO_2 controls when drafted in 2005, approximately 50 GW of new power generation capacity would be installed per year from 2006 to 2010, compared to the 10 GW of small thermal power units to be shut down each year. Figure 4.1 reports the actual capacity available during 2005–2010. This information is used to give an updated projection for 2010 of total capacity, total thermal electric power output (in billion kilowatt-hour), and the contribution of the small plants. These revised projections are given in table 9.5.

Given the higher fuel requirements of the small, inefficient plants, this change results in a lower cost share for fuel inputs and a lower average cost of electricity. We represent this reduction in inputs of coal and oil (per kilowatt-hour of electricity) by reducing the energy intensity parameter and shifting the power sector cost function down.[10] Table 9.4 shows that the average unit cost for small plants is 0.704 yuan per kilowatt-hour compared to the 0.286 yuan per kilowatt-hour average for

Table 9.5
The economics of small power plant shutdown policy

Year	Total Electricity Output (billion kWh)	Thermal Electricity Output (billion kWh)	Small-Plant Output (billion kWh)	Original Fuel Cost Share (percent)	Fuel Cost Share after Shutdown (percent)	Reduction in Cost per kWh (percent)
2005	2544	2047	400	37.0	37.0	
2006	2827	2374	350	36.3	36.2	−2.3
2007	3110	2698	300	35.7	35.4	−4.1
2008	3394	2862	250	35.0	34.6	−5.9
2009	3677	3026	200	34.4	33.9	−7.4
2010	3960	3190	150	33.8	33.2	−8.8

Source: Cao, Garbaccio, and Ho (2009), updated with information from chapter 4.

the larger plants. Given the small contribution of the small plants, this big gap in costs result in modest changes in fuel cost shares and unit costs, with the fuel cost share falling to 33.2% in 2010 compared to 33.8% in the base case. The average unit cost is estimated to fall by 8.8% in 2010 compared to the base case where there are no plant shutdowns.

The power sector is unusually complicated compared to the other industries; some of the features have been noted in section 9.2. Another feature is the uneven load demand over the course of the day and over the different seasons. The system has plants that are idle or operating below capacity much of the time. In some countries a market system leads to "base load plants" which operate all the time and "peaking plants" which operate only during the peak demand hours. The peaking plants are normally those with the lower capital costs but higher fuel costs such as gas-fired plants. Such a system would have made our average cost calculations, comparing small plants with large ones, more complicated. In China, however, the centralized system used a "San Gong" power dispatch model during the 11th FYP period. This dispatch model refers to the three principles of equity, fairness, and transparency—in effect having all plants share the total demand proportionally instead of trying to minimize total system costs.[11] The preceding simple cost comparisons are more consistent with such a dispatch system than they would be if China followed a Western-style least-cost market dispatch model.

9.3.2 FGD Installation Policy

At the end of 2005, flue-gas desulfurization equipment had been installed on 46.2 GW of the total 391 GW of thermal power capacity. In order to meet the SO_2

reduction target of the plan, another 167 GW of FGD equipment was scheduled to be installed on existing power generation units by 2010.[12] All new power generation units were mandated to have FGD equipment, and JES (2007) projected 250 GW of new capacity during the plan period. According to the original 11th FYP target, it was expected that by 2010, FGD would be installed on almost 85% of total coal-fired capacity. Section 4.2.2 gives data on the actual construction of power units during this period and puts the total installed coal-fired capacity in 2010 at 651 GW, estimating the actual installation of FGD equipment to have reached 80%. This estimate is remarkably close to the projected rate given the upheavals in the world economy during 2008–2010.

The requirement that all new plants must have FGD equipment was also in the 10th FYP, and so it is included in the base case. The difference between the base case and the policy case is the retrofit requirement.

The costs of the FGD installation policy can be divided into two types: direct and economy-wide. The direct costs of the FGD policy include the capital costs of the FGD equipment and operation and maintenance costs, which include additional electricity for the operation of the equipment and thus an increase in fuel inputs. The economy-wide costs are the indirect costs on different sectors in the rest of the economy, which are discussed in the next section.

Capital costs for FGD units manufactured in China have fallen by more than half since the 1990s. These costs now range from 150 yuan/kW for a 600 MW plant to 180 yuan/kW for a 100 MW plant (CAEP 2007). As the cost of constructing a 600 MW plant without FGD is approximately 4000 yuan/kW, the addition of FGD equipment represents about a 3.8% increase in capital costs. The unit operating cost of the FGD equipment (per ton of SO_2 removed) depends on the size of the plant and sulfur content of the coal used, and ranges from 1244 yuan per ton of SO_2 for a 100-MW plant to 800 yuan per ton for a 1000-MW plant using coal with a sulfur content of 1% (section 4.3.2 notes that the average sulfur content for coal used in the power sector is 1%, see also figure 4.2).

Low-sulfur coal raises the cost per ton removed, from 1020 yuan per ton average for all plants using 1% sulfur coal to 1840 yuan per ton average for 0.5% sulfur coal. The Chinese Academy for Environmental Planning also reported that coal with a sulfur content of less than 0.5% makes up 30% of coal combusted in the power sector, and coal with a sulfur content of 0.5–1% composes another 35% (CAEP 2007). Averaging over plant sizes and coal types, it estimated that running FGD equipment raises operating costs by 2.4%. In terms of the price of delivered electricity, which includes transmission costs, the additional cost of running FGD equipment is only 1.5%.

In 2006, 25.3% of total electricity output (in kilowatt-hours) was produced by generation units equipped with FGD, as shown in table 9.6. In keeping with the projected level of capacity and our estimate of total output, the amount of electricity produced by units with FGD installed and operating should have increased to 64.6% in 2010. The additional cost imposed by the 11th FYP is the cost of operating the retrofitted plants; recall that the requirement that new plants have FGD equipment is in both the base and policy cases. Since operating an FGD unit raises the delivered electricity cost by 1.5%, we estimate that the average cost of all electricity generated rose by approximately 0.03% in 2006, as shown in the second-to-last column of table 9.6. (The share of FGD electricity is 14.0% in the base case and 16.2% in the policy case, hence $0.0003 = 0.015 \times [0.162 - 0.140]$). By 2010 the cost shock is more noticeable, but still small, 0.35% ($1.5\% \times [0.76 - 0.52]$). We represent this change as an upward shift of the cost function, which is equivalent to a negative productivity shock.[13] That is, the installation and operation of the FGD equipment increased the inputs (capital, labor, and energy) required to generate the same amount of electricity.

9.4   The Costs and Benefits of the 11th Five-Year Plan SO$_2$ Policies

9.4.1   Methodology for the Analyses

The impacts of the 11th FYPlan SO$_2$ policies on the rest of the economy are estimated by simulating the economic model with the small unit shutdown and FGD policies and comparing the results to the "business-as-usual" (BAU), or base case described in section 9.2. As described in chapter 4, the BAU scenario includes existing environmental policies but not the specific additional policies targeting SO$_2$ of the 11th FYP. It assumes that the FGD units already installed in 2005 continue to operate and that FGD would be required on newly constructed plants, but that no retrofitting of FGD on existing plants occurred. It also assumes that no shutdown of small power plants was mandated. We report three comparisons to produce a complete accounting of the economic effects of the 11th FYP: shutdown policy only, FGD policy only, and both policies combined (note that the combined case is the only one analyzed in the assessment of benefits in chapters 4–8). At the end of this section, we discuss how the cost estimates of the policies described in the previous section are implemented in the simulations.

9.4.2   Impacts of the Small Power Plant Shutdown Policy

As we discussed in section 9.2, the higher-cost small generation units exist in part because of implicit and explicit subsidies from the government. In our simulation,

Table 9.6
Economics of FGD policy

Year	Projected in Chapter 4						Reported
	Total Electricity Output (billion kWh)	Thermal Electricity Output (billion kWh)	FGD Capacity (GW)	Total FGD Electricity Output (billion kWh)	FGD Electricity as Share of Total Output (percent)	Base Case (no retrofit): FGD Electricity (billion kWh)	Increase in Average Cost (percent)	FGD Capacity (GW)
2005	2544	2047	46	715	25.3	582	0.03	
2006	2827	2374	144	1186	38.1	922	0.11	
2007	3110	2698	242	1600	47.1	1218	0.19	
2008	3394	2862	340	1989	54.1	1497	0.27	
2009	3677	3026	438	2557	64.6	1762	0.35	
2010	3960	3190	536					560

Note: "FGD electricity" refers to power generated by plants with FGD equipment installed.

we recognize the reduction of 8.8% in unit costs by 2010 given in table 9.5, and at the same time assume a corresponding reduction in subsidies, leaving the price of electricity unchanged. In the policy simulation we hold all other government expenditures at the same level as in the base case. This is a common approach in such analysis to allow a straightforward welfare calculation. Social welfare depends on private goods such as food and public goods such as roads; the method of holding the supply of public goods constant allows us to focus on changes in private consumption to estimate the net welfare effect of the policy.

The reduction in total government expenditure due to the reduction in subsidies is recycled as reductions in taxes. A large number of studies have shown that the form of revenue recycling can influence the net cost of a policy (e.g., Bovenberg 1999). In our simulation, we reduce all tax rates proportionately based on the savings from the subsidy removal. Because much of the tax revenue in the base year is derived from taxes on enterprises, as opposed to income taxes (as is the case in the United States), the main beneficiaries of the tax reduction are enterprises. Since enterprises use retained earnings to finance some of their investment, the tax reduction leads to an increase in investment.

The effects of the policies on energy prices and consumption are reported in table 9.7 for all years; the effects on the main macro variables are given in table 9.8; and the effects on all sectors in 2010 are given in table 9.9. As noted, the electricity prices are unchanged in the shutdown-only case, with the withdrawal of subsidies offsetting the lower coal and oil consumption. That fall in the consumption of coal is considerable under this policy; by 2010 it is 5.5% lower than the base case. The price of coal is lower as a result of this smaller demand; by 2010 the price is 1.06% lower.

The shutdown policy results in reduced required inputs for the production of the average kilowatt-hour of electricity, and this reduction is equivalent to a small positive productivity shock to the economy. The effects on the main economic aggregates for this and the other two 11th FYP scenarios (FGD and combined) are given in table 9.8. In the shutdown case, aggregate GDP rises slightly in each year compared to the base case, which in turn results in higher investment. By the end of the Five-Year Plan period in 2010, the combined change in productivity and the larger capital stock results in an increase in GDP of 0.77% from the baseline. Household consumption rises by 0.50% and total investment by 1.1%. As discussed before, we assume that government expenditure is held constant. Since the effect of the tax reduction is larger for enterprises than for households, the percentage rise in investment is greater than the rise in consumption. This difference shifts the overall

Table 9.7
The effects of 11th FYP policies on energy use and prices (percent change from base case)

Year	Electricity Use	Electricity Price	Coal Use	Coal Price	Oil Use	Oil Price
Shutdown Policy						
2006	−0.29	0.00	−1.15	−0.36	−0.10	−0.19
2007	−0.54	0.00	−2.31	−0.65	−0.18	−0.40
2008	−0.76	0.00	−3.44	−0.85	−0.27	−0.55
2009	−0.92	0.00	−4.62	−0.92	−0.48	−0.60
2010	−1.05	0.00	−5.52	−1.06	−0.54	−0.74
FGD Policy						
2006	−0.04	0.04	−0.01	0.01	0.00	0.00
2007	−0.13	0.13	−0.02	0.02	−0.01	0.01
2008	−0.21	0.21	−0.03	0.03	−0.02	0.02
2009	−0.30	0.30	−0.04	0.04	−0.02	0.02
2010	−0.40	0.40	−0.05	0.05	−0.02	0.03
Combined Shutdown and FGD Policy						
2006	−0.32	0.04	−1.15	−0.35	−0.11	−0.19
2007	−0.68	0.13	−2.33	−0.63	−0.19	−0.39
2008	−0.98	0.22	−3.48	−0.82	−0.28	−0.54
2009	−1.22	0.30	−4.67	−0.88	−0.50	−0.58
2010	−1.44	0.40	−5.57	−1.01	−0.56	−0.71

composition of output slightly, with higher growth in the construction and nonmetallic mineral products (cement industries, for example) than in the services sector.

The small-plant shutdown reduces the use of diesel, but total oil use declines only slightly because part of the reduction in oil use by the electricity sector is offset by a small increase in consumption in other sectors, such as transportation. With the reduction in coal and oil use due to the small unit shutdown, CO_2 emissions fall by 4.4% versus the base case.

It is natural to ask why this policy was not implemented earlier because it has positive effects on both the environment and the economy. One answer is that the costs and benefits of the policy are distributed to different segments of the population. As noted earlier, in many cases small power plants were built in areas that were underserved by the main electrical grid and in response to past energy shortages. Closing them would also have negative impacts on local employment.

Table 9.8
Effects of 11th FYP SO_2 policies in 2010

Variable	Base Case in 2010	11th FYP SO_2 Effect (percent change)		
		Shutdown Policy	FGD Policy	Combined Shutdown and FGD Policies
GDP (billion yuan 2005)	30,933	0.77	−0.03	0.74
Consumption (billion yuan 2005)	11,883	0.50	−0.03	0.47
Investment (billion yuan 2005)	12,460	1.10	−0.03	1.07
Government consumption (billion yuan 2005)	3,624	0.05	−0.02	0.05
Fossil energy use (million tons of standard coal equivalent)	2,761	−4.24	−0.05	−4.29
Coal use (million tons)	2,799	−5.52	−0.05	−5.57
Oil use (million tons)	449	−0.54	−0.02	−0.56
Gas use (billion cubic meters)	89	−3.42	−0.06	−3.48
Electricity (billion kWh)	3,217	−1.05	−0.40	−1.45
CO_2 emissions (fossil fuel; million tons)	7,255			−4.6
PM_{10} emissions (1000 tons)	17,671			−3.6
SO_2 emissions (1000 tons)	32,472			−19.8
NO_X (1000 tons)	24,446			−1.3

Health and Crop Benefits of Combined Shutdown and FGD Policies of 11th FYP	Cases	Value (billion yuan)
$PM_{2.5}$		
Avoided acute mortality	12,300	6.35
Avoided chronic mortality	73,900	38.28
Other health benefits		1.57
Ozone		
Avoided acute mortality	120	0.06
Other health benefits		0.01
Crop benefits		0.00
Total health benefits (including acute mortality)		8.00
Total health benefits (including chronic mortality)		39.92

	Percent
Total health benefits (chronic mortality) / GDP	0.13
Total crop benefits / GDP	0.00

Table 9.9
Effects of 11th FYP policies on industry output and prices (percent change in 2010)

Sector	Changes in Output			Changes in Price		
	Combined	FGD	Shutdown	Combined	FGD	Shutdown
Agriculture	0.29	−0.01	0.31	−0.16	0.01	−0.17
Coal mining	−5.26	−0.05	−5.21	−1.01	0.05	−1.06
Oil mining	−0.15	−0.04	−0.11	−0.51	0.03	−0.53
Gas mining	−3.33	−0.06	−3.27	−0.91	0.06	−0.97
Nonenergy mining	0.62	−0.06	0.69	−0.47	0.04	−0.52
Food manufacturing	0.75	−0.02	0.78	−0.68	0.02	−0.70
Textile	0.44	−0.03	0.47	−0.49	0.03	−0.52
Apparel and leather	0.46	−0.02	0.48	−0.56	0.02	−0.58
Lumber	0.56	−0.03	0.59	−0.51	0.03	−0.55
Paper	0.45	−0.03	0.48	−0.53	0.03	−0.56
Petroleum refining and coal	−0.31	−0.03	−0.28	−0.71	0.03	−0.74
Chemicals	0.49	−0.05	0.55	−0.46	0.04	−0.51
Nonmetal mineral products	0.60	−0.05	0.66	−0.56	0.05	−0.60
Primary metals	0.52	−0.05	0.58	−0.54	0.05	−0.59
Fabricated metals	0.41	−0.05	0.46	−0.50	0.04	−0.54
Machinery	0.53	−0.04	0.57	−0.53	0.03	−0.57
Transportation equipment	0.75	−0.03	0.79	−0.62	0.03	−0.65
Electrical machinery	0.13	−0.03	0.17	−0.54	0.03	−0.58
Electronic and telecom equipment	0.48	−0.03	0.51	−0.39	0.02	−0.42
Instruments	0.12	−0.03	0.15	−0.35	0.02	−0.37
Other manufacturing	0.51	−0.02	0.54	−0.48	0.02	−0.50
Electric utilities	−1.44	−0.40	−1.05	0.40	0.40	0.00
Gas utilities	0.49	−0.04	0.53	−0.71	0.04	−0.76
Construction	0.99	−0.03	1.03	−0.51	0.03	−0.55
Transportation	0.14	−0.02	0.16	−0.61	0.02	−0.63
Communications	0.47	−0.03	0.49	−0.53	0.02	−0.56
Commerce	0.48	−0.02	0.51	−0.67	0.02	−0.70
Accommodation and restaurants	0.38	−0.02	0.40	−0.51	0.02	−0.54
Finance and insurance	0.22	−0.02	0.24	−0.63	0.02	−0.64
Real estate	0.28	−0.01	0.30	−0.22	0.01	−0.24
Business services	0.23	−0.02	0.26	−0.48	0.02	−0.50
Other services	0.15	−0.03	0.17	−0.40	0.03	−0.43
Administration	−0.03	−0.02	−0.04	−0.27	0.02	−0.29

Moreover, replacing small plants with larger ones would also require relatively large capital expenditures not easily made by some localities. In many cases there will also be a need for additional transmission lines to new plants after the small ones are shut down. However, both the economic and environmental benefits are more widely spread across the entire population.

9.4.3   Impacts of the FGD Installation Policy

The change in average cost due to the FGD requirement, given in table 9.6 for all years, is simulated and the impacts on energy prices and quantities are given in the second panel of table 9.7. This small increase in costs raises delivered electricity prices by only 0.04% in 2006 and by 0.40% in 2010. Given our unit elasticity assumption, the overall electricity use is reduced by approximately the same percentages. The higher cost of electricity leads to a small decline in the output of electricity-intensive industries such as chemicals, nonmetal mineral products, and primary metals, as shown in table 9.9. The use of FGD also increases the amount of coal required to generate a kilowatt-hour of deliverable electricity. However, this increase is offset by the reduction in the demand for electricity and the reduction in the demand for coal due to the lower output of the electricity-intensive industries. The result is a small, 0.05%, net decline in coal consumption in 2010.

This change is equivalent to a small negative productivity shock and results in a slight decline in GDP versus the base case, with corresponding reductions in the consumption and investment components of GDP (see table 9.8). The lower amount of investment in each period results in a smaller capital stock in the subsequent periods. By the end of the Five-Year Plan period, the smaller capital stock and lower productivity results in GDP being about 0.03% below the baseline. There is also a slight change in the composition of output, with the electricity-intensive sectors declining the most. The output of less electricity-intensive industries, such as agriculture and finance, falls by a smaller amount.

Because it is not electricity-intensive, transportation is only slightly affected by the FGD policy. The net effect of reductions in manufacturing and transportation, and other smaller changes, is a 0.02% decline in oil consumption in 2010 (table 9.7). The effect of the FGD policy on natural gas consumption is small, as most natural gas use is in industry such as chemical manufacturing.

9.4.4   Combined Impacts of the FGD and Shutdown Policies

Our final 11th FYP simulation combines the small unit shutdown and FGD installation policies to generate an estimate of the overall impacts of fully implementing

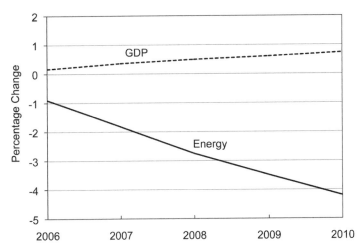

Figure 9.2
Change in GDP and fossil fuel use due to power sector policies (combined shutdown and FGD policies of 11th FYP).

the FYP's SO$_2$ reduction policies. As shown in tables 9.7–9.9, the impacts for 2010 are essentially additive. The effects on GDP and energy use for each year are plotted in figure 9.2. For the combined policy, GDP in the last year of the plan is 0.74% above the baseline, while primary fossil fuel use is 4.3% below it. This outcome is due mainly to the productivity improvement and increase in capital stock resulting from the small unit shutdown, which offsets the slight decline in GDP resulting from the costs of the FGD equipment.

Looking at the industry effects given in table 9.9, we see that the biggest increase in price is for electric utilities (0.4%), while the coal mining price falls by 1.0% and petroleum refining by 0.71%. These are accompanied by large reductions in the output of coal (−5.3%) and electricity (−1.4%) versus the base case, and a smaller reduction in petroleum refining (−0.31%). The higher income due to the small productivity gain leads to output increases in most sectors of about half a percent—for example, food manufacturing (0.7%), real estate (0.3%), and commerce (0.5%).

The 5.6% reduction in coal use is almost entirely due to the shutdown policy—that is, due to a big reduction in coal demand of the power sector that overwhelms a small increase in output of the manufacturing sector. The reductions in oil and gas use are also mostly due to this shutdown effect. Therefore, the reductions in PM$_{10}$ emissions (−3.6%) and carbon emissions (−4.6%) compared to the base case are also due mainly to the closure of inefficient small plants.

However, the reduction in 2010 SO_2 emissions versus the base case of 19.8% is due in large part to the FGD equipment requirements, with the reduction in coal use due to the small-plant closures making a smaller contribution. This reduction in SO_2 emissions translates to an 11.6% decline from 2005 levels, and thereby achieves the 11th FYP 10% reduction target. In addition, NO_X emissions fall by a mere 1.3% versus the base case (but recall that they are rising swiftly over time in both cases), with the reduction in coal and oil use dominating the small increase in NO_X emissions from expanded transportation output.

These emissions under the 11th FYP policies are then used to construct a new set of policy-case emissions in each grid cell as described in detail in section 6.3.4. Figure 6.8 gives the reduction in the emissions of seven chemical species for the major sectors. As noted in chapter 6, the geographical effects from the power sector alone are significant and noticeable; the changes in emissions from the other sectors are trivial in comparison.

As described in chapter 7, the gridded emissions of the combined shutdown-FGD policy of the 11th FYP are then used by the GEOS-Chem atmospheric model to estimate a new set of concentrations, for each location, throughout the year. The results are discussed in detail in chapter 7, and figure 7.5 shows the considerable improvement in air quality. The reduction in concentration of sulfate particles (in the $PM_{2.5}$ class) in some areas of Shandong, Henan, and Hebei is more than 4 micrograms per cubic meter ($\mu g/m^3$) as shown in figure 7.5B, and the reduction in SO_2 concentrations in some areas of those provinces, Shanxi, and Jiangsu exceeds 8 parts per billion by volume (ppbv), as shown in figure 7.5A. These changes should be interpreted with respect to the underlying average concentrations in the base case given in figure 7.4. For example, in October 2010, the estimated monthly average sulfate concentration ranged from 5 to $25\,\mu g/m^3$ in the eastern half of China, and the SO_2 concentration ranged from 8 to 40 ppbv; the reductions in those concentrations of $4\,\mu g/m^3$ and 8 ppbv, respectively, are thus considerable improvements. The changes in ozone concentration are small and more mixed in this policy case; overall there is a tiny net reduction in concentrations throughout the country, but the effects on health and crop productivity are inconsequential, especially compared to $PM_{2.5}$ impacts.

The reductions in fine particle pollution lead to a reduction in health damages that are described in detail in chapter 8. Figure 8.4B shows the regional distribution of the reduction in acute mortality. This distribution is the combination of the effects of the reductions in pollution levels and location of the dense population centers. A region centered on Henan and western Shandong, along with Beijing, Shanghai, and other major cities, exhibits large improvements in air quality combined with dense population, and thus has the largest reductions in health damages. For the country as a whole, table 9.8 reports the health effects due to $PM_{2.5}$ and ozone reductions. Considering only acute mortality from the two pollutants, the policies of the 11th FYP result in an estimated reduction of around 12,400 cases annually, which are valued at 6.4 billion yuan using willingness-to-pay estimates described in appendix B and table B.1.

As discussed in chapter 8, the estimate for avoided premature mortality from chronic exposure to $PM_{2.5}$ is more uncertain because it is based on epidemiological studies in the United States, but if we accept the lowest concentration-response coefficient from that literature as suggested in that chapter, we project an annual reduction of 73,900 cases of premature mortality. Using the valuations for a statistical case of premature mortality and for morbidity endpoints discussed in appendix B, the total yuan value of this health improvement would be 39.9 billion yuan, equivalent to 0.13% of GDP.

We should repeat an earlier caution that these are conservative estimates that might understate the effects, and note that some of the studies reviewed in chapter 1 employ assumptions that would lead to higher estimated health effects. Our conservative estimates are partly a result of our atmospheric modeling methodology discussed in chapter 7. The model uses a grid that is roughly 50 km by 66 km for each cell. Thus, it computes one number that represents the average concentration in each cell. The dispersion of primary particulate matter, however, falls quickly with distance; that is, the concentration within 3 km of the source is much higher than the concentration 20 km away. Moreover, very often the highest population density is closest to the factories that are the emission sources. The concentration of primary PM decreases rapidly with distance and thus a finer grid would have captured larger reductions in areas of high concentrations and estimated a larger health impact. This effect is illustrated in Ho and Nielsen (2007, chapter 5). On the other hand, we should also note that our atmospheric model covers the whole country and that SO_2 travels a long distance. The model thus includes such regional effects, which are only roughly estimated in Ho and Nielsen (2007).

In sum, the 11th FYP power sector policies were effective at reducing emissions not only of SO_2 but also of primary PM and carbon, and ultimately the concentrations of secondary PM and ozone. These reductions in local pollutants improved public health and, for the period examined (2006–2010), were accompanied by a positive effect on GDP due to the replacement of costly small power plants with more efficient large ones. The effect of the 11th FYP emission controls on grain productivity, however, was negligible. While we have not included the transition cost of relocating workers laid off from the plants that were shut down or the likely higher costs of transmission, it is fair to say that the benefits of the shutdown policy were, in principle, large enough to justify generous compensation to those adversely affected by the policy. The FGD requirement had a cost in terms of lower GDP available for consumption, but it was commensurate with the conservatively estimated local health benefits. We should note that we did not consider other benefits, such as those in other countries from reduced SO_2 and sulfate originating in China.

Finally, our model does not incorporate the endogenous feedback of damages to human health and ecosystems from exposure to pollution. If we could include the effects of better health on labor productivity and agricultural output, the 11th FYP policies would have had even greater positive effects on the economy.

9.5   The Costs and Benefits of a Carbon Tax in China

As described in chapter 2, at the time of writing the Ministry of Finance has been actively discussing environmental tax reform, and a commitment to imposition of unspecified environment taxes was included in the 12th FYP for 2011–2015 (Xinhua 2011). What form an environmental tax might take is subject to continuing debate; policies that have been discussed include higher gasoline taxes, reform of current resource taxes, taxes on air and water pollutants, and carbon taxes. It is thus an opportune time to bring our integrated analysis to bear on carbon pricing options. We should note that in many European countries, pricing of carbon is already implemented in some form; in the United States some states have active carbon markets as they wait for federal government action; and at the time of writing, China is preparing policy experiments in carbon trading on a local scale. While the current outlook for a substantial coordinated international carbon market is not bright, unilateral actions that go further than these current efforts are certainly possible.

9.5.1   The Economics of a Carbon Tax

We now discuss how we use the economic model described in section 9.2 to simulate the effects of a tax on fossil fuels based on their carbon content. While a comprehensive carbon price would also tax CO_2 from noncombustion sources, notably process emissions from cement manufacturing, we shall ignore these emissions here for simplicity. The simple policy considered is a direct unit tax on the consumption of fossil fuels—that is, a tax imposed per ton of coal, per ton of oil, and per cubic meter of natural gas.[14] Both domestically produced and imported fuels are taxed at the same rate.

Let the carbon tax rate be tx^u yuan per ton of carbon, and the carbon content per unit of fuel j be c_j. The carbon tax per unit of fuel j is then simply

$$t_j^c = tx^u c_j \quad \text{for} \quad j = \text{coal, oil, gas} \tag{9.1}$$

This tax appears on the right-hand side of the price equation (A.4) in appendix A.

In choosing a carbon tax rate to examine, we consider both the experience in other countries and the tax proposals discussed in reports of China's Ministry of Finance (MOF). Carbon prices in the European Trading Scheme were about €13 per ton of CO_2 in 2009, or around 125 yuan per ton (at the exchange rate in 2009). Earlier in 2007, the ETS price was up to an equivalent of 210 yuan per ton. The U.S. Environmental Protection Agency (U.S. EPA) analysis of the Waxman-Markey bill projected an initial carbon price of $13–$17 per ton of CO_2 (or 85–115 yuan per ton at 2010 exchange rates), equivalent to more than 100% of the price of coal at the mine mouth.

Moreover, it is important to note that CO_2 is a stock pollutant; that is, the damage that it causes depends on its accumulated amount, unlike particulate matter, which is produced and washed out in a relatively short time. Thus the damages due to CO_2 in 2050 will only be modestly affected by emissions in any given year. However, many carbon-reducing strategies involve new capital and new technologies, and a rapid reduction is very costly. That is, it is more expensive to reduce a ton of emitted CO_2 in 2012 than to reduce a ton in 2030 if both reductions are planned today; the more distant target date allows firms to change their investment plans gradually, whereas an immediate reduction requires abandoning still useful capital. The reduction in climate risk in 2050 is the same in either case. As others have pointed out, these factors mean that annual reductions in CO_2 may be gradually increased over time to achieve the same cumulative target by 2050 (e.g., Aldy and Pizer 2008; Newell and Pizer 2003). One should start with small reduction targets that exploit the easiest cuts and then impose bigger reduction targets that firms will be able to meet more easily by replacing normally depreciated capital with new technologies.

We thus believe that such a new tax should start at a low rate to allow the economy to adjust gradually, a point also made in a study by the MOF Research Institute of Fiscal Science (Su et al. 2009) where they suggest a tax that rises from 10 yuan per ton of CO_2 in 2012 to 40 yuan per ton in 2020.

We start with a tax of 100 yuan per ton of carbon, about 27 yuan per ton of CO_2. In 2005, the base year of the model, that would be worth about US$3.30 per ton of CO_2. With a mine mouth price of coal in 2005 of 360 yuan per ton, this would still be a nontrivial tax of about 14% on the price of China's primary energy source. For the purposes of comparing with the SO_2 policy in the 11th FYP, the simulation here consists of imposing this 100 yuan per ton of carbon tax on the use of coal, oil, and gas every year from 2006 to 2010.

The coal consumption in China was 2.6 billion tons in 2007. If such a tax were collected on this amount, the revenue would be 140 billion yuan, or more than 3% of total government revenues. As we have noted in section 9.4, how the new tax revenue is used has an impact on economic growth, and since this is such a large new tax, the effect on growth may be substantial. Our first scenario recycles the revenue as lump sums back to households to maintain the base-case level of government spending and to maintain revenue neutrality.[15] This is the carbon tax scenario for which emissions, air quality, and health and agricultural effects are assessed in chapters 4–8.

A second scenario uses the revenue to cut existing distortionary taxes, an approach that many argue is better for economic growth. For simplicity we assume that all the tax rates in the base case are reduced by the same fraction ξ_t; that is, in the policy case, the new tax rates on capital income, value added, and sales are

$$t_t^k = \xi_t t_{t0}^k, \quad t_t^V = \xi_t t_{t0}^{VAT}, \quad t_{jt}^t = \xi_t t_{jt0}^t \tag{9.2}$$

where t_{t0}^k, t_{t0}^{VAT}, and t_{t0}^S are the capital income tax, value-added tax, and sales tax rate in the base year 2005.[16]

The scaling factor ξ_t is endogenously determined so that real government consumption in period t is equal to the base-case level, GG_t^{base} (consumption being government expenditures on goods and services excluding transfers, subsidies and interest payments):

$$GG_t = GG_t^{base} \tag{9.3}$$

where GG_t is the quantity index of government purchases and is given in appendix A, equation (A.17b).

9.5.2   The Impacts of a Carbon Tax

The two carbon tax policy scenarios are simulated for each year during 2006–2010, beginning with the same initial capital stock and labor supply as in the base case. The results of comparing the two scenarios with the base case for 2010 are given in table 9.10 for the key aggregate variables. The changes over time for GDP and

Table 9.10
Effects of a carbon tax in 2010

Variable	Base Case 2010	Tax Effect (percent change)	
		Lump Sum Transfer	Reduced Other Taxes
GDP (billion yuan 2005)	30,955	−0.19	−0.03
Consumption (billion yuan 2005)	11,891	0.13	−0.14
Investment (billion yuan 2005)	12,468	−0.25	0.28
Government consumption (billion yuan 2005)	3,625	0.00	0.00
Fossil energy use (million tons of standard coal equivalent)	2,764	−11.5	−11.3
Coal use (million tons)	2,802	−14.6	−14.4
Oil use (million tons)	449	−2.6	−2.3
Gas use (billion cubic meters)	89	−8.1	−7.9
Electricity (billion kWh)	3,245	−4.1	−3.8
CO_2 emissions (fossil fuel; million tons)	7,255	−12.2	*
PM_{10} emissions (1000 tons)	17,671	−10.6	*
SO_2 emissions (1000 tons)	32,472	−14.0	*
NO_X (1000 tons)	24,446	−11.1	*
Carbon tax as percent of total tax revenue		3.07	3.09

Health and Crop Benefits of Carbon Tax with Lump Sum Transfer	Cases	Value (billion yuan)
$PM_{2.5}$		
Avoided acute mortality	17,200	8.89
Avoided chronic mortality	103,000	53.48
Other health benefits		2.20
Ozone		
Avoided acute mortality	1,380	0.72
Other health benefits		0.11
Crop benefits		5.11
Total health benefits (including acute mortality)		11.92
Total health benefits (including chronic mortality)		56.51

	Percent
Total health benefits (chronic mortality) as percent GDP	0.18
Total crop benefits as percent of GDP	0.02

*Emissions and air quality effects not modeled for this case.

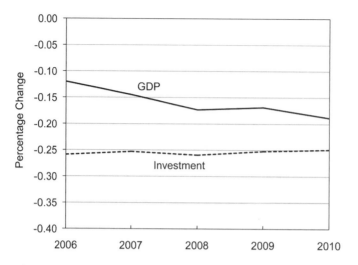

Figure 9.3
Change in GDP and investment due to 100 yuan per ton CO_2 tax (lump sum transfer revenue case).

investment, in the case where the new revenue is rebated in lump sums back to households, are plotted in figure 9.3, while changes in energy consumption and carbon emissions are plotted in figure 9.4. Finally, the changes in the prices of the 33 commodities for this case are given in table 9.11.

We first discuss the lump sum transfer case. The carbon tax raises the price of coal at the mine mouth by 14% and the price of oil by 2% in the first year. Given the simple form of our cost functions, the demands for these fuels are reduced in proportion to these changes in prices. The tax increase raises the costs of producing carbon-intensive products such as primary metals, nonmetallic mineral products (including cement), and transportation services. The producers consequently must raise their prices to cover these higher costs, which will reduce their sales and output. This reduction in the output of carbon-intensive products further contributes to the reduction in fossil fuel use. Total fossil energy consumption falls by 10.5% and carbon emissions by 11% in the first year (figure 9.4).

Over time there is an additional secondary effect on energy use beyond the direct price effect of the carbon tax: changes in the level of investment. Investment is determined by the level and composition of aggregate income as well as by the price of investment goods. The composition of income is affected by the disposition of the carbon tax revenue, which we previously noted is sizable, at 3.1% of total government revenue. In the lump sum transfer scenario, the transfer of new revenue

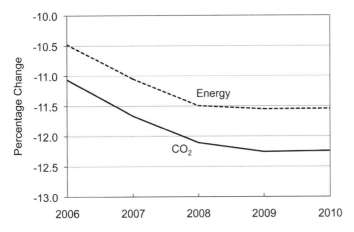

Figure 9.4
Change in energy use and CO_2 emissions due to 100 yuan per ton CO_2 tax (lump sum transfer revenue case).

back to households raises consumption at the expense of investment. In the second scenario where taxes on enterprises are cut, allowing higher retained earnings, investment rises at the expense of consumption, as shown in table 9.10.

However, the prices of investment goods such as buildings, transportation equipment, and machinery rise relative to the price of consumption goods because they have a high level of embodied carbon. (While the final assembly of machinery and transportation equipment may use relatively little energy, the steel, glass, and other materials used to produce them are energy-intensive goods, and the higher prices of these materials due to the carbon tax result in higher costs for machinery and equipment.) This statement is true for both the lump sum transfer case and the reduced-other-taxes case. In the lump sum transfer scenario, the changes in incomes and the more expensive capital goods both act to reduce investment in each period. This reduction is shown in figure 9.3, and the cumulative effect is a 0.1% smaller stock of capital by 2010. The smaller stock of capital means less income and less output, and thus a second channel for the reduction in fuel consumption.

The smaller stock of capital and the price wedges due to the carbon tax lead to a 0.19% fall in 2010 GDP. Coal use in 2010 is 14.6% lower because of the "income effect" of the lower GDP and the "price effect" of more expensive coal.[17] Electricity use falls by 4.1% as a result of the higher price of electricity and the reduced demand from lower output of electricity-intensive products. The output of the oil refining sector falls by only 2.3%, since petroleum products bear a much smaller carbon tax compared to coal. CO_2 emissions fall by 12.2%, a change that is less than the reduction in coal use because of the switch to other fuels.

From table 9.11 we see that, in the lump sum transfer case, the energy intensive manufacturing industries—chemicals, nonmetallic mineral products, primary metals—all have output declines in excess of 1% and price increases of close to 1%. Transportation services output falls by 0.7%. The aggregate shift away from investment spending toward consumption results in a small expansion of the agriculture and food manufacturing industries. We note again that there is a reduction of GDP by 2010, so the output losses at the industry level far exceed the output gains.

The industries producing the carbon intensive commodities are also the biggest emitters of local pollutants—primary PM, SO_2, and NO_x—and these emissions are thus sharply reduced with the contractions of these sectors and their reductions in fuel use. These effects are discussed in greater detail in chapter 6 and are shown in figure 6.8. To summarize them, by 2010 the annual PM_{10} and NO_x emissions are 11% lower than the base case while SO_2 emissions are 14% lower.

These emissions under the carbon tax policy are then used to construct a new set of emissions in each grid cell for the lump sum transfer case. As for the 11th FYP results, this emissions inventory is then taken into the GEOS-Chem atmospheric model to estimate a new set of concentrations, given in detail in chapter 7. In figure 7.6 we can see that the reduction in $PM_{2.5}$ concentration in some areas is more than $12 \mu g/m^3$ in fall and winter, and for ozone the peak reduction is more than 1.0 ppbv. Again, these changes should be compared to the concentrations in the base case given in figure 7.4. For example, in October 2010 the mean $PM_{2.5}$ concentration ranges from 20 to $140 \mu g/m^3$ in the eastern half of China through most of the year, although it can reach $180 \mu g/m^3$ in spring.

From the base case and policy case concentrations, we obtain the changes in concentration in each grid cell, which are then combined with the population map as described in chapter 8. Multiplying the concentration change with the concentration-response coefficient and the population produces the changes in health effects. Figure 8.4A shows the geographical distribution of the reduction in acute mortality; the biggest reductions are in Henan, Hebei, Shandong, and Jiangsu provinces. Table 8.4 shows the reduction in the incidence of all the health points in 2010: the number of cases of acute mortality (premature deaths) due to $PM_{2.5}$ falls by 17,200 a year, and hospital admissions for cardiovascular reasons by 30,000. The reduction in ozone concentrations reduces acute mortality by an additional 1400 cases. If we use the chronic mortality estimates derived from U.S. studies, the effect is estimated at 103,000 cases a year.[18]

Table 9.11

Effects of the carbon tax on industry output and prices in 2010, lump sum transfer case (percent change)

Sector		Output	Price
1	Agriculture	0.04	0.43
2	Coal mining and processing	−14.69	16.39
3	Crude petroleum mining	−2.28	2.35
4	Natural gas mining	−8.11	8.71
5	Nonenergy mining	−1.18	0.65
6	Food products, tobacco	0.06	0.52
7	Textile goods	−0.45	0.63
8	Apparel, leather	−0.15	0.48
9	Sawmills and furniture	−0.55	0.68
10	Paper products, printing	−0.51	0.65
11	Petroleum refining and coking	−2.33	2.19
12	Chemical	−1.04	0.94
13	Nonmetal mineral products	−1.11	1.13
14	Metals smelting and pressing	−1.35	1.08
15	Metal products	−0.86	0.76
16	Machinery and equipment	−0.76	0.65
17	Transport equipment	−0.49	0.64
18	Electrical machinery	−0.67	0.61
19	Electronic and telecom equipment	−0.37	0.39
20	Instruments	−0.49	0.32
21	Other manufacturing	−0.39	0.45
22	Electricity, steam, hot water	−4.07	3.72
23	Gas production and supply	−3.43	3.90
24	Construction	−0.32	0.67
25	Transportation	−0.73	0.70
26	Communications	−0.11	0.35
27	Trade	−0.27	0.47
28	Accommodation and food	−0.13	0.47
29	Finance and insurance	0.01	0.24
30	Real estate	−0.08	0.61
31	Business services	−0.45	0.52
32	Other services	−0.24	0.65
33	Public administration	0.11	0.40

These health effects are translated into economic terms using the valuations given in table B.1 and are reported at the bottom of table 9.10. The combined health effects of all health endpoints (including acute but not chronic mortality) in table 8.4 imply a reduction in total health damages of 11.9 billion yuan in 2010 (in 2005 yuan). Of this total, acute mortality from $PM_{2.5}$ is the biggest contributor at 8.9 billion yuan. Since the dominant factor for total health damages in each grid cell is acute mortality, the damages map is very similar to the acute mortality map in figure 8.4A.

The more uncertain chronic mortality effect is monetized at 53.5 billion yuan, and when we include the valuation for hospital admissions and outpatient visits for both PM and ozone, the total reduction in health damages is 56.5 billion. This figure is almost five times the value for the total that only includes acute effects and is equivalent to 0.18% of the projected base-case GDP in 2010.

Finally, the net national reduction of ozone levels due to the carbon tax (lump sum transfer case) leads to productivity gains in major grain crops, as described in section 8.2. The changes in agricultural output are estimated using exposure and dose-response assessment that are analogous to those of the health benefit analysis. The mean estimates of increased national yields are 174,000 metric tons of wheat, 993,000 tons of rice, and 1,450,000 tons of maize (table 8.7). Valuing these crops at market prices, the gain in output is worth 5.1 billion yuan in 2010, which is much greater than the health benefit from ozone reduction (details in section 8.2.4). This is equivalent to 0.02% of base-case GDP in 2010, implying a total benefit of the carbon tax, including health effects described earlier, of 0.20% of GDP.

9.5.3 Comparing the Lump Sum and Tax Cut Revenue Cases

We have already noted that if the new carbon tax revenue were used to cut existing taxes instead of being transferred back to households, it would significantly change the composition of final demand: consumption would be lower and investment higher as a result of the assumed higher retained earnings from the reduced taxes on enterprise income. We report this simulation to spotlight how the use of revenues matters. (Note that for this case we report only the effects on the economy, energy use, and carbon emissions; we do not discuss pollution emissions, air quality, and health and crop effects that are evaluated for other policy cases in chapters 4–8.)

The cumulative effect of this higher investment and lower consumption is a smaller GDP loss in 2010 compared to the base case, 0.03% versus 0.19% for the lump sum transfer case, as shown in the last column of table 9.10. This gives a slightly smaller reduction in energy use, −11.3% versus −11.5%, a drop in consumption that applies to all three fossil fuels.

The modest effects in comparison to other studies of carbon taxes (e.g., Aunan et al. 2007) result from the short horizon considered here. The cumulative effects over a longer period are suggested by a similar, but longer, simulation using an earlier version of our economic model, again analyzing use of revenues raised by environmental taxes to reduce other taxes (Ho and Jorgenson 2007, table 10.3). In the first year, the simulated fall in GDP was 0.04%, but by the 20th year the GDP was actually higher than the base case, by 0.18%. The fall in coal use due to the environmental taxes in the first year was 3.95%, but with the higher output in the 20th year, the fall in coal use was only 3.36%. Such a pattern would apply to the effects reported here in table 9.10: the initial GDP loss would become a GDP gain later, and the reduction in energy use and emissions would be substantially smaller than in the lump sum tax case.

Again we emphasize that these comparisons are intended to highlight the importance of the parallel revenue recycling choice when carbon taxes are imposed. They are not intended to argue for a policy to benefit enterprises at the expense of consumption. One could alternatively argue that a tax cut that more directly increases household incomes will lead to a more balanced change in consumption and investment that would minimize the discounted stream of any consumption losses. Economists often consider such a measure of consumption as the most appropriate for measuring welfare change. Net welfare change, of course, is not just this consumption change but also the change in health and other damages along with climate risks, which are not considered here.

Our finding of monetized health and agricultural benefits exceeding the loss in GDP is also given by Aunan et al. (2007) and Vennemo et al. (2009), as discussed in chapter 1. To our knowledge, this Norwegian-Chinese team represents the only other research group publishing in the peer-reviewed literature that explicitly estimates the value of the local benefits of carbon control in China.

9.6 Comparing the Carbon Tax with the 11th Five-Year Plan SO$_2$ Policies

The simulated results of the 11th FYP reported in section 9.4 indicate a small GDP effect, a modest CO$_2$ reduction, and a substantial SO$_2$ reduction. The actual records of economic growth and emissions in the 2006–2010 period are the result of the policies analyzed, other policies implemented but not considered here, the worldwide economic shocks that are not represented, and myriad other real-world factors not considered. Social science analysis cannot take place in a controlled laboratory, and such modeling cannot be validated in a way that readers with a scientific background may be familiar with. Economic analysis is necessarily an abstraction that focuses on a narrow set of interactions that ignore the countless secondary factors. Thus the analysis using the model simulations here cannot be directly validated, and one cannot observe a real economy where the counterfactual base case occurred. However, if interpreted cautiously these simulation results provide useful insights.

The official record of SO_2 emissions shows a reduction from 25.5 million tons in 2005 to 22.1 in 2009 (NBS and MEP 2010, table 4.1); GDP growth between 2005 and 2008 was 11.2% per annum, somewhat faster than that envisaged in 2005 when the FYP was drafted. The first year of the economic recession in the richer countries, following the global financial crisis, was 2008, and the Chinese government responded with macroeconomic measures that resulted in a much smaller slowdown than originally feared; GDP growth between 2007 and 2008 was still 9.0%. While the complex interactions between government policy and world economic effects are difficult to parse, let alone the impact of the SO_2 policies of the 11th FYP within this interaction, we believe it is reasonable to say that whatever effects the SO_2 policies on the macro economy may have been, they were small compared to those of other economic shocks at the time. That is, the simulated macroeconomic results in table 9.8 are not contradicted by the actual experience through 2008.

In other words, a government official wishing to ask, at the time of writing in 2011, what the impact of the FYP SO_2 policy was, is not asking an impossible or pointless question even though the actual economic history during this period was too complex to represent precisely. We believe that analyses such as the one presented here provide a useful guide to some of the impacts of the policy, following the example of U.S. EPA (1997).[19] We estimated a tiny impact on overall economic growth and modest effects at the industry level, and the actual data that officials can now see do not contradict these results. We thus believe that a comparison of the simulated carbon tax effects with those of the 11th FYP SO_2 policy—that is, a comparison of an untried policy with a policy that was actually implemented—will provide useful insights for the formulation and implementation of future policies.

The comparison of the carbon tax and SO_2 policy of the 11th FYP is given in table 9.12, that is, the difference between the policy cases and the base case in 2010. Chiefly because of the FGD component, the policy reduces SO_2 emissions by 19.8%

Table 9.12
Comparison of the effects of a 100 yuan per ton carbon tax with the 11th FYP SO_2 policy in 2010

Variable	11th FYP—Combined Shutdown and FGD Policies (percent change)	Carbon Tax Lump Sum Transfer Case (percent change)
GDP	0.74	−0.19
Consumption	0.47	0.13
Investment	1.07	−0.25
Government consumption	0.05	0.00
Fossil energy use	−4.29	−11.5
Coal use	−5.57	−14.6
Oil use	−0.56	−2.58
Gas use	−3.48	−8.08
Electricity	−1.45	−1.03
CO_2 emissions (fossil fuel combustion)	−4.6	−12.2
PM_{10} emissions	−3.6	−10.6
SO_2 emissions	−19.8	−14.0
NO_X	−1.3	−11.1

Health and Crop Benefits	Cases	Value (billion yuan)	Cases	Value (billion yuan)
$PM_{2.5}$				
Avoided acute mortality	12,300	6.35	17,200	8.89
Avoided chronic mortality	73,900	38.28	103,000	53.48
Other health benefits		1.57		2.20
Ozone				
Avoided acute mortality	120	0.06	1380	0.72
Other health benefits		0.01		0.11
Crop benefits		0.00		5.11
Total health effects (including acute mortality)		8.00		11.92
Total health effects (including chronic mortality)		39.92		56.51

	Percent	Percent
Total health benefits (chronic mortality) as percent of GDP	0.13	0.18
Total crop benefits as percent of GDP	0.00	0.02

versus the base case and 11.6% versus 2005, achieving the target of the 11th FYP. In contrast, the reduction of fuel use due to the carbon tax, without any direct SO_2 controls, reduces it by a still-sizable, if smaller, 14.0%.

However, in terms of coal use, the carbon tax generates a 14.6% reduction versus the base case, compared to the 5.6% reduction in the 11th FYP (which is mostly due to the shutdown policy). The reduction in oil and gas use is also larger despite some fuel switching from coal to oil and gas. These reductions in fuel consumption reduce the PM_{10} emissions by 10.6% in the carbon policy case, compared to only 3.6% in the SO_2 policy case. The reduction in NO_X emissions is a large 11.1% versus the base case, compared to 1.3% in the 11th FYP case.

The atmospheric concentration of $PM_{2.5}$ depends on the emissions of many pollutants, in particular, both primary PM emissions and SO_2 emissions. The net effect of the larger reduction in primary PM and smaller reduction in SO_2 in the carbon tax case is a larger improvement in air quality. The map in figure 7.6D gives the effect of the carbon tax on $PM_{2.5}$ concentrations and may be compared to figure 7.5B, which gives the effect of the SO_2 policy on concentrations of sulfate (a $PM_{2.5}$ species). The effects on avoided acute mortality due to changes in $PM_{2.5}$ and ozone for both policies are mapped in figure 8.4; the larger areas of dark shading are clear in the carbon tax map. The reduction in the cases of acute mortality due to $PM_{2.5}$ and ozone is 18,600 in the carbon policy, compared to 12,400 in the 11th FYP policy (12,300 + 120 in the bottom section of table 9.12). If we consider the estimates for chronic mortality, the results are 103,000 avoided cases for the carbon tax versus 73,900 for the 11th FYP, all due to $PM_{2.5}$ exposures.

The differences in fuel consumption between the two policies correspondingly result in differences in CO_2 emissions. Under the direct pressure of the carbon tax, CO_2 emissions are reduced by 12.2%, whereas with the more modest coal savings in the 11th FYP the emissions are reduced by only 4.6%.

Thus we can conclude that for the same five-year period, a tax of 100 yuan per ton of carbon could have generated greater reductions in all air pollutants except SO_2 and sulfate fine particles than the ambitious 11th FYP SO_2 policy, and the resulting reduction in health damages would have been 40–50% bigger. The carbon tax would also reduce CO_2 emissions by more than twice the savings generated by the small-plant shutdown policy in the FYP. The effect on GDP versus the base case is negative in the five-year window for the CO_2 tax case but positive for the 11th FYP because of the savings from the plant shutdowns. We should note again that the small-plant shutdown is a one-off policy; by the end of the 11th FYP, there were relatively few inefficient plants left to close. The impact on GDP of an FGD-only policy would have been a small negative effect, like the carbon tax policy.

We end this comparison by reemphasizing that the time frame of the policies considered is important. Over a longer horizon the carbon tax would generate large revenues that can be used for higher investment that may lead to positive GDP effects later, or alternatively lead to smaller consumption costs in the immediate term. The FGD policy, if sustained to retrofit all thermal power plants or extended to other industries, will continue to generate large health benefits, but with a continuing small cost to GDP and consumption.

9.7 Conclusion

This chapter provides the last step in our integrated approach to estimating the cost and benefits of emission control policies in China. The preceding chapters discuss how emissions are related to industry activity, how air quality is affected by these emissions, and how the health and agricultural effects of reduced pollution are estimated. Those earlier chapters thus were about the *benefits* of policies to reduce air pollution. This chapter describes how the policies affect the economy, how the aggregate GDP is changed, and how individual industries are affected—that is, the *costs* of the policies. We also describe here the effect on carbon emissions, without trying to value the benefits to China and the rest of the world of CO_2 reductions.

We applied this methodology to estimate the costs and benefits of SO_2 control elements of the 11th FYP, a major environmental policy that is providing valuable experience for Chinese policy makers. There is currently also official interest in environmental taxes and other options in carbon control; the official goal of the government, announced prior to the 2009 Copenhagen meetings of the United Nations Framework Convention on Climate Change, is to reduce the aggregate carbon intensity in 2020 by 40–45% compared to the 2005 level. We thus also applied our integrated analysis to a counterfactual carbon tax of 100 yuan per ton of carbon (or roughly 27 yuan per ton of CO_2) implemented from 2006 to 2010 instead of the 11th FYP SO_2 controls. This is a level of taxation that is a reasonable starting point for such a policy and that also provides a suitable comparison in terms of the general scale of effects to the small-plant shutdown and FGD policies in the 11th FYP.

We conclude that the 11th FYP policy demonstrates the effectiveness of well-targeted technology mandates for pollution control: a substantial improvement in air quality achieved at modest cost to GDP even if we were to ignore the negative GDP costs of the shutdown policy. The reduced pollution from the installation of FGD equipment and shutdown of small plants leads to a substantial reduction in health damages, even when a conservative assessment method is used. If Western estimates of chronic mortality effects are used instead, the avoided premature deaths come to tens of thousands a year.

The small-plant shutdown policy also has the added benefit of reducing energy consumption and raising GDP slightly. This finding suggests that extending such a policy of closing small inefficient units to other sectors, while compensating workers affected by the shutdowns, should be high on the lists of policies to consider as part of the reform agenda.

This experience with SO_2 control is encouraging for the prospect of effective NO_x-control technology mandates that are being developed for the 12th FYP, although the cost structure and technical removal efficiency of NO_x-control devices could be less advantageous than FGD and the potential health benefits have not yet been examined. This is an obvious avenue of future research.

While reducing CO_2 emissions is not the top priority for the Chinese government, the use of a carbon tax may serve other purposes that are within the nation's leading interests. Reducing energy consumption and local pollution are very urgent national priorities, and our simulation of a tax of 100 yuan per ton of carbon projects substantial energy conservation and improvements in air quality at a modest cost to GDP. We estimate that the reduced combustion of fossil fuels would lead to a reduction in premature mortality that is greater than that estimated for the 11th FYP. The policy's effect on ozone could yield sizable additional benefits in terms of improved productivity of major grain crops.

Such a tax would raise government revenues equal to 3% of total current revenues and would even allow a corresponding cut in existing taxes. Such a parallel recycling of revenue would limit the cost of a lower GDP that would occur under the other revenue recycling option that we considered. The benefits to public health alone would be bigger than the reduction in aggregate consumption due to the carbon tax. In addition, annual CO_2 emissions would be reduced by 12%, a likely substantial long-term benefit to the global environment.

Estimating these benefits involves estimating various quantities including the concentration of particulates, the health impact, and the valuation of various health endpoints. As noted in the chapters describing the methodologies to estimate these quantities, there is uncertainty about these estimates given the statistical methods used. The uncertainty for valuations, for example, is large. The numerical results of such exercises should thus be read with these uncertainties in mind. We have taken a conservative approach in selecting these uncertain parameters and believe that our general conclusions about the size of the benefits in relation to the costs are sound. The 11th FYP policies might not have been the most efficient pollution reduction policies possible, but we can say with some confidence that they provided a substantial net benefit. Similarly, although the estimated net benefits of a carbon tax may be uncertain, we believe they are positive.

Given these benefits of carbon control, it is natural to ask next what the smallest net cost option is. An obvious policy to consider, as recommended by many economists, would be to start with a low level of carbon tax and gradually raise it as enterprises and households adapt to the higher energy prices. Machinery and buildings are gradually replaced during the natural course of economic activity, and a gradual scaling up of the tax would allow this replacement process to include more energy-efficient options as they become available throughout the period of rising carbon prices.

The size of the reduction in CO_2 emissions points to a substantial gain for the rest of the world. As we and others have argued, this reduction in global climate risks is a strong reason for richer countries to support such efforts at energy conservation and shifting to lower-carbon energy sources (Ho and Nielsen 2007). Support could be in the form of policy design, energy efficiency plans such as building energy standards, and technology transfer. However, regardless of international support, our results suggest that it would be in China's own domestic national interest to price carbon in order to encourage such energy transitions. The reduction in health damages due to the improved air quality alone would justify the costs of such a tax.

Acknowledgment

This chapter draws on our earlier work with Richard Garbaccio and Hongwei Yang in JES (2007) and Cao, Garbaccio, and Ho (2009).

Notes

1. The base case is a counterfactual simulation without the policies that were actually put in place. For the policy case we do not take the actual data for those years but instead simulate the model based on the same initial year as the base case. This step is taken to maintain comparability between the two cases; the model is an abstraction of the actual complex economy focusing on key relationships, and in order to avoid distracting detail, many actual events and interactions are not represented. Furthermore, the actual data for 2009 and 2010 were only available after the completion of this report.

2. In this model there is no forward-looking behavior where expected changes in future policies can affect behavior today. This approach is in contrast to models with intertemporal equilibrium and such forward-looking behavior, see, e.g., Fisher-Vanden and Ho (2007). Such

models are much more complex and difficult to solve, and we have chosen the simpler approach here.

3. A variable is said to be "exogenous" if it is predetermined before any solution of the model, e.g., population. A variable is "endogenous" if it is determined within the model simultaneously with other variables, e.g., prices.

4. By inelastic labor supply we mean that the total hours worked is a predetermined number not affected by economic events. The alternative formulation of having hours worked dependent on the wage rate seems to us to be too elaborate to implement sensibly for the current Chinese economy, with its large pool of underemployed workers. A more elaborate model with heterogenous workers might change the details but not likely the overall comparisons of costs and benefits.

5. For example, Jorgenson, Ho, and Stiroh (2005, chapter 6) show how more college-educated workers are hired even as their relative wages rise. In the economics literature, the phenomenon of skills-biased technical change is a much-discussed topic. Jorgenson, Gollop, and Fraumeni (1987) give a prominent discussion of energy-saving technical change in the United States.

6. A 17-sector version of the 2005 input-output table is published by the NBS in the *China Statistical Yearbook 2008* (tables 2-26 and 2-27). Shantong Li and Jianwu He of the Development Research Center obtained unpublished detailed data from the NBS and estimated an IO table with 42 industries, which they kindly shared with us. Their methods are described in Zhai and Li (2000).

7. The information in this section, and the next, is drawn from Cao, Garbaccio, and Ho (2009), which relies on JES (2007). This made use of projections made in 2005 and is not identical to the more recent data on actual installation of generating capacity and FGD equipment given in chapter 4. These figures are, however, very close.

8. The NDRC's Energy Research Institute estimates that in 2006 there were about 115 GW of capacity provided by coal- and oil-fired units under 100 MW, out of a total of 391 GW of thermal-fired capacity. This figures come from the studies summarized in JES (2007).

9. Because the replacement plants are large, they may be sited further away from load centers than the retired plants, which may mean higher transmission costs. This factor is not considered in our analysis.

10. That is, the α_{Ej} in equation (A.3) is lowered, offset by a corresponding increase in the capital share parameter. The unit cost reduction is represented by a reduction of the $g(t)$ term.

11. The standard power dispatch model in China during the 11th FYP was the "San Gong" model ["three 'gongs'": "gong1 ping2" (common level); "gong1 zheng4" (fair); "gong1 kai1" (transparent)]. This dispatch model, still in effect in many provinces of China at the time of writing, uses monthly and yearly contracts to guarantee utilization rates for each plant in a system. It is designed to share the market "fairly," not to minimize cost, fuel use, or emissions. Peak load is met by ramping up generation at individual plants in large part proportionately, scheduled to achieve the contractual commitments across the system. Unexpectedly high or low load is met by proportional increases or decreases in the contracted utilization rates at all power plants. Our assumptions that the power lost by retirement of small plants is made up in the operating hours at existing and new plants (those hours in fact declined over the period), as well as our use of average coal use and emission rates, are thus reasonable representations of the power dispatch protocol.

12. This 167 GW of FGD includes 39 GW carried over from the previous five-year plan and 128 GW of FGD installations newly mandated.

13. That is, we raise the $g(t)$ term in the unit cost equation (A.3) for the electricity, steam, and hot water sector.

14. In contrast to a unit tax, an ad valorem tax would be based on the yuan value of the fuel, for example, a sales tax of x percent on the purchase value. Given that we are concerned with the quantity of CO_2, we use the unit tax policy here.

15. In the detailed model description in appendix A, we are adding the carbon tax revenue on the right-hand side of the private income equation (A.5).

16. See the revenue equation (A.11).

17. Economists view the main determinants of household demand for a commodity as household income (or wealth), and the price of the commodity; as income rises, the demand for most goods should rise, and as the price rises, demand falls. The former effect is called the income effect, and the latter is the price (or substitution) effect.

18. The issues surrounding the distinction between acute and chronic mortality are discussed in section 8.1.2.

19. In the U.S. environmental community, a well-known analysis of a policy already implemented is the U.S. EPA's analysis of the Clean Air Act of 1970 published in 1997, U.S. EPA (1997). This retrospective look was requested by Congress and served as an important source of analysis for discussions of revisions to that law.

References

Aldy, J. E., and W. A. Pizer. 2008. Issues in Designing U.S. Climate Change Policy. Resources for the Future Discussion Paper 08-20, Washington, DC.

Aunan, K., T. Berntsen, D. O'Connor, T. Hindman Persson, H. Vennemo, and F. Zhai. 2007. Benefits and costs to China of a climate policy. *Environment and Development Economics* 12:471–497.

Bovenberg, A. L. 1999. Green tax reforms and the double dividend: An updated reader's guide. *International Tax and Public Finance* 6(3):421–443.

Chinese Academy for Environmental Planning (CAEP). 2007. Development of Power Industry and Cost for Sulfur Dioxide Emission Control in China. Chinese Academy for Environmental Planning, State Environmental Protection Administration, Beijing.

Cao, J., R. Garbaccio, and M. Ho. 2009. China's 11th Five-Year Plan and the environment: Reducing SO_2 emissions. *Review of Environmental Economics and Policy* 3(2):231–250.

Cao, J., M. Ho, D. Jorgenson, R. Ren, L. L. Sun, and X. M. Yue. 2009. Industrial and aggregate measures of productivity growth in China, 1982–2000. *Review of Income and Wealth* 55(s1):485–513.

Fisher-Vanden, K., and M. S. Ho. 2007. How do market reforms affect China's responsiveness to environmental policy? *Journal of Development Economics* 82(1):200–233, January.

Ho, M. S., and D. J. Jorgenson. 2007. Policies to control air pollution. Chapter 10 in *Clearing the Air: The Health and Economic Damages of Air Pollution in China*, ed, M. S. Ho and C. P. Nielsen. Cambridge, MA: MIT Press.

Ho, M. S., and C. P. Nielsen, eds. 2007. *Clearing the Air: The Health and Economic Damages of Air Pollution in China*. Cambridge, MA: MIT Press.

Holz, C. 2006. China's Economic Growth, 1978–2025: What we know today about China's economic growth tomorrow. Mimeo, Hong Kong University of Science and Technology, December.

International Energy Agency (IEA). 2007–2008. *World Energy Outlook (2007 and 2008 editions)*. Paris: OECD/IEA. Available at www.worldenergyoutlook.org/.

Joint Economic Study (JES) of the U.S.-China Strategic Economic Dialogue. 2007. U.S.-China Joint Economic Study: Economic analyses of energy saving and pollution abatement policies for the electric power sectors of China and the United States (Summary for Policymakers). Washington, DC, and Beijing: U.S. Environmental Protection Agency and China State Environmental Protection Administration.

Jorgenson, D. W., F. M. Gollop, and B. M. Fraumeni. 1987. *Productivity and U.S. Economic Growth*. Cambridge, MA: Harvard University Press.

Jorgenson, D. W., M. S. Ho, and K. J. Stiroh. 2005. *Information Technology and the American Growth Resurgence*. Cambridge, MA: MIT Press.

Li, S. T. 2009. 2030 nian woguo jingji shehui fazhan qianjing zhangwang (Outlook of China's economic and social development through 2030). Development Research Center of the State Council, Beijing. In Chinese.

National Bureau of Statistics (NBS). 2008. *2005 Input-Output Tables of China*. Beijing: China Statistics Press.

National Bureau of Statistics (NBS) and Ministry of Environmental Protection (MEP). 2010. *China Statistical Yearbook on Environment 2010*. Beijing: China Statistics Press.

Newell, R. G., and W.A. Pizer. 2003. Regulating stock externalities under uncertainty. *Journal of Environmental Economics and Management* 45(2, Suppl.): 416–432.

Su, M., Z. H. Fu, W. Xu, Z. G. Wang, X. Li, and Q. Liang. 2009. Woguo kai zheng tanshui wenti yanjiu (Study on levying a carbon tax in China). Research Institute for Fiscal Science, Ministry of Finance, Beijing. In Chinese.

United States Energy Information Administration (U.S. EIA). 2009. *Annual Energy Outlook 2009*. Available at http://www.eia.gov/oiaf/archive/aeo09/index.html, last accessed July 19, 2011.

U.S. Environmental Protection Agency (U.S. EPA). 1997. *The Benefits and Costs of the Clean Air Act, 1970 to 1990*. U.S. Environmental Protection Agency, Washington, DC, October.

Vennemo, H., K. Aunan, J. W. He, T. Hu, and S. T. Li. 2009. Benefits and costs to China of three different climate treaties. *Resource and Energy Economics* 31:139–160.

Xinhua News Agency. 2011. Authorized release: National Economic and Social Development Twelfth Five-Year Plan. 16 March. Available at http://news.xinhuanet.com/politics/2011-03/16/c_121193916.htm, last accessed July 22, 2011. In Chinese.

Zhai, F., and S. T. Li. 2000. The Implications of Accession to WTO on China's Economy. Third Annual Conference on Global Economic Analysis, Melbourne, Australia. 27–30 June.

III

Appendixes

A

Economic-Environmental Model of China

Jing Cao, Mun S. Ho, and Dale W. Jorgenson

In this appendix we describe the model for China in some detail, beginning with the modeling of each of the main economic agents. Then in section A.2 we describe the data and parameters underlying the model. A previous version of this model of the Chinese economy is used in Ho and Nielsen (2007), and here we describe the updates to it. This is a multisector model of economic growth where the main drivers of growth are population, total factor productivity growth, and changes in the quality of labor and capital. It has a dynamic recursive structure—that is, a structure in which investment is determined by fixed savings rate as in the Solow model.

A.1   Structure of the Model

We discuss the five main actors in the economy in turn—producers, households, capital owners, government, and foreigners. For easy reference, table A.1 lists variables that are referred to with some frequency. In general, a bar above a symbol indicates that it is a plan parameter or variable, while a tilde indicates a market variable. Symbols without markings are total quantities or average prices. To reduce unnecessary notation, we drop the time subscript t from our equations whenever possible.

A.1.1   Production

Each of the 33 industries is assumed to produce its output using a constant-returns-to-scale technology. For each sector j the output at time t, QI_{jt}, is expressed as

$$QI_j = f(KD_j, LD_j, TD_j, A_{1j}, \dots, A_{nj}, t) \tag{A.1}$$

where KD_j, LD_j, TD_j, and A_{ij} are capital, labor, land, and intermediate inputs, respectively.[1] In sectors for which both plan and market allocation exists, output is

Table A.1
Selected parameters and variables in the economic model

Parameters	
s_i^e	Export subsidy rate on good i
t_i^c	Carbon tax rate on good i
t^k	Tax rate on capital income
t^L	Tax rate on labor income
t_i^r	Net import tariff rate on good i
t_i^t	Net indirect tax (output tax less subsidy) rate on good i
t^x	Unit tax per ton of carbon

Endogenous Variables	
G_I	Interest on government bonds paid to households
G_INV	Investment through the government budget
G_IR	Interest on government bonds paid to the rest of the world
$G\_transfer$	Government transfer payments to households
P_i^{KD}	Rental price of market capital by sector
PE_i^*	Export price in foreign currency for good i
PI_i	Producer price of good i
PI_i^t	Purchaser price of good i including taxes
PL	Average wage
PL_i	Wage in sector i
PM_i	Import price in domestic currency for good i
PM_i^*	Import price in foreign currency for good i
PS_i	Supply price of good i
PT_i	Rental price of land of type i
QI_i	Total output for sector i
QS_i	Total supply for sector i
$r(B^*)$	Payments by enterprises to the rest of the world
$R\_transfer$	Transfers to households from the rest of the world

made up of two components, the plan quota output ($\overline{QI}_j$) and the output sold on the market ($\widetilde{QI}_j$). The plan quota output is sold at the state-set price ($\overline{PI}_j$) while the output in excess of the quota is sold at the market price ($\widetilde{PI}_j$). The PI and QI names are chosen to reflect that these are domestic industry variables, as opposed to commodities (PC) or total supply (PS), the sum of domestic output and imports.

A more detailed discussion of how this plan-market formulation is different from standard market economy models is given in Garbaccio, Ho, and Jorgenson (1999).

In summary, if the constraints are not binding, then the "two-tier plan/market" economy operates at the margin as a market economy with lump sum transfers between agents. The capital stock in each industry consists of two parts—the fixed capital, $\bar{K}_j$, that is inherited from the initial period, and the market portion, $\widehat{KD}_j$, that is rented at the market rate. The before-tax return to the owners of fixed capital in sector j is

$$profit_j = \overline{PI_j}\,\overline{QI}_j + \widehat{PI_j}\,\widehat{QI}_j - \widehat{P_j^{KD}}\,\widehat{KD}_j - PL_jLD_j - PT_jTD_j - \sum_i \overline{PS_i}\,\bar{A}_{ij} - \sum_i \widehat{PS_i}\,\tilde{A}_{ij} \tag{A.2}$$

For each industry, given the capital stock $\bar{K}_j$ and prices, the first-order conditions from maximizing equation (A.2), subject to equation (A.1), determine the market and total input demands. For example, the demand for input i by industry j is given by

$$\widehat{PS_i} = \widehat{PI_j}\,\frac{\partial f(.)}{\partial A_{ij}}$$

Note that this involves only the market prices; the infra-marginal plan prices do not enter into the first-order condition.

Given the lack of reliable time-series data for estimating substitution elasticities, we use simple Cobb-Douglas production functions. We write the production function as a series of nested functions, where the top tier expresses output as a function of capital, labor, land, energy inputs (E), and nonenergy intermediates (M). Equation (A.1) for the output of industry j at time t then becomes

$$QI_{jt} = g(t)KD_{jt}^{\alpha_{Kj}}\,LD_{jt}^{\alpha_{Lj}}\,TD_{jt}^{\alpha_{Tj}}\,E_{jt}^{\alpha_{Ej}}\,M_{jt}^{\alpha_{Mj}} \tag{A.3}$$

where

$$\log E_{jt} = \sum_k \alpha_{kj}^E \log A_{kjt}$$

with $k = \{$coal, oil mining, gas mining, petroleum refining, electricity, gas distribution$\}$, and

$$\log M_{jt} = \sum_k \alpha_{kj}^M \log A_{kjt}$$

with $k = \{$nonenergy intermediate goods$\}$. Here α_{Ej} is the cost share of all energy inputs into industry j (share of the total costs of producing good j), and α_{kj}^E is the cost share of energy of type k within the aggregate energy input. Similarly, α_{Mj} is the cost share of total nonenergy intermediate inputs in industry j, and α_{kj}^M is the share of intermediate nonenergy input of type k within the aggregate nonenergy intermediate input.

To allow for biased technical change—that is, changes in demand for inputs over time which are independent of price changes—the α_{Ej} coefficients are indexed by time and are updated exogenously. We set α_{Ej} to fall gradually over the next 40 years while the labor coefficient α_{Lj} rises correspondingly. The composition of the aggregate energy input E_j (i.e., the α_{kj}^E coefficients) is also allowed to change over time. The coefficient $g(t)$ in equation (A.3) represents technical progress, and the change in $g(t)$ is assumed to be of the exponential form: $\dot{g}_j(t) = A_j \exp(-\mu_j t)$. This implies technical change that is rapid initially but gradually declines toward zero.

The price to buyers of industry output includes the indirect tax on output, the externality ad valorem tax, and the carbon tax per unit output:

$$PI_i^t = (1 + t_i^t + t_i^x)PI_i + t_i^c \tag{A.4}$$

A.1.2 Households

The household sector derives utility from the consumption of commodities, is assumed to supply labor inelastically, and owns a share of the capital stock. It also receives income transfers from the government and foreigners and receives interest on its holdings of public debt. Private income, Y^p, after taxes and the payment of various nontax fees (*FEE*) can then be written as

$$Y^p = YL + DIV + G\_I + G\_transfer + R\_transfer - FEE \tag{A.5}$$

where *YL* denotes labor income from supplying *LS* units of effective labor, less income taxes:

$$YL = (1 - t^L)PLLS \tag{A.6}$$

The relationship between labor demand and supply is given in equation (A.33). *LS* is a function of the working age population, average annual hours, and an index of labor quality:

$$LS_t = POP_t^w hr_t q_t^L \tag{A.7}$$

DIV denotes dividends from the household's share of capital income and is explained in equation (A.23). *G_I* and *G_transfer* represent interest and transfers from the government, and *R_transfer* is transfers from the rest of the world.

Household income is allocated between consumption (VCC_t) and savings. In this model we use a simple Solow growth model formulation with an exogenous savings rate (s_t) to determine private savings (S_t^p):

$$S_t^p = s_t Y_t^p = Y_t^p - VCC_t \tag{A.8}$$

Household utility is a linear logarithmic function of the consumption of goods such that

$$U_t = U(C_{1t}, \ldots, C_{nt}) = \sum_i \alpha_{it}^C \log C_{it} \tag{A.9}$$

The households are allocated some proportion of goods at the plan prices, which we denote by $\bar{C}_i$. Assuming that the plan constraints are not binding, then given market prices and total expenditures, the first-order conditions derived from equation (A.9), as in the producer problem above, determine household demand for commodities, C_i, where $C_i = \bar{C}_i + \tilde{C}_i$. $\tilde{C}_i$ is household purchases of commodities at market prices. The household budget can be written as

$$VCC = \sum_i (\widetilde{PS_i}\tilde{C}_i + \overline{PS_i}\bar{C}_i) \tag{A.10}$$

We use such a Cobb-Douglas utility function because we currently lack the disaggregated data to estimate an income elastic functional form. However, one would expect demand patterns to change with rising incomes, and this is implemented by allowing the α_{it}^C coefficients to change over time.

A.1.3 Government and Taxes

In the model, the government has two major roles. First, it sets plan prices and output quotas and allocates investment funds. Second, it imposes taxes, purchases commodities, and redistributes resources. Public revenue comes from direct taxes on capital, value-added taxes, indirect taxes on output, tariffs on imports, the externality tax, and other nontax receipts:

$$Rev = \sum_j t^k (P_j^{KD} KD_j - D_j) + t^V \sum_j (P_j^{KD} KD_j + PL_j LD_j + PT_j TD_j) + \sum_i t_i^t PI_i QI_i \tag{A.11}$$
$$+ R\_EXT + \sum_i t_i^r PM_i^* M_i + \sum_i t_i^c (QI_i - X_i + M_i) + FEE$$

where D_j is the depreciation allowance and X_i and M_i are the exports and imports of good i.

In particular, the revenue from the externality, or green, tax on output is

$$R\_EXT_t = \sum_j t_{jt}^x PI_{jt} QI_{jt} \tag{A.12}$$

In one application of the model described in Ho and Nielsen (2007, chapter 10), the externality tax rate is set proportional to the marginal air pollution damages from output j:

$$t_{jt}^x = \lambda MD_{jt-1}^O \tag{A.13}$$

When we consider a tax on fossil fuels based on the carbon content, the externality tax per unit of fuel j is

$$t_j^x = tx^u c_j \tag{A.14}$$

where c_j is the carbon content per unit of fuel of type j.

The nontax payments to the government are set as a fixed share of household income:

$$FEE_t = \gamma^{NHH} Y_t^p \tag{A.15}$$

Total government expenditure is the sum of commodity purchases and other payments:

$$Expend = VGG + G\_INV + \sum s_i^e PI_i X_i + G\_I + G\_IR + G\_transfer \tag{A.16}$$

Government purchases of specific commodities are allocated as shares of the total value of government expenditures, VGG. For good i,

$$PS_i G_i = \alpha_i^G VGG \tag{A.17a}$$

We construct a price index for government purchases as $\log PGG = \sum_i \alpha_i^G \log PS_i$. The real quantity of government purchases is then

$$GG = \frac{VGG}{PGG} \tag{A.17b}$$

Transfers are set equal to a fixed rate of the population multiplied by the wage rate:

$$G\_transfer = \gamma^{tr} PL_t POP_t \tag{A.17c}$$

The difference between revenue and expenditure is the deficit ΔG, which is covered by increases in the public debt, both domestic (B) and foreign (B^{G^*}):

$$\Delta G_t = Expend_t - Rev_t \tag{A.18}$$

$$B_t + B_t^{G^*} = B_{t-1} + B_{t-1}^{G^*} + \Delta G_t \tag{A.19}$$

The deficit and interest payments are set exogenously, and equation (A.18) is satisfied by making the level of total nominal government expenditure on goods, VGG, endogenous in the base case. In simulating policy cases we would often set the real government expenditures in the policy case equal to those in the base case. In the policy case we would use some endogenous tax variable to satisfy equation (A.18).

A.1.4   Capital, Investment, and the Financial System

We model the structure of investment in a fairly simple manner. In the Chinese economy, some state-owned enterprises receive investment funds directly from the state budget and are allocated credit on favorable terms through the state-owned banking system. Nonstate enterprises get a negligible share of state investment funds and must borrow at competitive interest rates. There is also a small but growing stock market that provides an alternative channel for private savings. We abstract from these features and define the capital stock in each sector j as the sum of two parts, which we call plan and market capital:

$$K_{jt} = \bar{K}_{jt} + \tilde{K}_{jt} \tag{A.20}$$

The plan portion evolves with plan investment and depreciation:

$$\bar{K}_{jt} = (1 - \delta)\bar{K}_{jt-1} + \psi_t^I \bar{I}_{jt} \; t = 1, 2, \ldots, T \tag{A.21}$$

The rate of depreciation is δ, and ψ_t^I is an aggregation that converts the investment units to capital stock units.[2] In this formulation, $\bar{K}_{j0}$ is the capital stock in sector j at the beginning of the simulation. This portion is assumed to be immobile across sectors. Over time, with depreciation and limited government investment, it will decline in importance. Each sector may also rent capital from the total stock of market capital, $\tilde{K}_t$:

$$\tilde{K}_t = \sum_j \tilde{K}_{jt} \text{ where } \tilde{K}_{ji} > 0 \tag{A.22}$$

The allocation of market capital to individual sectors, $\tilde{K}_{jt}$, is based on sectoral rates of return. As in equation (A.2), the rental price of market capital by sector is $\tilde{P}_j^{KD}$. The supply of $\tilde{K}_{jt}$, subject to equation (A.22), is written as a translog function of all of the market capital rental prices, $\tilde{K}_{jt} = K_j(\tilde{P}_1^{KD}, \ldots, \tilde{P}_n^{KD})$.

In the three sectors of agriculture, crude petroleum, and gas mining, land is a factor of production. We have assumed that agricultural land and oil fields are supplied inelastically, abstracting from the complex property rights issues regarding land in China. After taxes, income derived from plan capital, market capital, and land is either kept as retained earnings by the enterprises, distributed as dividends, or paid to foreign owners:

$$\sum_j profits_j + \sum_j \tilde{P}_j^{KD}\tilde{K}_j + \sum_j PT_jT_j = tax(k) + RE + DIV + r(B^*) \tag{A.23}$$

where $tax(k)$ is total tax on capital and value added (the first two terms on the right-hand side of equation A.11).[3]

As discussed later, total investment in the model is determined by savings. This total *VII* is then distributed to the individual investment goods sectors through fixed shares, α_{it}^I:

$$PS_{it}I_{it} = \alpha_{it}^I VII_t \tag{A.24}$$

A portion of sectoral investment $\bar{I}_t$ is allocated directly by the government, while the remainder $\tilde{I}_t$ is allocated through other channels.[4] The total, I_t, can be written as

$$I_t = \tilde{I}_t + \overline{I}_t = I_{1t}^{\alpha_1^I} I_{2t}^{\alpha_2^I} \cdots I_{nt}^{\alpha_n^I} \tag{A.25}$$

As in equation (A.21) for the plan capital stock, the market capital stock $\tilde{K}_{jt}$ evolves with new market investment:

$$\tilde{K}_{jt} = (1-\delta)\tilde{K}_{jt-1} + \psi_t^I \tilde{I}_{jt} \tag{A.26}$$

A.1.5   Foreign Sector

Trade flows are modeled using the method followed in most single-country models. Imports are considered to be imperfect substitutes for domestic commodities, and exports face a downward-sloping demand curve. We write the total supply of commodity i as a CES function of the domestic (QI_i) and imported good (M_i):

$$QS_i = A_0 \left[\alpha^d QI_i^\rho + \alpha^m M_i^\rho \right]^{\frac{1}{\rho}} \tag{A.27}$$

where $PS_i QS_i = PI_i^t QI_i + PM_i M_i$ is the value of total supply. The purchaser's price for domestic goods PI_i^t is discussed in section A.1.1. The price of imports to buyers is the foreign price plus tariffs (less export subsidies), multiplied by a world relative price, e:

$$PM_i = e(1+t_i^r)PM_i^* \tag{A.28}$$

Exports are written as a simple function of the domestic price relative to world prices adjusted for export subsidies (s_{it}^e):

$$X_{it} = EX_{it} \left(\frac{\widetilde{PI}_{it}}{e_t(1+s_{it}^e)PE_{it}^*} \right)^{\eta_i} \tag{A.29}$$

where EX_{it} is base case exports that are projected exogenously.

The current account balance is equal to exports minus imports less net factor payments plus transfers:

$$CA = \sum_i \frac{PI_i X_i}{(1+s_i^e)} - \sum_i PM_i M_i - r(B^*) - G\_IR + R\_transfer \tag{A.30}$$

Like the government deficits, the current account balances are set exogenously and accumulate into stocks of net foreign debt, both private (B_t^*) and public (B_t^{G*}):

$$B_t^* + B_t^{G^*} = B_{t-1}^* + B_{t-1}^{G^*} - CA_t \tag{A.31}$$

A.1.6   Markets

The economy is in equilibrium in period t when the market prices clear the markets for the 33 commodities and the three factors. The supply of commodity i must satisfy the total of intermediate and final demands:

$$QS_i = \sum_j A_{ij} + C_i + I_i + G_i + X_i, i = 1, 2, \dots, 33 \tag{A.32}$$

For the labor market, we assume that labor is perfectly mobile across sectors so there is one average market wage that balances supply and demand. As is standard in models of this type, we reconcile this wage with the observed spread of sectoral wages using wage distribution coefficients ψ_{jt}^L. Each industry pays $PL_{jt} = \psi_{jt}^L PL_t / (1 - t_j^V)$ for a unit of labor. The labor market equilibrium is then given as

$$\sum_j \psi_{jt}^L LD_{jt} = LS_t \tag{A.33}$$

For the nonplan portion of the capital market, adjustments in the market price of capital $\tilde{P}_j^{KD}$ clear the market in sector j:

$$KD_{jt} = \psi_{jt}^K K_{jt} \tag{A.34}$$

where ψ_{jt}^K converts the units of capital stock into the units used in the production function. The rental price PT_j adjusts to clear the market for land:

$$TD_j = T_j \tag{A.35}$$

where $j = $ "agriculture," "crude petroleum," and "gas mining."

In this model without foresight, investment equals savings. There is no market where the supply of savings is equated to the demand for investment. The sum of savings by households, businesses (as retained earnings), and the government is equal to the total value of investment plus the budget deficit and net foreign investment:

$$S^P + RE + G\_INV = VII + \Delta G + CA \tag{A.36}$$

The budget deficit and current account balance are fixed exogenously in each period. The world relative price (e) adjusts to hold the current account balance at its exogenously determined level.

The model is a constant-returns-to-scale model and is homogeneous in prices; that is, doubling all prices leaves the economy unchanged. We are free to choose any price normalization.

A.2   Parameters, Exogenous Variables, and Data Sources

The key input into the model is the Social Accounting Matrix (SAM) for 2005. This traces the flow of commodities and payments among the producers, household, government, and rest of the world. The SAM is assembled from the 2005 input-output table which was derived from the 2002 benchmark input-output (IO) table.[5] This SAM is constructed by the authors jointly with the Development Research Center in Beijing.[6] A summary of this SAM is given in figure A.1, and the actual matrix used is disaggregated to the 33 sectors and commodities. From this we derive the labor and capital incomes, the tax revenues for each type of tax, the expenditures on specific commodities by the household, government, and foreign sectors, and government payments of all types in equation (A.16).

These payments are combined with employment and capital input data to give the compensation rates for labor and capital for each sector. The estimates for employment and capital stocks by sector are taken from a productivity study of China (Cao et al. 2009) that supplements the official data with labor force surveys. The various tax and subsidy rates are not statutory rates but are implied average rates derived by dividing revenues by the related denominator—value of industry output, capital income, total value added, and imports.

The exogenous variables in the model include total population, working age population, saving rates, dividend payout rates, government taxes and deficits, world prices for traded goods, current account deficits, rate of productivity growth, rate of improvement in capital and labor quality, and work force participation. These variables may of course be endogenous (i.e., they interact among each other), but we ignore this possibility and specify them independently.

The assumption that affects the growth rate the most is the household savings rate s_t. Our assumption is to have s_t beginning at the observed 35.6% for 2005 and gradually falling to 21% in 2020 and 16% in 2030. National private savings is household savings plus the retained earnings of enterprises. There is no official policy for dividend payouts, and we assume that the share of retained earnings will fall and dividend payouts rise to reflect the diminishing role of state enterprises in the economy in the long run. The dividend rate—that is, the share not used for retained earnings—was 40.6% in 2005, and we project it to rise to 67% by 2020. It should be pointed out that national savings and investment in the Chinese data includes capital such as roads and other public infrastructure, items that are excluded from the "gross fixed private investment" item in most other countries' national accounts.

The total government sector (national and local governments combined) ran a deficit of about 300 billion yuan in 2005, and this is assumed to fall gradually to zero by 2020 to give a smooth transition to the long run.

In the labor supply expression equation (A.7), we have the product of the working-age population, annual average hours, and quality. Population projections

Receipts \ Expenditures	Commodity	Industry	Labor	Capital	Land	Households	Enterprise	VAT, Business Tax	Govt	ROW	Capital account	Total
Commodity	0	36051	0	0	0	7122	0	0	2661	6850	7934	60616
Industry	55204	0	0	0	0	0	0	0	0	0	0	55204
Labor	0	6986	0	0	0	0	0	0	0	0	0	6986
Capital	0	8190	0	0	0	0	0	0	0	0	0	8190
Land	0	915	0	0	0	0	0	0	0	0	0	915
Households	0	0	6986	0	787	0	3058	0	418	21	0	11270
Enterprise	0	0	0	8190	128	0	0	0	0	0	0	8319
VAT, Business Tax	0	1457	0	0	0	0	0	0	0	0	0	1457
Government	0	1604	0	0	0	209	789	1457	0	0	300	4360
ROW	5412	0	0	0	0	0	0	0	11	0	0	5423
Capital account	0	0	0	0	0	3939	4472	0	1270	-1447	0	8234
Total	60616	55204	6986	8190	915	11270	8319	1457	4360	5423	8234	

Addendum: GDP=19153

Figure A.1
Summary Social Accounting Matrix for China, 2005 (billion yuan).

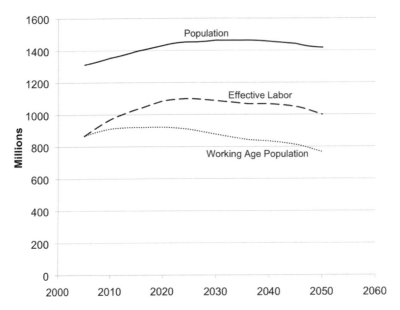

Figure A.2
Population and labor projections.

by age groups are taken from projections made by the Population Division of the Department of Economic and Social Affairs of the United Nations Secretariat.[7] The composition of the work force changes over time with a bigger portion of educated workers, bigger or smaller portion of more experienced workers, and an older average age. This quality of labor input index q_t^l is estimated in Cao et al. (2009) to have grown at 0.9% per year for the period 1983–2000. Given the expectation of continued higher educational attainment in the future, we assume that China's aggregate labor quality will continue to rise but at a diminishing rate. By 2040 the quality index is assumed to grow at only 0.2% per year. For comparison, the U.S. labor quality growth peaked at 0.5% during the 1960s, and fell to 0.3% per year during 1995–2000 (Jorgenson, Ho, and Stiroh 2005, table 6.5).

Total labor hours depend also on the participation rate and annual working hours. There are no comprehensive data on the number of hours worked, and based on comparisons to other countries, we project it to rise as a result of improvements in the functioning of the labor market—lower underemployment, seasonal unemployment, and other labor market frictions. We assume that hours worked per capita rise at 0.5% per year initially but slow down over time. The results are plotted in figure A.2.

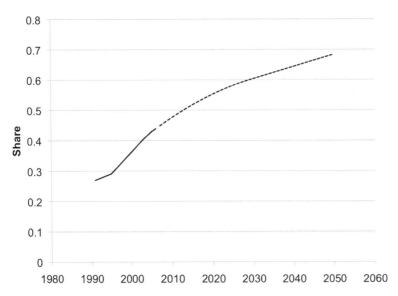

Figure A.3
Urban share of population, sample period (solid) and projections (dashed).

The health effects of air pollution depend to a large degree on the size of the urban population, that is, the population exposed to the high levels of pollution concentration. Garbaccio, Ho, and Jorgenson (2000) note that the urbanization rate of the Chinese population was quite similar to the U.S. experience at a comparable period and argue that it is reasonable to project a rapidly rising rate for China. Liang and O'Neill (2009) made a projection of urbanization rates for various countries in a report for the International Institute for Applied Systems Analysis (IIASA), and we use their projections for China. The result is plotted in figure A.3.

An adjustment for improvements in future capital "quality," or composition, is also made (the ψ_t^I coefficient in equation A.21). Cao et al. (2009) note how the composition of the capital stock in China has shifted toward assets with shorter life, that is, toward a smaller share of structures and a larger share of equipment such as computers. They explain how assets that have shorter useful lives generate higher annual capital services per dollar of capital stock, and hence are of a higher quality in the terminology of Jorgenson, Ho, and Stiroh (2005). We project that capital quality rises by 2% per year initially, then gradually decelerates. The supply of land for agriculture, oil mining, and gas mining is simply set fixed for all periods equal to the base year value.

Table A.2 Miscellaneous tax rates and coefficients

Rate	Symbol	Coefficient
Tax rate on capital income	tk	0.0948
Indirect tax rate on output	tt	−0.004 to 0.074
VAT rate	tv	0 to 0.189
Import tax rate	tr	0 to 0.14
Nontax payment share	γ^{NHH}	0.0109
Government transfer rate	γ^{tr}	0.2576
Household savings rate (2005)		0.3562
Dividend payout rate (2005)		0.4061

Tax rates are set equal to those for 2005 derived from the SAM. These are summarized in table A.2. For the government deficit, ΔG, we set it at the base year 1.57% of GDP initially, declining steadily towards zero in the long run. These deficits are accumulated into the stocks of domestic and foreign debt, B_t and B_t^{G*}, assuming a constant division between domestic and foreign borrowing. Data for the stock of debt and interest paid on it comes from the *China Statistical Yearbook* (NBS 2008a, table 8-13, 8-19) and the 2005 Social Accounting Matrix. Government transfers, *G_transfer*, are set to rise in proportion with population and average wage. The nontax fees paid by households are set to be a fixed share of GDP equal to the base year's share (table A.2).

The current account balance has swung to a huge surplus in the recent years. There is no simple consensus about the future evolution of this variable, and after setting it as a share of GDP at the observed sample period values, we set it to decline rapidly to zero. This CA_t deficit is also the assumed rate of borrowing from the world. Import prices PM^{i*} are assumed fixed at the base year value for every period with one important exception. World oil price forecasts are taken from the U.S. Energy Information Administration (2009) and shown in figure A.4.[8] The model also requires estimates of world demand for Chinese exports, EX_{it}. In line with recent Chinese experience, we project a rather high initial rate of growth for exports, beginning at a 7% growth rate and falling gradually.

The base year input-output data for 2005 were constructed by the National Bureau of Statistics in 2008 (NBS 2008b), and when we incorporated those into our model in 2010, the macroeconomic variables for 2006–2008 were released. These include the GDP, investment, and current account surplus. The current account has swung into a large surplus together with a high share of investment in GDP. As a result, the private savings rate is unusually high for 2006 and 2007. We

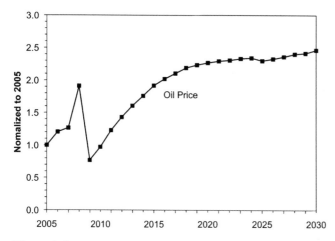

Figure A.4
Oil price projected in base case, normalized to 1.0 in 2005. Oil price is world price projected by the U.S. Energy Information Administration.

take these figures into account in setting the savings rate and current account balance as separate shares of GDP for these years.

A.2.1 Parameters

The rate of productivity growth is another factor that has a large effect on the base case growth rate of the economy but has little impact on the difference between cases. Total factor productivity (TFP) growth at the industry level in the 1982–2000 period shows a very wide range of performance as estimated by Cao et al. (2009), ranging from –10% to 5% per year. The Domar-weighted productivity growth for all industries was 2.7% for 1982–2000. To keep the base case as simple as possible we ignore this wide range of observed TFP growth, and in our projections of sector productivity terms in equation (A.3), we initially set all the μ_j's to the same value, 0.018. These are then adjusted to match actual GDP growth rates in the initial years for which we have actual data.

The value share parameters of the production functions (α_{Kj}, α_{Lj}, etc.) are set to the values in the 2005 IO table in the first year of the simulation. For future periods we change most of these parameters such that they gradually resemble the shares found in the U.S. IO table for 1997. The exceptions to this approach are the coal inputs for all the sectors, which are set to converge to a value between current Chinese and 1997 U.S. shares.[9] The rate of reduction in energy use is set at a modest level relative to the rapid improvements in recent Chinese history. We assume that the share of energy in industry output is reduced gradually to 60% of the 2005 levels in 40 years. This assumption is conservative compared, for example, to the performance of the electric power industry during the 1990–1999 period. In that time the thermal output grew 88%, whereas coal input only rose 61%, a rate of improvement of approximately 1.5% per year (NBS 1999, tables 4-5 and 4-15).

The α_{it}^C parameters of the consumption function are set in a similar way. That is, for the first period they are equal to the shares in the 2005 Social Accounting Matrix, and for the future periods they gradually approach 1997 U.S. shares except for coal. This projection implies a higher projected demand for private vehicles and gasoline than that assumed in most other models of China. The coefficients determining demand for different types of investment goods (α_{it}^I) and different types of government purchases (α_{it}^G) are projected identically.

Given the lack of estimates for trade elasticities for China, we simply set them at some conservative values. The ρ coefficients in the import demand functions are set at 0.2, and the η coefficients in the export function (A.29) are set to -1.2. The base share of exports and imports are taken from the SAM.

Notes

1. QI_j denotes the quantity of industry j's output. This is to distinguish it from QC_j, the quantity of commodity j. In the actual model each industry may produce more than one commodity, and each commodity may be produced by more than one industry. In the language of the input-output tables, we make use of both the USE and MAKE matrices. For ease of exposition, we ignore this distinction here.

2. Both K and I are aggregates of many asset types, ranging from computer equipment to structures. The compositions of total investment and total capital stock are different, and an aggregation coefficient is needed to reconcile the historical series.

3. In China, a substantial part of the "dividends" is actually income from agricultural land.

4. It should be noted that the industries in the Chinese accounts include many sectors that would be considered public goods in other countries. Examples include local transit, education, and health.

5. The input-output table is given in NBS (2008b). The benchmark IO table for 2002 is derived from detailed enterprise census data. The 2005 IO table is extrapolated by the NBS using less detailed data.

6. We are grateful to Shantong Li and Jianwu He of the Department of Development Strategy and Regional Economy, Development Research Center, State Council of the PRC, for sharing their data and ideas with us. The procedure for constructing this Social Accounting Matrix is described in Zhai and Li (2000).

7. The demographic projections are from *World Population Prospects: The 2008 Revision* downloaded from the U.N. web site: http://esa.un.org/unpp/.

8. The projections for crude oil prices are taken from U.S. EIA (2009, table 112).

9. We have chosen to use U.S. patterns in our projections of these exogenous parameters because they seem to be a reasonable anchor. While it is unlikely that China's economy in 40 years' time will mirror the U.S. economy of 1997, it is also unlikely to closely resemble any other economy. Other projections, such as those by the World Bank (1994), use the input-output tables of developed countries including the United States.

References

Cao, J., M. Ho, D. Jorgenson, R. Ren, L. L. Sun, and X. M. Yue. 2009. Industrial and aggregate measures of productivity growth in China, 1982–2000. *Review of Income and Wealth* 55(s1):485–513.

Garbaccio, R., M. Ho, and D. Jorgenson. 1999. Controlling carbon emissions in China. *Environment and Development Economics* 4(4):493-518.

———. 2000. The Health Benefits of Controlling Carbon Emissions in China. Proceedings of Workshop on the Ancillary Benefits and Costs of Greenhouse Gas Mitigation. Organisation for Economic Co-operation and Development. Washington, DC. 27-29 March.

Ho, M. S., and C. P. Nielsen, eds. 2007. *Clearing the Air: The Health and Economic Damages of Air Pollution in China*. Cambridge, MA: MIT Press.

Jorgenson, D. W., M. S. Ho, and K. J. Stiroh. 2005. *Information Technology and the American Growth Resurgence*. Cambridge, MA: MIT Press.

Liang, L. W., and B. O'Neill. 2009. Household Projections for Rural and Urban Areas of Major Regions of the World. International Institute for Applied Systems Analysis Report, IR-09-026.

National Bureau of Statistics (NBS). 1999. *China Energy Statistical Yearbook, 1997–1999*. Beijing: China Statistics Press.

———. 2008a. *China Statistical Yearbook, 2007*. Beijing: China Statistics Press.

———. 2008b. *2005 Input-Output Tables of China*. Beijing: China Statistics Press.

United States Energy Information Administration (U.S. EIA). 2009. *Annual Energy Outlook 2009*. Available at http://www.eia.gov/oiaf/archive/aeo09/index.html, last accessed July 19, 2011.

World Bank. 1994. *China: Issues and Option in Greenhouse Gas Emissions Control*. Washington, DC: World Bank.

Zhai, F., and S. T. Li. 2000. The Implications of Accession to WTO on China's Economy. Third Annual Conference on Global Economic Analysis, Melbourne, Australia. 27–30 June.

B

The Valuation of Health Damages

Yu Lei and Mun S. Ho

B.1   Introduction

In order to compare the costs and benefits of policies, economists often monetize health damages. For example, the U.S. Environmental Protection Agency's *Guidelines for Preparing Economic Analyses* provides a discussion of the various methods of estimating the value of statistical life for use in valuing the reduction in mortality rates due to environmental policies (U.S. EPA 2000, chapter 7). Our previous integrated study, Ho and Nielsen (2007), used valuations derived from a survey reported in the same book by our colleagues Zhou and Hammitt (2007), along with evidence from an earlier Harvard China Project study (Wang and Mullahy 2006) and from the international literature, to monetize the benefits of green taxes. The yuan values of the health benefits give us a common metric to compare to the yuan costs of lower consumption or lower GDP. The valuations of the damage reduction due to the policies examined in this book are given in chapter 9. In this appendix we explain how we derived our valuation parameters.

The methodology and reasoning behind the valuation of environmental damages is discussed comprehensively in many places, including the previously cited U.S. EPA (2000) and Viscusi and Aldy (2003). Some specific studies for China are discussed in Ho and Nielsen (2007, chapter 9). We summarize this discussion briefly here for those who may not be familiar with this literature.

Two different measures have been used to monetize health effects; one is the "human capital" approach, which simply counts the wages lost and medical costs of illnesses or premature death. This method ignores pain, loss of leisure time, and other losses, and economists prefer a more comprehensive measure of the losses when available. We thus have the second measure, the "willingness-to-pay" (WTP) approach that is closer to the preferred measure of welfare. The WTP for, say, a reduction in the risk of contracting chronic bronchitis during the year is the maximum amount an individual is willing to pay in exchange for such a reduction. The two main approaches used to estimate the WTP, discussed next, will give an idea of the concepts involved.

One common method used to assess the WTP is by comparing the wages of different occupations with different risks. Jobs that have a higher risk of injury, illness, or death such as those in coal mining, construction, or emergency services pay a higher wage relative to less risky jobs with otherwise equivalent attributes, such as skill level. Economists ascribe this gap to the amount needed to compensate workers for these higher risks. From information about wages for different occupations and the probability of fatal accidents or other health risks, analysts derive a measure of the WTP to avoid those risks. The method is known as a "revealed preference" approach, because values of the worker (WTP to avoid risk) are revealed by his or her market preferences (in this case in the labor market, choosing a job with known risks at a particular wage). We know of only one revealed preference study in China, Guo and Hammitt (2009), which was conducted earlier in the Harvard China Project.

The second method, the "contingent valuation" (or "stated preference") approach, involves surveying people about their valuation of risks in a hypothetical setting. For example, Zhou and Hammitt (2007) asked randomly selected adults in three parts of China questions like the following: (1) how much they were willing to pay to prevent a minor illness like a recent cold; (2) how much they were willing to pay to reduce their lifetime chance of developing chronic bronchitis by 5 percent from an initial risk of 18 percent; and (3) how much they were willing to pay to reduce their probability of death in the next year by 1 in 1000, from an initial level of 7 in 1000.

The value of statistical life (VSL) is essentially calculated by dividing the mean amount people were willing to pay by the reduction in mortality risk:

$$\text{VSL} = \left(\frac{WTP}{1/1000} \right) \tag{B.1}$$

From their survey results, Zhou and Hammitt estimated a mean VSL between 230,000 and 500,000 yuan in 1999.

In this study we use the results from the contingent valuation surveys for the VSL to value the mortality effects. Given the lack of estimates for morbidity WTP values in China, we use a simple cost of illness (COI) method. We describe these in turn in the following sections.

B.2   Mortality Valuation

We surveyed the Chinese VSL literature and found the studies that are summarized in table B.1. Note first that with one exception these studies are not based on national samples; each study surveys people in particular areas of China, and obviously the results from high-income Shanghai would be different from rural Anqing, Anhui. Second, the surveys were conducted in different years, and there were substantial changes in real and nominal incomes during that period. The VSL estimates

Table B.1
Comparison of value of statistical life (VSL) estimates for China

Study[a]	Location of survey or study data	Year of survey[b]	VSL (1000 yuan in survey year yuan)	Mean income in survey year (yuan)	VSL/ Income	VSL (1000 yuan in year 2000 yuan)	Method[c]
Wang and Mullahy (2006*)	Chongqing	1998	286	5467	52.3	288	CV
Zhou and Hammitt (2007*)	Beijing	1999	514	9183	56.0	524	CV
Zhou and Hammitt (2007*)	Anqing	1999	235	5320	44.2	240	CV
Guo and Hammitt (2009*)	Nationwide	2000	360	6208	58.0	360	Wage
Guo, Haab, and Hammitt (2006*)	Chengdu	2005	190	11,359	16.7	163	CV
Krupnick et al. (2006)	Shanghai/ Chongqing	2005	1,400	14,912	93.9	1,198	CV
National mean		2005	341	6367	53.5		

[a]Studies marked with an asterisk were conducted under the Harvard China Project.
[b]Guo and Hammitt (2009*) used data from various years, but adjusted to 2000 prices.
[c]CV denotes a contingent valuation (stated preference) study; "Wage" denotes a wage differential (revealed preference) study.

from the studies are given in the fourth column and inflated or deflated to year 2000 prices in the seventh column.

The obvious feature of the estimates is that the VSL differs widely across the studies, and it can differ by region in the same study (i.e., using the same survey methods), as in Zhou and Hammitt (2007). While there is no more than one study for any region in the same year, we note that Chongqing is covered by both Wang and Mullahy (2006) and Krupnick et al. (2006) and that these show a big differ-ence; this may be due either to the differing survey years or the methods used, or to both. The estimated VSLs range from 190,000 yuan for Chengdu in Guo, Haab, and Hammitt (2006) to 1.4 million yuan for Shanghai and Chongqing in Krupnick et al. (2006); the latter thus is more than seven times the former.

These studies, as well as others, have noted how the WTP depends on age, edu-cation, and obviously, income. Wang and Mullahy (2006), for example, note how sensitive WTP is to income. The income effect on VSLs is a very important issue because income elasticity is used to translate estimates from one region to another, or from one year to another.

To give a quick idea of the possible income effect, we collected the disposable income per capita for the cities and regions in those studies from the regional sta-tistical yearbooks or government announcements. These mean incomes are given in the fifth column of table B.1. In the next column we compute the ratio of the mean VSL to mean income. We can see that this ratio ranges from 16.7 in the Chengdu study by Guo, Haab, and Hammitt (2006) to 93.9 in the study of Shanghai and Chongqing by Krupnick et al. (2006); a factor of 5.6. Of course, this range of values is not necessarily the result of the income elasticity of WTP for risk reduction; it may simply reflect the noncomparability of the different study methods. The differ-ence for the mean VSLs between the Guo, Haab, and Hammitt (2006) and Krupnick et al. (2006) studies is huge, despite the surveys being carried out in the same year and in cities featuring relatively high mean incomes. Nevertheless, each of the first four studies listed in table B.1, all carried out around the year 1999, generated a VSL-to-income ratio that is remarkably similar.

Widely ranging VSL estimates are typical across CV studies, and they pose an unfortunately familiar research quandary: how should one distill this evidence for a VSL estimate to use in our cost-benefit analysis? A formal meta-analysis is beyond the scope of our study, and thus, given the similarity in the VSL-to-income ratios, we choose to take the simple mean ratio of the studies and apply it to the mean national income. This mean ratio for the first four studies is 52.6, and the mean for all six studies is a remarkably similar 53.5. We therefore use 53.5 to estimate the national VSL. The disposal income in 2005, averaged over the urban mean and rural mean, was 6367 yuan per capita. Multiplying this by 53.5 gives 340,000 yuan (in 2005 currency) as the national VSL.[1]

In Ho and Nielsen (2007, chapter 9), we discussed how we arrived at a central estimate of 370,000 yuan (in 1997 currency) for the VSL, based on Zhou and Hammitt (2007). That this figure is so close to 340,000 yuan is unsurprising given that the late 1990s was a period of low inflation in China, that two of the six estimates in table B.1 are from the same Zhou and Hammitt study, and that the VSL-to-income ratios are quite similar across different studies.

The economic model described in chapter 9 projects a rapidly rising level of income in China, in line with most forecasts. Thus the VSLs estimated for the base year would rapidly be inapplicable, as we need to adjust the VSL in subsequent years for rising incomes. As we have discussed in Ho and Nielsen (2007, section 9.3.4), the estimates for income elasticities from cross-sectional studies and cross-country studies of WTP are very difficult to reconcile and to apply. We choose to use the most transparent alternative and scale the VSLs proportionately to per capita incomes, that is, to assume an income elasticity of one.

Let V_{ht} denote the value of health effect h in period t, using base year 2005 prices. The various health effects are listed in table B.2. Thus the value of mortality effects is given by multiplying the mean VSL-to-income ratio by y_t, the per capita income in year t:

$$V_{ht} = \bar{v}y_t \tag{B.2}$$

where $\bar{v} = 53.5$ and $h = $ mortality.

The projections of the base case income per capita are given in table B.2, and the resulting projected VSLs are in the last column. Over the 10-year period, the VSL would rise by more than 230 percent.

Table B.2
Projected value of statistical life (VSL) in 2005 yuan

Year	Per Capita Income Index	VSL
2005	1	340,277
2010	1.52	517,221
2015	2.32	789,443

B.3   Morbidity Valuation

Our first preference is to use WTP estimates to value morbidity risks. However, there are few WTP studies for morbidity in China, and the ones that exist have focused on just a few endpoints such as chronic bronchitis and asthma. One of these is Zhou and Hammitt (2007) that, as noted earlier, asked respondents to value risks of chronic bronchitis in a 1999 CV survey. They estimated a value of 13,360 yuan for chronic bronchitis in Anqing, and 27,440 for Beijing.

The Guo, Haab, and Hammitt (2006) survey of Chengdu in 2005 covered asthma, yielding an estimated median value of a statistical case of asthma at 13,685 yuan. Peng and Tian (2003) is not a conventional WTP study, in that they only surveyed 356 patients with respiratory diseases in Shanghai, estimating a value of 21,739 yuan from this sample of people already experiencing this illness.

Given this paucity of evidence from Chinese studies, one can alternatively transfer WTP estimates for all the health endpoints from other countries or use the simpler cost-of-illness (COI) measure. Given the uncertainties about using estimates from rich countries for China, we choose to use the more transparently estimated COIs. For this we include direct medical expenses: medicines, laboratory tests, treatment and overnight stays in health care facilities, and indirect costs of lost wages during hospital stays. There are no estimates of other costs such as transportation to health facilities or costs of lodging during treatment and recovery, and these have to be left out.

The direct medical expenses for cardiovascular hospital admission, respiratory hospital admission, and outpatient visits in 2005 are taken from the *Chinese Health Statistics Yearbook 2005* (MOH 2006). These are given in the first row of table B.3. Average durations of cardiovascular and respiratory hospital admissions are also obtained from the same source. The wages lost are estimated simply by multiplying the duration of hospital admissions by the average daily wage in 2005. For outpatient visits, each one is assumed to have lost one day of wages. The estimated lost wages are given in the third row of table B.3.

To gain further insight, we collected historical data on medical expenses. Since the Chinese health care reforms in the 1990s, there have been dramatic changes, moving from a state-provided system to a market system. Medical care–related costs and behaviors changed dramatically with these reforms. We thus concentrate on the data for 2000–2005. There was less information about expenses differentiated by diseases prior to 2005, and we computed an average over all of the available disease categories. We assume that the trend of these average expenses applies also to the

Table B.3

Cost of illness (COI), base year 2005 and projections

	Cardiovascular Hospital Admissions	Respiratory Hospital Admissions	Outpatient Visits
Costs in 2005			
Direct medical expenses (yuan 2005)	6627	3006	127
Duration (days)	12.0	7.9	1.0
Wages lost (yuan 2005)	209	138	17
Total COI	6836	3144	144
Projections of COI (yuan 2005)			
2005	6836	3144	144
2010	10,402	5203	220
2015	15,873	8379	335

Source: MOH (2006).

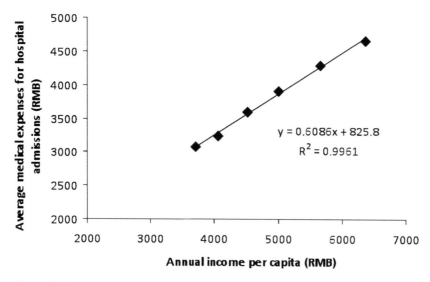

Figure B.1

Medical expenses for hospital admissions and annual per capita income, 2000–2005.

trend for cardiovascular and respiratory diseases. These average medical expenses are compared to average personal income in figure B.1.

The plot of the points lies remarkably close to a straight line, and we thus use it to project future medical costs. That is, as in the VSL projections given in table B.2, we assume that the value of morbidity risks, V_{ht}, rises proportionally with income per capita. The projections, in 2005 yuan, are given in the bottom half of table B.3.

B.4   Valuation of Policy Effects

Given the preceding estimates of the VSL and COI for hospital admissions and outpatient visits, we can now put a yuan value to the health effects estimated in chapter 8. The value of health damages is the number of cases of health effect h (HE_{ht}) multiplied by its valuation, and then summed over all h. The value of avoided health damages in year t due to policy p, compared to the base case, is thus simply

$$V_t^p = \sum_h \left(HE_{ht}^b - HE_{ht}^p \right) V_{ht} \tag{B.3}$$

We should point out that while the policy may change incomes compared to the base case, we do not change the valuations. They remain at the V_{ht}'s given in tables B.2 and B.3, valuations that are projected from the base case simulation of the economic model.

Note

1. Table 10-1 of the *China Statistical Yearbook 2007* (NBS 2008) gives the mean disposable incomes of urban and rural households, and table 4-1 gives the population by residence. These concepts correspond to the incomes for the regions given in the fifth column of table B.1.

References

Guo, X. Q., T. C. Haab, and J. K. Hammitt. 2006*. Contingent valuation and the economic value of air-pollution-related health risks in China. American Agricultural Economics Association Annual Meeting. Long Beach, California, 23–26 July.

Guo, X. Q., and J. K. Hammitt. 2009*. Compensating wage differentials with unemployment: Evidence from China. *Environmental and Resource Economics* 42:187–209.

Ho, M. S., and C. P. Nielsen, eds. 2007*. *Clearing the Air: The Health and Economic Damages of Air Pollution in China.* Cambridge, MA: MIT Press.

Krupnick, A., S. Hoffmann, B. Larsen, X. Z. Peng, R. Tao, and C. Yan. 2006. *The Willingness to Pay for Mortality Risk Reductions in Shanghai and Chongqing, China.* World Bank. Washington, DC.

Ministry of Health of China (MOH). 2006. *China's Health Statistics Yearbook 2006.* Beijing: Peking Union Medical College Press. In Chinese.

National Bureau of Statistics (NBS). 2008. *China Statistical Yearbook 2007*. Beijing: China Statistical Press.

Peng, X., and W. Tian. 2003. WTP Study on the economic loss of the air-pollution-related diseases in Shanghai. *World Economic Forum* 2:32–43. In Chinese.

U.S. Environmental Protection Agency (U.S. EPA). 2000. *Guidelines for Preparing Economic Analyses*. U.S. EPA report 240-R-00-003. Washington, DC, September.

Viscusi, W. K., and J. E. Aldy. 2003. The value of a statistical life: A critical review of market estimates throughout the world. *Journal of Risk and Uncertainty* 27(1):5–76.

Wang, H., and J. Mullahy. 2006*. Willingness to pay for reducing fatal risk by improving air quality: A contingent valuation study in Chongqing, China. *Science of the Total Environment* 367:50–57.

Zhou, Y., and J. K. Hammitt. 2007*. The economic value of air-pollution-related health risks in China: A contingent valuation study. Chapter 8 in *Clearing the Air: The Health and Economic Damages of Air Pollution in China*, ed. M. S. Ho and C. P. Nielsen. Cambridge, MA: MIT Press.

C

New Assumptions and Methods for the 2013–2020 Policy Cases

Yu Zhao, Yuxuan Wang, Yu Lei, and Chris P. Nielsen

C.1 Assumptions of the Emission Estimates

C.1.1 Base Year 2007

The methodology of estimating emissions for the base year of 2007 is generally the same as that for 2005 described in chapters 4–6, with some new assumptions and adjustments as follows.

The unit-based database of coal-fired power plants for 2007 is adjusted from that originally compiled for 2005, using information about newly built and retired units between 2005 and 2007 that was already used in emission estimation for the 2006–2010 policy analyses.

New on-road vehicles are assumed to have met China's stage I and II standards (equivalent to Euro I and II) beginning in 2000 and 2004, respectively. The fleet compositions of different control levels by vehicle type for 2007 are calculated based on an average vehicle age of 15 years.

While emission factors and removal efficiencies of air pollution control devices for detailed technologies are assumed unchanged from 2005 to 2007, the penetration of newer and larger plants with advanced technologies is estimated to have reduced the sector-average emission factors for given sources in such sectors as power generation, on-road transportation, and cement production.

C.1.2 Base Case 2020

The base case for 2020 must include national emission control policies after 2010, especially with regard to China's new NO_x control strategies; the results are thus an educated guess of China's emissions.

The activity levels (i.e., energy consumption or industrial production) by sector are obtained from the projections of the economic model for these new policy cases, as described in chapter 3.

For the coal-fired power sector, the removal efficiency of wet flue gas desulfurization (FGD) for units built before or in 2010 is assumed to be 80%, while that for units built after 2010 is assumed to be 90%. The units in national "key areas" (i.e., those with large populations, concentrated economic activity, and high emissions, including eastern, south-central, and north-central China) are assumed to have installed selective catalytic reduction (SCR) systems, as introduced in chapter 4, or selective noncatalytic reduction (SNCR) systems. The NO_x removal efficiency is assumed to be 60% (a typical value for SNCR) for units built before or in 2010, and 70% (a typical value for SCR) for units after 2010.

For cement production, as a result of penetration of precalciner kilns with advanced emission control devices discussed in chapters 5 and 6, the sector-average emission factors are estimated to decline from 8.7 to 4.4 kg SO_2 per ton of coal, and 33 to 9 kg particulate matter (PM) per ton of coal, from 2007 to 2020. Only a tiny improvement in the emissions factor of NO_x (from 10.0 to 9.9 kg NO_x per ton of coal) is projected for the same period (Lei et al. 2011).

For iron and steel production, FGD systems are assumed to be installed, and the removal efficiency of SO_2 is assumed to be 80%. Because of the side benefits of wet FGD on PM control, 80% of PM emissions are assumed to be reduced simultaneously.

For other fossil fuel combustion in the industrial and combined residential/commercial sectors, 15% and 20% reductions of emission factors are assumed, respectively, for SO_2 and NO_x from 2007 to 2020, attributed to implementation of current emission control legislation. The NO_x emission factors for those sources, however, are assumed unchanged.

For transportation, new on-road vehicles have been required since 2007 to meet China's stage III standard (equivalent to Euro III) and off-road sources to meet the stage I standard. We assume that stage IV and V standards will be required for on-road vehicles after 2010 and 2012, respectively, and stage II and III standards for nonroad sources in the same years. Based on a projection of fleet composition, the average NO_x emission factors for on-road and nonroad sources are estimated to decline 55% and 28% from 2007 to 2020, respectively, and the analogous emission factors for PM are 82% and 55%, respectively.

For all the carbon tax cases in 2020, the emission factors are assumed unchanged from those for base case, and the changes of activity levels are obtained from the estimates of the economic model.

C.2   Assumptions of the Atmospheric Simulations

The 2020 simulations are conducted using the same version of the nested-grid GEOS-Chem model as in chapter 7. The meteorological fields used are from June 2006 to December 2007. Model simulations from June to December 2006 are for spin-up purposes, and the results from January 2007 to December 2007 are used for the actual analysis. This is different from the simulations of chapter 7, in which we used the 2006 meteorology.

Except for the anthropogenic emissions of species provided for the 2020 cases by the research described in the prior section (i.e., SO_2, NO_X, $PM_{2.5}$, PM_{10}, BC, and OC for the base case, scenarios 2/4 and scenario 3), the model uses the 2007 base year emissions for all other species, including volatile organic compounds (VOCs).

C.3   Assumptions of the Health and Agricultural Benefit Assessments

The recommendations from chapter 8 on the concentration-response functions for the health assessment, as well as on the crop distributions, exposure indices, and exposure-relative yield functions for the agricultural assessment, are applied in the 2013–2020 tax cases.

To scale up the population distribution to 2020 for the health assessment, we first assume that China's population will reach 1.43 billion by that year. We assume that the geographical distribution is constant with the 2010 simulations of chapter 7, because no systematic information on nationwide migration is available for us to adjust it.

The health valuation function is changed from that developed in appendix A and applied in chapter 9 according to output of the economic model described in chapter 3 and methodologies built into BenMAP. The value of a statistical life (VSL) and costs of hospital admissions and outpatient visits are estimated to rise along with growth of average income and GDP. Our estimate of VSL in 2020 is 1.33 million yuan in 2007 yuan, and the costs in 2020 of hospital admissions due to cardiovascular disease, hospital admissions due to respiratory disease, and outpatient visits are estimated to be 32,700 yuan, 15,000 yuan, and 600 yuan, respectively (all in 2007 yuan).

The average market prices of wheat, rice, and maize in China in the base year of 2007 are reported by FAO (2010) at 1575, 1762, and 1471 yuan per ton, respectively. The market prices in 2020 will increase according to the projection by the economic model of an increase in price of all agricultural output relative to the average of all commodities. This projection yields price estimates of 2294, 2514, and 2142 yuan per ton (in 2007 yuan), respectively, in 2020, which are used to value the increased grain yields resulting from net ozone reductions under the carbon taxes. These prices are within 20% of alternative estimates for 2020 derived by linear extrapolation of historic prices of these grains in China from 1991 to 2009, as reported in FAO (2010).

References

Food and Agriculture Organization of the United Nations (FAO). 2010. *FAO PriceSTAT database*. Available at http://faostat.fao.org/site/570/default.aspx#ancor, last accessed on December 18, 2011.

Lei, Y, Q. Zhang, C. P. Nielsen, and K. B. He. 2011. An inventory of primary air pollutants and CO_2 emissions from cement production in China, 1990–2020. *Atmospheric Environment* 45(1):147–154.

Contributors

Jing Cao is Associate Professor, Department of Economics, School of Economics and Management, Tsinghua University. She earned her Ph.D. in public policy from the Harvard Kennedy School of Government.

Kebin He is Professor, School of Environment, Tsinghua University, where he also received his Ph.D. in environmental engineering.

Mun S. Ho is Visiting Fellow at the Institute for Quantitative Social Science, Harvard University, and Visiting Scholar at Resources for the Future, Washington, DC. He earned a Ph.D. in economics from Harvard University.

Dale W. Jorgenson is Samuel W. Morris University Professor, Harvard University. He received his Ph.D. in economics from Harvard University.

Yu Lei is Associate Research Fellow and Assistant Director of the Atmospheric Environmental Institute, Chinese Academy for Environmental Planning, Beijing. He earned a Ph.D. in environmental science and engineering from Tsinghua University and was a postdoctoral researcher at the Harvard China Project and Harvard School of Engineering and Applied Sciences.

Chris P. Nielsen is the Executive Director of the Harvard China Project, School of Engineering and Applied Sciences, Harvard University. He received an S.M. in technology and policy from the Massachusetts Institute of Technology.

Yuxuan Wang is Associate Professor, Center for Earth System Science, Tsinghua University, Beijing. She earned a Ph.D. in atmospheric science from the Department of Earth and Planetary Sciences, Harvard University, where she also served as a postdoctoral researcher, research associate, and lecturer.

Wei Wei is Assistant Professor, School of Environmental Science and Technology, Beijing University of Technology. She received a Ph.D. in environmental science and engineering from Tsinghua University.

Qiang Zhang is Associate Professor, Center for Earth System Science, Tsinghua University. He received a Ph.D. in environmental science and engineering from Tsinghua University.

Yu Zhao is Professor, School of Environment, Nanjing University. He earned a Ph.D. in environmental science and engineering from Tsinghua University and was a postdoctoral researcher at the Harvard China Project and Harvard School of Engineering and Applied Sciences.

Index

Note: Page numbers in italics refer to figures, tables, and boxed material.

ACE-Asia (Asian Pacific Regional Aerosol Characterization Experiment), 225
Acetaldehyde (C_2H_4O) emissions, 243, *244*
Activity levels
  for cement production, 206, *207*, 231
  for coal-fired power plant, 166–168, *167*, *168*
  for emission inventory
    in 2005, 229, *229–233*, *230*
    in 2010, *233–235*, *234*, *235*
Acute premature mortality
  under 11th FYP SO_2 controls and past carbon tax scenario, *92*, 92–95, *94*, *96*, 296–300, 302–303, 306–308, *307*, *309–310*
  under future carbon tax scenarios on, 138–143, *144–147*
  ozone and, 302–304, 306, *306–307*, *310*
  $PM_{2.5}$ emissions and, 297–300, *299*, 306–308, *307*, *309*
Ad valorem tax, 371n14
AEEI (autonomous energy efficiency improvement), 335, 370n5
Aerosol optical depth (AOD), 32–34, *33*
Agricultural benefits
  assessment of, 311–323
  assumptions and methods for future, 405–406
  concentration-response relationships in, 317–319, *319*
  main grain crops in, 311–313, *312*, *314–315*
  ozone exposure indices for, 313–316
  rationale for, 311
  of future carbon tax scenarios, 138, 143, *144–147*
  of past policy scenarios, 88–95
    under carbon tax scenario, 94–95, *96*, 319–323, *320–323*, *357*, *362*
    under 11th FYP SO_2 controls, 94, *94*, 319–323, *320–323*, *348*
    framework for, 93–94
Agricultural sector, activity level of, 231
Aircraft-based measurement campaigns, 30
Air Pollution Index (API), 50n11
Air Quality Index (AQI), 50n11
Akimoto et al. (2006), 191, 225
Aldehyde emissions, 243, *244*
Alkane emissions, 243, *243*, *244*
Alkene emissions, 243, *243*, *244*
Alkyne emissions, *243*
Ammonia (NH_3) emissions
  in emission inventory, 249–255, *252–253*
  gridded, 248, *248*, *250–251*
  national, regional, and sector distribution in 2005 of, *239*, 242
  role as $PM_{2.5}$ precursor of, 5–6, 33–34
  sources of, *23*, 227, 231
Anthropogenic emission inventory. *See* Emission inventory
Anthropogenic emissions, from cement production, *213*, 213–214
AOD (aerosol optical depth), 32–34, *33*

AOT40 index, 315–316
AP-42 database, 183–185, *184*, 198n4, 235, 238
API (Air Pollution Index), 50n11
AQI (Air Quality Index), 50n11
Aromatic emissions, *243*
Ash content, from coal-fired power plants, *170*, *178*, 182–183, 237
Asian Pacific Regional Aerosol Characterization Experiment (ACE-Asia), 225
Atmospheric concentrations
  under 2005 base-year simulation, 270–274, *272*, *274*
  under 2010 base case, 274–278, *276–277*
  effects of future carbon taxes on, 132–137, *134–137*
  effects of past policy scenarios on, 84–88
    under carbon tax, 86–88, *89*, *90*, *280–281*, 280–285, *284*
    under 11th FYP SO_2 controls, 86, *87*, 278–280, *279*
    GEOS-Chem atmospheric model for, 84–86, *85*
  official statistics on, 19–22, *20*
  scientific research on, 28–40
    of CO_2, 38–40, *39*
    data sources for, 28–30
    of NO_X, 30–32, *31*
    of ozone, 35–38, *36–37*
    of particulate matter, 32–34, *33*
    of SO_2, 34–35
Atmospheric dispersion models, 264–265
Atmospheric environment, 3–49, 263–288
  official perspectives on, 14–21
  overview of, 3–9
  review of other integrated assessments of, 40–48, *41–43*
  scientific research perspectives on, 22–40
  sources of data on, 11–14
Atmospheric modeling, 263–288
  conclusions based on, 286–288
  methods for, 267–270
    GEOS-Chem model in, 267–269, *268*
    model setup in, 269–270

  new assumptions under future carbon tax scenarios for, 405
  overview of numerical models for, 264–267
    atmospheric dispersion models as, 264–265
    Eulerian gridded chemical transport models as, 265–267
  rationale for, 263–264
Aunan and Pan (2004), 296
Aunan et al. (2007), *41*, 45–46, 48, 363
Automobiles. *See* Vehicles
Autonomous energy efficiency improvement (AEEI), 335, 370n5

Baghouse filters, for cement production, 207
Balloon-based measurement campaigns, 30
Base case
  for coal-fired power plant emissions, 165–166
  for economic model, 330, 337–340, *339*, 369n1
  for future carbon tax scenarios, 109–111, *110*, *111*
  effects on economy and energy use of, *115*, *116*, 117
  effects on emissions of, 128–129, *129*
  new assumptions and methods for, 403–404
  for past policy scenarios, 63–64
    economy and energy use under, 69, *70*
    emissions under, 77–79, *78*, *80*
"Base load plants," 342
BAU ("business-as-usual") scenario, in economic model, 344
BC. *See* Black carbon (BC)
Beijing Olympics, ozone levels at, 38
Benefit assessment model, 66
Benefit Mapping and Analysis Program (BenMAP), 88–91, 291–293
  flow diagram for, *291*, *292*
  geographical distribution with, 306–308, *309*
  in He et al. (2010), 47
  key inputs to, 88–91, *91*
  relative risk in, 305, *306*

Benefits. *See* Costs and benefits
Biased technical change, in economic
  model, 335, 370n5, 378
Biofuel
  emission factors of, *236*, 238
  by region, 231, *232*
Black carbon (BC), definition of, 256n1
Black carbon (BC) emissions
  emission factors of, *236*, 237
  in emission inventory
    gridded, 248, *250–251*
    national, regional, and sector distribution
      in 2005 of, *239*, *240*, 241
    by scenario and sector in 2010, 243–247,
      244–245
    trends in, 249–255, *252–253*
Boden, Marland, and Andres (2010),
  26–27, *27*, *28*
Boiler type, for coal-fired power plants,
  169, *170*, *180–181*, 180–182
Bootstrap simulation method, for coal-fired
  power plant emissions, 176
Bottom-up methodology
  for cement production emissions,
    204–206
  for CO2 emissions, 114, 155n13, 256n5
  defined, 22, 163–164
Burner pattern, for coal-fired power plants,
  169, *170*, 181–182
"Business-as-usual" (BAU) scenario, in
  economic model, 344

Calcination, of carbonates, 203, 205, 208,
  212, 214
Calcium carbonate (CaCO3), calcination of,
  203, 212
CALPUFF model, 265
Cao, Garbaccio, and Ho (2009), 370n7
Cao, Zhang, and Zheng (2006), 237,
  253
CaO (quicklime), in clinker, 208
Cao and Ho (2009), 15–16
Cao et al. (2009), 386, 387
Capacity projections, 154n5
Cap-and-trade policy, 104–105
Capital, in economic model, 380–382
Capital market, in economic model, 383

Capital stock
  in economic model, 335, 377, 380–382
  past carbon tax effect on, 359
Carbonates, calcination of, 203, 205, 208,
  212, 214
Carbon dioxide (CO2), as stock pollutant,
  355
Carbon dioxide (CO2) concentrations,
  scientific research on, 38–40, *39*
Carbon dioxide (CO2) emissions
  in 2005–2009, 18, 26–28, 238, 242, 247,
    339
  bottom-up research on, 256n5
  from cement production
    from 1990 to 2008, *211*, 212, *213*
    emission factors of, 205, 208–209
    estimates for 2010 of, 218, *218*, *219*
  effects of 11th FYP SO2 controls on, 81,
    247, *348*
  effects of future carbon taxes on, 127–132
    annual change, 117, *117*, 118–119
    vs. base case, 128–129, *129*
    cumulative, 118, *118*
    economic model of, *357*, *358*
    in end year, *116*
    in first year, *115*
    with various carbon tax cases, 129–132,
      130–131, *148*
  effects of past carbon taxes on, 81, *365*
  major contributors to global, 3
  official statistics on, 18–19
  scientific research on, 26–28, *27*
Carbon Dioxide Information Analysis
  Center (CDIAC), of Oak Ridge
  National Laboratory, 26–27, 212
Carbon intensity, 107, 154n4
  under future carbon tax scenarios,
    118–119, *119*
Carbon monoxide (CO), in ozone
  formation, 283
Carbon monoxide (CO) concentrations,
  official statistics on, 21
Carbon monoxide (CO) emissions, from
  cement production
  from 1990 to 2008, *211*, 213, *213*
  emission factors of, 207–208, *208*
  estimates for 2010 of, 218, *218*, *219*

Carbon prices. *See also* Carbon taxes;
   Carbon trading
in Emission Trading System, 113–114
in U.S., 113–114
Carbon taxes
future scenarios (2013–2020) for, 103–153
   base case for, 109–111, *110, 111*
   comparison of effects of various options
      for, 143–150, *148*
   conclusions based on, 151–153
   effects on economy and energy use of,
      114–127, *115*
   effects on emissions of, 127–132
   effects on health and agricultural
      damages of, 138–147
   effects on pollution concentration of,
      132–137
   framework for assessing, 106, *106*
   health and agricultural benefits of, *318,*
      319–323, *320–323*
   multilateral, 108, *125–126*
      effects on economy and energy use of,
         115–118
   new assumptions and methods for, 404
   overview of, 103–109
   as percent of total tax revenue, *115,*
      116
   revenue options in, 108, *112,* 114,
      123–126, 356
   tax rate options for, *112,* 112–114
past scenario (2006–2010) for, 59
   atmospheric modeling for, *280–281,*
      280–285, *284*
   vs. cap-and-trade policy, 104–105
   and CO_2 accumulation, 355
   for coal-fired power plant emissions,
      165–166
   direct unit, 354–355, 371n14
   economic impact of, 356–362, *357–359,*
      361
   vs. 11th FYP SO_2 controls, 363–367,
      365
   with lump sum vs. tax cut revenue
      cases, 362–363
   economics of, 354–356
   effects on economy and energy use of,
      71–74, *73, 75*
effects on industry output and prices of,
   358, *361*
and emission inventory, 254–255
emissions under, 81–84, *83*
health and agricultural benefits of,
   94–95, *96,* 306–311, *307, 309, 310,*
   357, 360–362
integrated framework to assess, 64–67,
   65
overview of, 61–63
ozone exposure changes due to, 316–317,
   318
pollution concentrations under, 86–88,
   89, 90
with revenues used to cut existing taxes,
   74, *75*
vs. SO_2 emissions reduction and base
   case, 95–100, *97–99*
rationale for, 103–104
Carbon trading, 104–105
Carbonyl emissions, *243*
Carousel tax fraud, 154n2
CDIAC (Carbon Dioxide Information
   Analysis Center), of Oak Ridge
   National Laboratory, 26–27, 212
Cement production, 203–220
activity rates for, 206, *207,* 231
bottom-up methodology for, 204–206
conclusions on, 219—220
emission factors for, 207–211
   of CO_2, 205, *208–209*
   in emission inventory, 238
   estimates for 2010 of, 217–218, *218*
   of particulate matter, 205–206, *209,*
      209–211, 210
   of SO_2, NO_x, and CO, 207–208, *208*
emissions from 1990 to 2008 for
   spatial distribution of, 214, *215–216*
   total, *211,* 211–214, *213*
estimates for 2010 for, 216–219, *217–219*
historical emissions from 1990 to 2008
   for, 211–214
new assumptions and methods in future
   carbon tax scenarios for, 404
overview of, 203–204
CFB (circulating fluidized bed) boiler, for
   coal-fired power plant, *170, 181*

CFBC (circulating fluidized bed combustion) system, for coal-fired power plant, *180*, 183, *184*
CGE (computable general equilibrium) model, 45
CH$_2$O (formaldehyde) emissions, 243, *244*
C$_2$H$_4$ (ethylene) emissions, 243, *244*
C$_2$H$_4$O (acetaldehyde) emissions, 243, *244*
C$_2$H$_6$ (ethane) emissions, 243, *244*
C$_3$H$_8$ (propane) emissions, *244*
Chemical transport models (CTMs), 3-D atmospheric, 265–267. *See also* Global atmospheric transport and chemistry (GEOS-Chem) model
China Animal Industry Yearbook (2006), 231
China Automotive Industry Yearbook (2006), 233
China Cement Association, 214
China Chemical Industry Yearbook (2006), 231
China Energy Statistical Yearbook (2006), 231
China Industry Economy Statistical Yearbook (2006), 231
China Rural Statistical Yearbook (2006), 231
China State Bureau of Technology Supervision (CSBTS), 169
China Statistical Yearbook (1991–2011), 206, 388
China Statistical Yearbook for Regional Economy (2006), 233
Chinese Health Statistics Yearbook (2005), 398
Chinese Journal Full-Text Database (CJFD), 175–176
Chinese National Environmental Monitoring Center (CNEMC), 19
Chronic premature mortality
  under 11th FYP SO$_2$ controls, 306, *307*
  under future carbon tax scenarios, *148*
  ozone exposure and, 306, *307*
  under past carbon tax scenario, *92*, 92–93, 306, *307*, 362
  PM$_{2.5}$ emissions and, 300–301, 306, *307*
Circulating fluidized bed (CFB) boiler, for coal-fired power plant, *170*, *181*

Circulating fluidized bed combustion (CFBC) system, for coal-fired power plant, *180*, 183, *184*
CJFD (Chinese Journal Full-Text Database), 175–176
Clean Air Act (U.S. 1970), 371n19
Clean Development Mechanism, 153
Cleaner Production Standard, 206
Clearing the Air: The Health and Economic Damages of Air Pollution in China. *See* Ho and Nielsen (2007)
Climate treaty designs, costs and benefits to China of, 46
Clinker
  in bottom-up methodology, 205
  in cement production, 206
  defined, 203
  emission factors for, 207, 208
  and spatial distribution of emissions, 212
CMAQ air quality model, 45, 47
CNEMC (Chinese National Environmental Monitoring Center), 19
CO. *See* Carbon monoxide (CO)
CO$_2$. *See* Carbon dioxide (CO$_2$)
Coal consumption
  in base case, 109
  by coal-fired power plants, 161, *162*, *163*
    operating hours and, 166–167, *167*
    trend in, 249, *254*
    unit size and, 167–168, *168*
  in economic model, *333*
    base case projection of, 338
  effects of carbon tax on, 71, 73, *115*, *116*, *117*, *148*, *357*, *359*
  effects of 11th FYP SO$_2$ controls on, 71, *347*, *348*, *351*
  level of particular matter emissions from, 228
  misleading official statistics on, 12
  by power sector, 161, *162*, *163*
  by production sector, *70*
Coal-fired power plants, 161–197
  base case for, 165–166
  coal consumption by, 161, *162*, *163*
    operating hours and, 166–167, *167*
    unit size and, 167–168, *168*

Coal-fired power plants (cont.)
  conclusions for, 194–197
  11th FYP SO₂ control case for, 165–166
  improving estimates of emission factors of,
    168–187
    vs. AP-42, 183–185, *184*, 198n4
    database for, 175–183, *177–182*
    discussion of, 183–187, *184*, *186*
    field measurements for, 169–174, *170*,
      172–174
    parameters contributing to variance in,
      185–187, *186*
  integrated emission factor database for,
    237
  methods for studying, 163–168
    activity levels as, 166–168, *167*, *168*
    scenarios as, 165–166
    unit-based methodology as, 163–165,
      164
  new assumptions in future carbon tax
    scenarios for, 404
  overview of, 161
  past carbon tax scenario for, 165–166
  policy for, 162–163
  results of, 187–194
    for emissions from 2000 to 2005,
      187–192, *188*, *191*
    for emissions in 2010, 190–194
    spatial allocation in, 194, *195–196*
  share of power sector of, 161, *162*
Coal price
  carbon tax and, 71, 355–356, 358, *361*
  effects of 11th FYP SO₂ controls on, *347*,
    349
Cobb-Douglas production functions, 377
Cobb-Douglas utility function, 379
COI (cost-of-illness) measure, 394, 398–
  399, *399*
Coking plants, energy consumption by,
  231, *232*
Column density, of atmospheric pollutants,
  30–32
Combusting fuel
  in emission inventory, 228–229
  level of SO₂ emissions from, 228
Commodity markets, in economic model,
  336

Commodity prices
  in economic model, 336
  under 11th FYP SO₂ control and past
    carbon tax scenarios, *73–74*
  under future carbon tax scenarios,
    120–122, *121*
Commodity supply, in economic model,
  336
"Common but differentiated
  responsibilities" principle, 103
Computable general equilibrium (CGE)
  model, 45
Concentration-response (C-R) functions,
  91–94, *92*, 138
  for agricultural benefits study, 317–319,
    318
  for health benefits study, 292, 296–306
    construction of applicable, 305, *306*
    of ozone, 302–304
    of PM₂.₅, 297–302, *299*, *302*
    sources of uncertainty in, 304–305
Constant-returns-to-scale model, 375, 383
Consumption
  in base cases, 68, 109–111, 154n8, *339*,
    340
  effects of carbon tax on, 74, *75*, *99*, *115*,
    116, *117*, *119*, *120*, *148*, *357*,
    359–360, *362–363*, *365*
  effects of 11th FYP SO₂ controls on, *99*,
    100, 346–347, *348*, *350*, *354*, *365*
"Contingent valuation" approach, to health
  damages, 394
Conventional air pollutants, scientific
  research on, 23–26, *24–26*
Corn production. *See* Maize production
Cost-of-illness (COI) measure, 394,
  398–399, *399*
Costs and benefits. *See also* Economic
  impact
  of 11th FYP SO₂ controls, 98–100, *99*,
    344–354
  of future carbon tax scenarios, 106, *106*,
    143–150, *148*
  in integrated framework assessment,
    64–65, *65*
  of past carbon tax scenario, *99*, 99–100,
    354–363

C-R functions. *See* Concentration-response (C-R) functions
Crop productivity. *See* Agricultural benefits
CSBTS (China State Bureau of Technology Supervision), 169
CTMs (chemical transport models), 3-D atmospheric, 265–267. *See also* Global atmospheric transport and chemistry (GEOS-Chem) model
Cui and Liu (2008), 208, 209
Current account balance, in economic model, 335, 382, 388

Data sources, 11–14
  official statistics as, 12–13, 14–22
  for scientific observations, 28–30
  scientific research as, 13–14, 22–40
Deadweight loss, 122
Department of Economic and Social Affairs (U.N.), Population Division of, 386
Developmental Research Center (China), 338, 384
Direct costs, in integrated framework assessment, 64
Direct unit carbon tax, 354–355, 371n14
Dispersion models, 264–265
Domar-weighted productivity growth, 389
Domestic use (residential and commercial) sector
  emission factors for, *236*
  energy consumption for
    in 2005, *232*
    in 2010, *235*
Downwind rural sites, 29
Dry deposition, in 3-D atmospheric chemical transport models, 266
Durban Platform, 103, 107
"Dust," 50n9
Dust collectors, for coal-fired power plants
  discussion of, 187, 192
  for emission factor database, 176, 178–180, 182
  field measurements from, 169, 171–174, *174*
  low-efficiency vs. high-efficiency, 193
Dust mobilization scheme, in atmospheric model, 278

"Dynamic recursive" model, 331, 369–370n2

ECM (Environmental Cost Model), 44
Economic growth, official statistics on, 14–16, *14–16*
Economic impact, 329–369. *See also* Costs and benefits
  base case for, 69, *70*, 330, 369n1
  conclusions on, 367–369
  economic model for, 331–340
  of 11th FYP SO_2 controls, 344–354
    vs. carbon tax, 363–367, *365*
    combined FGD and shutdown, *347–349*, 350–354, *351*
    of FGD installation, *347–349*, 350
    methodology for analyses of, 344
    for small power plant shutdowns, 344–350, *347–349*
  of future carbon tax scenarios, 114–127
    with alternative revenue uses, 122–123
    in end year, 114, *116*
    in first year, 114, *115*
    with various carbon tax rates, 114–122, *117–121*
  overview of, 329–331
  of past carbon tax scenario, 71–74, *73*, *75*, 356–362, *357–359*, *361*
    vs. 11th FYP SO_2 controls, 363–367, *365*
    with lump sum vs. tax cut revenue cases, 362–363
Economic index, 50n8
Economic model, 331–336, 375–390
  base case projection in, 330, 337–340, *339*, 369n1
  central features of, 68, 68–69
  data sources for, 383–389, *385–389*
  as "dynamic recursive" or Solow type, 331, 369–370n2
  emissions in, *334*
  exogenous variables in, *375*, *376*, 384
  "general equilibrium" effects in, 331–332
  limitations of, 67
  output, employment, and fuel use in, *333*
  parameters of, *375*, *376*, 389–390
  production sectors in, 68, 69, *70*

Economic model (cont.)
  special treatment of electric power
    industry in, 337
  structure of, 375–383
    capital, investment, and financial system
      in, 380–382
    foreign sector in, 382
    government and taxes in, 379–380
    households in, 378–379
    markets in, 382–383
    production in, 375–378
Economics
  of 11th FYP SO₂ controls, 340–344
    of FGD installation, 342–344, *345*
    for small power plant shutdown,
      340–342, *341, 342*
  of past carbon tax scenario, 354–356
EDGAR data set, 189–191
EEA (European Environment Agency),
  316
Effective labor, in economic model, 335
EFs. *See* Emission factors (EFs)
EIA (Energy Information Agency, U.S.), 26,
  109, 335, 388
EITE (energy-intensive, trade-exposed)
    industries, subsidies to, *119–121*
  effects on economy and energy use of,
    115–119
Electricity consumption
  in base case, 109–111
  effects of carbon tax on, *115, 116, 357,*
    359
  effects of 11th FYP SO₂ controls on, *347,*
    348, 351
Electricity growth, in economic model,
  338–340
Electricity price, effects of 11th FYP SO₂
    controls on, *347*
Electric power industry, in economic model,
  337
Electrostatic precipitator (ESP)
  for cement production, 212
  for coal-fired power plants
    conclusions on, 194–197
    discussion of, 185
    in emission factor database, 176, *178,*
      181, 182

  in field measurements, 169, *170,*
    173–174
  results of, 193
11th Five-Year Plan (11th FYP, 2006–2010)
    SO₂ controls, 4–5, 59
  agricultural benefits of, 94
  atmospheric modeling for, 278–280, *279*
  vs. carbon tax and base case, 95–100,
    97–99
  for coal-fired power plant emissions,
    165–166
  economic impact of, 69–71, *72,* 344–354
    vs. carbon tax, 363–367, *365*
    for combined FGD and shutdown,
      347–349, 350–354, *351*
    for FGD installation, *347–349, 350*
    methodology for analyses of, 344
    for small power plant shutdowns,
      344–350, *347–349*
  economics of, 340–344
    for FGD installation, 342–344, *345*
    for small power plant shutdown,
      340–342, *341, 342*
  effects on emissions of, 79–81, *82*
  effects on energy use of, 69–71, *72, 347*
  and emission inventory, 249–254
  health benefits of, 94, *94,* 306–311, *307,*
    309, 310
  integrated framework to assess, 64–67,
    65
  pollution concentrations under, 86, *87*
  targets of, 61
Emberson et al. (2009), 319
EMEP/CORINAIR Atmospheric Emission
    Inventory Guidebook, 235, 238
Emission factors (EFs)
  for cement production, 207–211
    2010 estimates of, 217–218, *218*
    of CO₂, 205, 208–209
    of particulate matter, 205–206, *209,*
      209–211, 210
    of SO₂, NOₓ, and CO, 207–208, *208*
  for coal-fired power plants, 168–187
    database of, 175–183, *177–182*
    discussion of, 183–187, *184, 186*
    field measurements of, 169–174, *170,*
      172–174

for economy-wide emission inventory, 235–238, *236*
Emission intensity, 105, 154n3
Emission inventory, 66, 225–255
  for cement production, 203–220
  for coal-fired power plants, 168–187
    database for, 175–183, *177–182*
    discussion of, 183–187, *184, 186*
    field measurements for, 169–174, *170, 172–174*
  for entire economy
    conclusions on, 255
    data sources for, 229–238
    emission sources for, 227–229
    geographic domain of, 226–227
    methodology for, 226–229, *230*
    rationale for, 225–226
    results and discussion of, 238–255
      for emissions in 2010, 243–247, *244–245*
      for gridded emissions, 247–251, *248, 250–251*
      implications for policy of, 249–255, *252–254*
      for national, regional, and sector emissions in 2005, 238–242, *239, 240*
      for speciation of NMVOC, 242–243, *243, 244*
Emissions
  effects of future carbon tax scenarios on, 127–132
  vs. base case, 128–129, *129*
  new assumptions for estimates of, 403–404
  with various carbon tax cases, 129–132, *130–131*
  effects of past policy scenarios on, 74–84
  under base case, 77–79, *78, 80*
  under carbon tax, 81–84, *83*
  under 11th FYP SO$_2$ controls, 79–81, *82*
  emission estimation for, 75–77, *77*
  estimation of, 75–77, *77*, 127–128, 155n17
  official statistics on, *17*, 17–19

scientific research on, 22–28
  for CO$_2$, 26–28, *27*
  for conventional air pollutants, 23–26, *24–26*
  uncertainties in, 28
sources of, 18, 76, 227–229
Emissions Trading System (ETS), 62, 113–114, 153–154n2, 355
Employment, in economic model, *333, 335*
Energy consumption
  in economic model, 332, *333, 334*
  base case projection of, 338, *339, 340*
  effects of future carbon tax scenarios on, 114–127
    with alternative revenue uses, 122–123
    in end year, 114, *116*
    in first year, 114, *115*
    with various carbon tax rates, 114–122, *117–121*
  effects of past policy scenarios on, 68–74
    under base case, 69, *70*
    under carbon tax, 71–74, *73, 75, 357, 358*
    economic growth and environment model for, 68, *68*
    under 11th FYP SO$_2$ controls, 69–71, *72, 347, 351, 351*
  official statistics on, 14–16, *14–16*
  by region and fuel, 231, *232*
  trends in, 249, *254*
Energy Information Agency (EIA, U.S.), 26, 109, 335, 388
Energy intensity, 15, 50n8
Energy-intensive, trade-exposed (EITE) industries, subsidies to, *119–121*
  effects on economy and energy use of, *115–119*
Energy Research Institute, 340, 370n8
Energy sector, in economic model, 336
Energy-to-GDP ratio, in base case, 109
Enterprise income tax rate, revenues used to cut, *112*, 114
  agricultural benefits of, 143, *144–147*
  comparing results of, *148, 150*
  effects on economy and energy use of, *117–119*, 122–127

Enterprise income tax rate, revenues used to cut (cont.)
  effects on emissions of, 129–132, *130–131*
  effects on pollution concentration of, 133–135, *134–135*, 137
  in end year, 114, *116*
  in first year of, 114, *115*
  health benefits of, 138–143, *139*, *141*, *142*
Environmental Cost Model (ECM), 44
Environmental Protection Agency (EPA, U.S.)
  analysis of Clean Air Act by, 371n19
  BenMAP tool of, 88, 291
  on CALPUFF model, 265
  on carbon tax, 355
  and Joint Economic Study, 45
  on NO$_X$ exposure and disease, 294
  ozone exposure index of, 313, 316
  on valuation of health damages, 393
Environmental Protection Bureaus (EPBs, China), 19, 28
EOTC (European Open Top Chamber) Project, 317
EPA. *See* Environmental Protection Agency (EPA, U.S.)
ESP. *See* Electrostatic precipitator (ESP)
Ethane (C$_2$H$_6$) emissions, 243, *244*
Ethylene (C$_2$H$_4$) emissions, 243, *244*
ETS (Emissions Trading System), 62, 113–114, 153–154n2, 355
EU (European Union), Emissions Trading System of, 62, 113–114, 153–154n2, 355
EUA (European Union Allowance), 62
Eulerian gridded chemical transport models, 265–267. *See also* Global atmospheric transport and chemistry (GEOS-Chem) model
European Environment Agency (EEA), 316
European Open Top Chamber (EOTC) Project, 317
European Union (EU), Emissions Trading System of, 62, 113–114, 153–154n2, 355
European Union Allowance (EUA), 62

Exogenous variables, in economic model, 336, 375, 384
  base case projection of, 338
  defined, 370n3
  listed, *376*
Exports
  in economic model, 335, 382
  under future carbon tax scenarios, *115*, *116*, 117
Exposure–relative yield functions, for agricultural crops, 317–319, *318*
Externality tax rate, 379

Fabric filter (FF) systems, for coal-fired power plants, 169, *170*, *181*, 182, 185, 187, 193
FAO (Food and Agriculture Organization, U.N.), 311
Fertilizer use, 231
Field measurements, for coal-fired power plants, 169–174, *170*, *172–174*
Financial system, in economic model, 380–382
Flue gas desulfurization (FGD)
  adverse effects of, 6
  for coal-fired power plants
    conclusions on, 197
    discussion of, 185, 187
    in emission factor database, 176, *180*, 181, *181*, 182
    field measurements of, 169–174, *170*
    policy on, 163, 165–166
    removal efficiencies of, 198n5
    results of, 192, 193
    wet-, 171, 173, 176, 182, 193
  in 11th FYP SO$_2$ controls, 61, 71, 163, 165–166
  installation policy for
    combined with small power plant shutdown, 347–349, *350–354*, *351*
    economics of, 342–344, *345*
    impacts of, 347–349, *350*
  Joint Economic Study on, 45
Flue gas volume per unit of fuel consumption, for coal-fired power plants, 175

Food and Agriculture Organization (FAO, U.N.), 311
Foreign sector, in economic model, 382
Formaldehyde (CH₂O) emissions, 243, *244*
Forward-looking behavior, in economic model, 369–370n2
Fossil fuel combustion, new assumptions and methods in future carbon tax scenarios for, 404
Fossil fuel energy use
effects of 11th FYP SO₂ controls on, *348, 351, 351*
effects of future carbon tax scenarios on, 114, *115, 116*
effects of past carbon tax scenario on, *357*
Fuel economy standards, for new cars, 4
Fuel use
in economic model, *333*
effects of SO₂ controls on, 71
by production sector, *70*
Fugitive dust, 278
Future carbon tax scenarios, 103–153
agricultural benefits of, 143, *144–147*
base case for, 109–111, *110, 111*
effects on economy and energy use of, *115, 116, 117*
effects on emissions of, 128–129, *129*
choice of time frame for, 107
comparison of effects of, 143–150, *148*
conclusions based on, 151–153
effects on economy and energy use of, 114–122, *115*
effects on emissions of, 127–132
effects on pollution concentration of, 132–137
framework for assessing, 106, *106*
health benefits of, 138–143, *139, 141, 142*
overview of, 103–109
Scenario F1 (tax of 30 yuan/ton), *112, 113*
comparing results of, 143–149, *148*
effects in end year of, 114, *116*
effects in first year of, 114, *115*
effects on economy and energy use of, 117–118, *117–119*
effects on emissions of, 129–132, *130–131*

Scenario F2 (tax of 10 yuan/ton rising to 50 yuan/ton), *112, 113*
agricultural benefits of, 143, *144–147*
comparing results of, *148, 149*
effects in end year of, 114, *116*
effects in first year of, 114, *115*
effects on economy and energy use of, *117–121, 118–122*
effects on emissions of, 129–132, *130–131*
effects on pollution concentration of, 133–135, *134–135,* 137
health benefits of, 138–143, *139, 141, 142*
Scenario F3 (tax of 10 yuan/ton rising to 100 yuan/ton), *112,* 113–114
agricultural benefits of, 143, *144–147*
comparing results of, *148,* 149–150
effects in end year of, 114, *116*
effects in first year of, 114, *115*
effects on economy and energy use of, *117–119,* 122
effects on emissions of, 129–132, *130–131*
effects on pollution concentration of, 135–137, *136–137*
health benefits of, 138–143, *139, 141, 142*
Scenario F4 (revenues used to cut enterprise income tax rate), *112,* 114
agricultural benefits of, 143, *144–147*
comparing results of, *148,* 150
effects in end year of, 114, *116*
effects in first year of, 114, *115*
effects on economy and energy use of, *117–119,* 122–127
effects on emissions of, 129–132, *130–131*
effects on pollution concentration of, 133–135, *134–135,* 137
health benefits of, 138–143, *139, 141, 142*
Scenario F5 (revenues used to protect vulnerable industries), 108, *123–125*
effects in end year of, 114, *116*
effects in first year of, 114, *115*

Future carbon tax scenarios (cont.)
  effects on economy and energy use of,
    117–119
  Scenario F6 (effect of carbon pricing in
    rest of world), 108, *125–126*
  effects in end year of, 114, *116*
  effects in first year of, 114, *115*
  effects on economy and energy use of,
    117, *118*
  size and timing of, 112–113

Garbaccio, Ho, and Jorgenson (1999), 376
Garbaccio, Ho, and Jorgenson (2000), 387
Gaseous pollutant emissions, from coal-
    fired power plants, 169–171, *170*
Gas use
  in economic model, *333*
  effects of 11th FYP SO₂ controls on, *348,*
    351
  effects of future carbon tax scenarios on,
    115, 116
  effects of past carbon tax scenario on, *357*
  by production sector, *70*
GDP. *See* Gross domestic product (GDP)
"General equilibrium" effects, in economic
    model, 331–332
Geographic information system (GIS)-based
    data set, 294
GEOS (Goddard Earth Observing System),
    267
GFED-2 inventory, 269
GHG (greenhouse gas) emissions, official
    statistics on, 18–19
Global atmospheric transport and chemistry
    (GEOS-Chem) model
  for atmospheric modeling of pollutant
    concentrations, 84–86, *85*, 132,
    267–269, *268*
  data sources for, 267–269
  for gridded emissions, 247–251, *248,*
    250–251
  in integrated framework to assess policies,
    66
  meteorological data for, 267–268,
    269–270
  model setup for, 269–270
  nested-grid version of, 267, *268*

for NMVOC emissions by species, 240, *243*
previous applications of, 269
for speciation of NMVOC emissions, 240,
    243, *244*
Global Modeling and Assimilation Office
    (GMAO), of NASA, 267
Goddard Earth Observing System (GEOS),
    267
GOSAT satellite, 52n32
Government consumption
  effects of 11th FYP SO₂ controls on, *348*
  effects of future carbon tax scenarios on,
    115, 116
  effects of past carbon tax scenario on, *357*
Government deficit, in economic model,
    335
Government expenditure, in economic
    model, 335, 380
Government purchases, 380
  quantity index of, *356*
Government revenue, in economic model,
    335
Government sector, in economic model,
    379–380
Government taxes, in economic model, 335
Grain production. *See also* Agricultural
    benefits
  annual yield, 311, *312*
  effects of future carbon tax scenarios on,
    138, 143, *144–148*
  effects of past carbon tax scenario on,
    319–323, *320–323*
  main crops in, 311–313, *312, 314–315*
  spatial distribution of, 312–313, *314–315*
Grate boiler, for coal-fired power plant,
    170, 180, 181, 189
Greenhouse gas (GHG) emissions, official
    statistics on, 18–19
Gregg et al. (2008), 28
Gridded emissions
  in economic model, *352, 360*
  in emission inventory, 247–248, *248,*
    250–251
Gross domestic product (GDP), 14–16, *15,*
    16
  in base case, 109
  in economic model, 338, *339, 340*

effects of alternative revenue uses on, *115*, *116*, 123

effects of 11th FYP SO₂ controls on, 71, 348, 351, *351*

effects of future carbon tax scenarios on, *115*, *116*, 117–119, *120*, 148

effects of past carbon tax scenario on, 74, 356–358, *357*, *358*

Gross output

in economic model, *333*, *334*

by production sector, *70*

Guidelines for Preparing Economic Analyses (EPA), 393

Guo, Haab, and Hammitt (2006), *395*, 396, 398

Guo and Hammitt (2009), 394, *395*

Halocarbon emissions, *243*

Handbook of Industrial Pollution Emission Rates (SEPA), 238

Hao et al. (2002), 237, 252

He, K. B., et al. (2005), 233

Health benefits

of future carbon tax scenarios, 138–143, *139*, *141*, *142*

of past policy scenarios, 88–95, 306–311, *307*, *309*, *310*

under carbon tax, 94–95, *96*, 306–311, *307*, *310*, *357*, 360–362

under 11th FYP SO₂ controls, 94, *94*, *348*, 352–353

framework for, 88–93, *91*, *92*

Health benefits assessment, 291–311

concentration-response relationships for, *292*, 296–306

construction of applicable functions for, 305, *306*

of ozone, 302–304

of PM₂.₅, 297–302, *299*, *302*

sources of uncertainty in, 304–305

methodology and data for, 291–296

air pollutants of concern in, 293–294

baseline incidence of health endpoints in, 295–296, *296*

framework of BenMAP in, 291–293, *292*

population exposure in, 291–292, 294, *295*

new assumptions in future carbon tax scenarios for, 405

results of, 306–311, *307*, *309*, *310*

Health damage valuation, 353, 362, 368, 393–400

of morbidity, 398–400, *400*

of mortality, *395*, 395–397, *397*

policy effects in, 400

Health Effects Institute (HEI, U.S.), 296, 303

He et al. (2010), 42, 46–47, 48

Ho and Nielsen (2007), 12, 40–44, *41*, 48, 263, 265, 294, 353, 375, 379, 393, 397

Hospital admissions

baseline incidence of, 295

effects of past carbon tax scenario and 11th FYP SO₂ controls on, 306, *307*

in framework of BenMAP, 292

ozone exposure and, 303–304

PM₂.₅ emissions and, 301, *302*

sources of uncertainty on, 304

Household demand, 371n17

Household savings rate, 332, 384

Household sector, in economic model, 332, 378–379

"Human capital" approach, to valuation of health damages, 393

Hydroxyl radical (OH), in ozone formation, 283–285, *284*

ICD (International Classification of Diseases), 295

IEA (International Energy Agency), 63, 111, 154n5, 225, 338, 340

Imports, in economic model, 335, 382

Income, per capita, 5

Income effect, in economic model, 332, 359, 371n17

Indirect costs, in integrated framework assessment, 64

Indoor air pollution, 47–48

Industrial sector

emission factors for, *236*

energy consumption by

in 2005, *232*

in 2010, *235*

Industrial Source Complex (ISC) model, 265
Industry output, 71–74, *73*
  under 11th FYP SO₂ controls, 71, 72, *349*
  under future carbon tax scenarios, 120–122, *121*
  under past carbon tax scenario, 360, *361*
Industry prices
  effects of carbon tax on, 71–74, *73*, 360, *361*
  under 11th FYP SO₂ controls, 71, 72, *349*
Inelastic labor supply, in economic model, 335, 370n4
Input-output (IO) tables, for economic model, 332, *333*, *334*, 336, 370n6, 384, 390n5
Institute of Geographical Sciences and Natural Resources Research (China), 294
Intake fraction, 40–44
Integrated assessments
  for analysis of national emission control policies
    application of, 7–8
    development of, 6–7
    overview of, 64–67, *65*
  of atmospheric environment, literature review of, 40–48, *41–43*
Intergovernmental Panel on Climate Change (IPCC), 205
International Classification of Diseases (ICD), 295
International Energy Agency (IEA), 63, 111, 154n5, 225, 338, 340
INTEX-B emission inventory, 213
Investments
  in economic model, 380–382
  under 11th FYP SO₂ controls, *348*
  under future carbon tax scenarios, *115*, *116*, 117–118, *119*, *120*, *148*
  under past carbon tax scenario, *357*, *358*, 358–359
IO (input-output) tables, for economic model, 332, *333*, *334*, 336, 370n6, 384, 390n5
IPCC (Intergovernmental Panel on Climate Change), 205

Iron and steel industry
  activity levels for, 231
  emission factors for, 238
  in emission inventory, 229
  new assumptions in future carbon tax scenarios for, 404
ISC (Industrial Source Complex) model, 265

Jiang (2004), 237
Jiang, Yang, and Liu (2002), 294
Joint Economic Study (JES 2007), 45, 48, 340, 341, 343, 370n7
Jorgenson, Ho, and Stiroh (2005), 387

Ketone emissions, 243, *244*
Kiln types, 203–205
  cement production by, 206, *207*
    estimates for 2010 of, 217, *217*
  emission factors by
    of particulate matter, *209*, 209–210, *210*
    of SO₂, NOₓ, and CO, 207–208, *208*
  and spatial distribution of emissions, 214
Klimont et al. (2001), *252–253*
Klimont et al. (2002), 242, *253*
Klimont et al. (2009), 189, 191, *191*, 249, *252–253*
Krupnick et al. (2006), *395*, 396
Kyoto Protocol (1997), 103, 153

Labor, in economic model, 335, 370n4
Labor market, in economic model, 383
Labor projections, in economic model, 386, *386*
Large point sources (LPS)
  activity levels of, 231
  defined, 227
LEAP (Long-Range Energy Alternatives Planning) model, 47
Lei (2008), 238
Lei, Y., et al. (2008), 209
Lei, Zhang, He, et al. (2011), *252–253*
Lei, Zhang, Nielsen, et al. (2011), 238
Lei et al. (2011), 191
Levy and Greco (2007), 294, 296, 298–299
Li, Duan, et al (2007), 238
Li, Wang, et al. (2007), 238

Liang, Fan, and Wei (2007), 105
Li et al. (2010), 34–35
Lin and McElroy (2011), 31, *31*
Lin et al. (2010), 32–34, *33*
Literature review, of other integrated
  assessments of atmospheric
  environment, 40–48, *41–43*
Livestock production, 231
Log-linear model, for population health
  response, 305
Long-Range Energy Alternatives Planning
  (LEAP) model, 47
Low-NO$_X$ burner (LNB), for coal-fired
  power plant, 169, 171, *180*, 182, 189,
  192–193
Low-pressure impactors (LPIs), 171
LPS (large point sources)
 activity levels of, 231
 defined, 227
Lu, Tong, and Liu (2011), 106
Lu et al. (2010), *252*
Lump sum transfer case, past carbon tax
  scenario with, 356
 economic impact of, 356–362, *357*, *358*,
  361
 vs. tax cut revenue case, 75, 362–363

M7 index, 316, 317, 319
M12 index, 316, 317, *318*, 319
Magnesium carbonate (MgCO3),
  calcination of, 212
Maize production
 annual yield, 311, *312*
 effects of future carbon tax scenarios on,
  138, *144*, *147*
 effects of past carbon tax scenario on,
  320, *323*
 exposure–relative yield functions for, *319*
 spatial distribution of, 313, *314–315*
Market-based emission controls, 9, 62, 67
Markets, in economic model, 382–383
Mass balance principle, 266
Matus et al. (2012), *43*, 47, 48
Medical expenses, 398–400, *399*
MEGAN inventory, 269
MEP. *See* Ministry of Environmental
  Protection (MEP, China)

Meteorological data, for GEOS-Chem
  model, 267–268, 269–270
MgCO3 (magnesium carbonate),
  calcination of, 212
Ministry of Environmental Protection
  (MEP, China), 44
 on activity levels of large point sources,
  231
 on coal-fired power plants, 167, 181, 192
 on emission trends, *252*
 on particulate matter emissions, *299*
 TSP estimates of, 332
Ministry of Finance (MOF, China), 105,
  354, 355
Ministry of Health (MOH, China),
  295–296
Monsoonal circulation, and ozone
  concentration, 272
Monte Carlo simulations, 51n19
 for coal-fired power plant emissions, 176,
  185
Morbidity
 effects of future carbon tax scenarios on,
  140
 effects of past carbon tax scenario and
  11th FYP SO$_2$ controls on, 306, *307*
 ozone and, 303–304
 PM$_{2.5}$ emissions and, 301–302, *302*
Morbidity, valuation of, 398–400, *400*
Mortality
 premature. *See also* Acute premature
  mortality; Chronic premature
  mortality
  effect of 11th FYP SO$_2$ controls on, 94,
   96, 306–308, *309–310*
  effect of future carbon tax scenarios on,
   138–143, *144–148*
  effect of past carbon tax scenario on,
   94–95, 96, 306–308, *309–310*
  methods of estimating, 88, 91–93,
   292–303
  other estimates of air pollution effects on,
   40, *41–43*, 44–47
 valuation of, *395*, 395–397, *397*
Multilateral carbon tax, 108, *125–126*
 effects on economy and energy use of,
  115–118

National Ambient Air Quality Standards (NAAQS), 297

National Bureau of Statistics (NBS, China)
as data source for economic model, 12, 13, 388–389
and emission inventory, 225
on emissions from coal-fired power plants, 189, 191, *191*

National Crop Loss Assessment Network (NCLAN) studies, 313, 317

National Development and Reform Commission (NDRC, China), 340, 370n8

National emission control policies, pre-existing
in base case and 11th FYP SO_2 control case, 165–166, 216–218, 234
in base case for future carbon tax scenarios, 403

National savings, in economic model, 332, 384

NBS. *See* National Bureau of Statistics (NBS, China)

NCLAN (National Crop Loss Assessment Network) studies, 313, 317

NDRC (National Development and Reform Commission, China), 340, 370n8

Netherlands Environment Assessment Agency (PBL), 26–27, *27*

NH_3 emissions. *See* Ammonia (NH_3) emissions

Nitrogen dioxide (NO_2) concentrations. *See also* Nitrogen oxide (NO_X) concentrations
official statistics on, *20*, 21–22
scientific research on, 30–32, *31*

Nitrogen dioxide (NO_2) emissions
from cement production, 218, *218*, 219, *219*
in economy-wide emission inventory
emission factors of, 236, *237*
gridded, 248, *250–251*
by scenario and sector in 2010, 243–247, *244–245*
trends in, 249–255, *252–253*
scientific research on, 25–26, *26*

Nitrogen oxide (NO_X), in ozone formation, 283–285, *284*

Nitrogen oxide (NO_X) concentrations
atmospheric modeling of
in 2010 base case, 275, *276–277*
in 2010 carbon tax case, *280–281*, 280–282
effects of future carbon tax scenarios on, 132–137, *134–137*, 148
official statistics on, *20*, 21–22
scientific research on, 30–32, *31*

Nitrogen oxide (NO_X) emissions
from cement production
from 1990 to 2008, 211, 213, *213*
emission factors of, 207–208, *208*
spatial distribution of, 213, *215–216*
from coal-fired power plants
from 2000 to 2005, *188*, 189
in 2010, 192–193
vs. AP-42, 183, *184*
database of, 175–183, *177*, *180*
field measurements of, 169–171, *170*
vs. other studies, 191, *191*
parameters contributing to variance in, 185, *186*
effects of future carbon tax scenarios on, 127–132
vs. base case, 128–129, *129*
with various carbon tax cases, *130–131*, 132
effects of past policy scenarios on
under base case, 77, 79, *80*
under carbon tax, 81–84, *83*, 357
under SO_2 controls, 79–81, *82*, *348*, 352
in emission inventory, 239, 240, *240*
health effects of, 294
official statistics on, 17, *18*
scientific research on, 25–26, *26*

NMVOC. *See* Non-methane volatile organic compounds (NMVOC)

NO_2. *See* Nitrogen dioxide (NO_2)

Noncombustion emissions, 76
from industry, 231

Non-methane volatile organic compounds (NMVOC), defined, 256n1

Non-methane volatile organic compounds
    (NMVOC) emissions
  emission factors of, 237
  in emission inventory, 249–255, *252–253*
  gridded, 248, *248*, *250–251*
  national, regional, and sector distribution
    in 2005 of, *239*, 242
  by scenario and sector in 2010, 243–247,
    244–245
  sources of, 227
  speciation of, 242–243, *243*, *244*
NO$_X$. *See* Nitrogen oxide (NO$_X$)

O$_3$. *See* Ozone (O$_3$)
Oak Ridge National Laboratory, Carbon
    Dioxide Information Analysis Center
    of, 212
OC. *See* Organic carbon (OC)
OCO (Orbiting Carbon Observatory),
    52n32
OCO2 (Orbiting Carbon Observatory 2),
    52n32
Official statistics, 14–22
  on atmospheric concentrations, 19–22, *20*
  on economy and energy use, 14–16,
    14–16
  on emissions, *17*, 17–19
  errors in, 12–13
OH (hydroxyl radical), in ozone formation,
    283–285, *284*
Ohara et al. (2007), 189, *191*, 249,
    252–253
Oil consumption
  in economic model, *333*
  under 11th FYP SO$_2$ controls, *347*, *348*,
    351
  under future carbon tax scenarios, *115*,
    116, *117*
  for on-road vehicles, 249, *254*
  under past carbon tax scenario, *357*
  by production sector, *70*
Oil prices
  in economic model, *388*, *389*
  effects of 11th FYP SO$_2$ controls on, *347*
  effects of past carbon tax scenario on, *71*
Oil refineries, energy consumption by, 231,
    232

Open-burned biomass
  emission factors of, *236*, 238
  by region, 231, *232*
Orbiting Carbon Observatory (OCO),
    52n32
Orbiting Carbon Observatory 2 (OCO2),
    52n32
Organic carbon (OC), 46
  defined, 256n1
  oxidation of, 212
Organic carbon (OC) emissions
  emission factors of, *236*, 237
  in emission inventory
    in 2010, 243–247, *244–245*
    gridded, 248, *250–251*
    national, regional, and sector distribution
      in 2005 of, *239*, *240*, 241–242
    trends in, 249–255, *252–253*
Outpatient visits
  baseline incidence of, 295
  effects of past carbon tax scenario and
    11th FYP SO$_2$ controls on, 306, *307*
  in framework of BenMAP, 292
  PM$_{2.5}$ emissions and, 301, *302*
Output, quantity of industry, *376*
Oxidation, of organic carbon, 212
Ozone (O$_3$)
  formation and removal of, 275, 283–285,
    284
  "good," 35, 52n30
Ozone (O$_3$) concentrations
  atmospheric modeling of
    in 2005 base-year simulation, 271–273,
      272
    in 2010 base case, 275–277, *276–277*
    in 2010 carbon tax case, 280–281,
      280–285
  daily maximum, 285–286, *286*
  effects of future carbon tax scenarios on,
    132–137, *134–137*
    agricultural benefits of, 138, *143*,
      144–147
    health benefits of, 138–143, *139*, *141*,
      142
  effects of past policy scenarios on
    under carbon tax, 86–88, *89*, *90*
    under SO$_2$ controls, 86, *87*, *352*

Ozone (O₃) concentrations (cont.)
  in GEOS-Chem model, 85–86
  official statistics on, 21
  scientific research on, 35–38, *36–37*
  seasonality of, 271–273, *272*, 282–285
  as secondary pollutant, 64
  in stratosphere ("good ozone"), 35,
    52n30
  in troposphere, 35
Ozone (O₃) exposure
  agricultural effects of, 311–323
    concentration-response relationships for,
      317–319, *319*
    framework for evaluating, 93–94
    under future carbon tax scenarios, 143,
      144
    indices for, 313–316
    main grain crops and, 311–313, *312*,
      314–315
    under past carbon tax scenario, 93–95,
      316–317, *318*
    rationale for studying, 311
    results for, 319–323, *320–323*
  health effects of, 293, 302–304
    construction of C-R functions for,
      306
    under 11th FYP SO₂ controls, *348*
    framework for evaluating, 88, 91–92
    under past carbon tax scenario vs. 11th
      FYP SO₂ controls, 94, *94*, 95, 306–
      307, 308–311, *310*
    indices for, 313–316
  "Ozone weekend effect," 88

Pan, Xiaochuan, 291
PAPA (Public Health and Air Pollution in
    Asia) program (U.S.), 303
Parameters, of economic model, 375, *376*,
    389–390
"Partial equilibrium" analysis, 331
Particulate matter (PM) concentration,
    atmospheric modeling of
  in 2005 base-year simulation, 273–274,
    273–274
  in 2010 base case, 276–277, *278*
  in 2010 carbon tax case, *280–281*,
    280–282

Particulate matter (PM) emissions
  from cement production
    from 1990 to 2008, *211*, 211–214,
      213
    emission factors of, *209*, 209–211, *210*
    estimates for 2010 of, 218, *218*, 219,
      219
    methodology for estimating, 205–206
    spatial distribution of, 214, *215–216*
  from coal combustion, 228
  from coal-fired power plants
    from 2000 to 2005, *188*, 189
    in 2010, 193–194
    vs. AP-42, 183–185, *184*
    database of, 175–183, *178*, *179*, *181*
    field measurements of, 171–174,
      172–174
    mass size distributions of, 171–173, *172*,
      173
    vs. other studies, *191*, 191–192
    parameters contributing to variance in,
      185–187, *186*
    penetration of, 174, *174*
    probability distribution bands for,
      176–178, *178*, *179*
    removal efficiencies for, 173–174
  effects of future carbon tax scenarios on,
    127–137, *134–137*
    vs. base case, 128–129, *129*
    health benefits of, 138–143, *139*, *141*,
      142
    with various tax cases, *130–131*, 132,
      148
  effects of past policy scenarios on
    under base case, 77, *78*, *79*, *80*
    under carbon tax, 81–84, *83*, 86–88, *89*,
      90
    under SO₂ controls, 79, 81, *82*, 86, *87*,
      348
  in emission inventory
    emission factors of, *236*, *237*
    gridded, *248*, *250–251*
    national, regional, and sector distribution
      in 2005 of, *239*, *240*, 240–241
    by scenario and sector in 2010, 243–247,
      244–245
    trends in, 249–255, *252–253*

health effects of, 293, 297–302
  acute premature mortality as, 298–300,
    299
  chronic premature mortality as, 300–301
  construction of C-R functions for, *306*
  under 11th FYP SO₂ controls, *348*
  morbidity as, 301–302, *302*
  under past carbon tax scenario vs. 11th
    FYP SO₂ controls, 306, *307*, 308, *309*
  sources of uncertainty in, 304
official statistics on, 17, 18, 21
penetration rate of
  from cement production, *210*, 210–211
  from coal-fired power plants, 174, *174*
scientific research on, 32–34, *33*
Particulate matter less than 2.5 microns in
    diameter (PM₂.₅), 4
precursors for, 33–34
as primary vs. secondary pollutant, 64
scientific research on, 32–34, *33*
Particulate matter less than 2.5 microns in
    diameter (PM₂.₅) concentrations
atmospheric modeling of
  in 2005 base-year simulation, 273–274,
    273–274
  in 2010 base case, 276–277, *278*
  in 2010 carbon tax case, *280–281*,
    280–282
causal factors for, 6
effects of future policy scenarios on,
    132–137, *134–137*
health effects of, 293, 297–302
  acute premature mortality as, 298–300,
    299
  chronic premature mortality as, 300–301
  construction of C-R functions for, *306*
  morbidity as, 301–302, *302*
  under past carbon tax scenario vs. 11th
    FYP SO₂ controls, 306, *307*, 308, *309*
  sources of uncertainty in, 304
official statistics on, *20*, 21
Particulate matter less than 2.5 microns in
    diameter (PM₂.₅) emissions
from cement production
  from 1990–2008, *211*, 213, *213*
  estimates for 2010 of, *218*, *219*
  spatial distribution of, 214, *215–216*

from coal-fired power plants
  database of, 175–183, *178*, *179*, *181*
  probability distribution bands for,
    176–178, *178*, *179*
  removal efficiencies of, 173–174
effects of future carbon tax scenarios on,
    127–129
  vs. base case, 128–129, *129*
effects of past policy scenarios on
  under carbon tax, 86–88, *89*, *90*
  under SO₂ controls, 86, *87*
in emission inventory
  emission factors of, *236*, 237
  gridded, 248, *250–251*
  national, regional, and sector
    distribution in 2005 of, *239*, *240*,
    240–241
  by scenario and sector in 2010, 243–247,
    244–245
  trends in, 249–255, *252–253*
Particulate matter less than 10 microns in
    diameter (PM₁₀) concentrations
estimating PM₂.₅ concentrations from, 293,
    297–299, 304
official statistics on, 20, 21
scientific research on, 32–34, *33*
Particulate matter less than 10 microns in
    diameter (PM₁₀) emissions
from cement production
  from 1990 to 2008, *211*, 213, *213*
  estimates for 2010 of, *218*, *219*
from coal-fired power plants
  field measurements of, 171–174,
    172–174
  mass size distributions of, 171–173, *172*,
    173
  penetration of, 174, *174*
  removal efficiencies for, 173–174
effects of future carbon tax scenarios on,
    127–132
  vs. base case, 128–129, *129*
  with various carbon tax cases, *130–131*,
    132, *148*
effects of past policy scenarios on
  under base case, 77, *79*, *80*
  under carbon tax, 81–84, *83*, *357*
  under SO₂ controls, 79, 81, *82*, *348*

Particulate matter less than 10 microns in diameter (PM₁₀) emissions (cont.)
  in emission inventory
    emission factors of, *236*, *237*
    gridded, 248, *250–251*
    national, regional, and sector distribution in 2005 of, *239*, *240*, 240–241
    by scenario and sector in 2010, 243–247, *244–245*
    trends in, 249–255, *252–253*
  health effects of, 293
  official statistics on, 18
  scientific research on, 23–25, *24*
Past emission control policy scenarios. *See* Carbon tax, past scenario for; 11th Five-Year Plan (11th FYP, 2006–2010) SO₂ controls
PBL (Netherlands Environment Assessment Agency), 26–27, *27*
PC (pulverized combustion) boiler, for coal-fired power plant, *170*, *180*, 180–181, *181*, 182
"Peaking plants," 342
Peng and Tian (2003), 398
Per capita income, 5
Permanent atmospheric measurement sites, 29–30
Pine forests, volatile organic compounds emitted by, 23, 76–77
Plant-by-plant databases, 66–67
PM. *See* Particulate matter (PM)
Pollutant plumes, atmospheric dispersion models for, 264–265
Pollution concentrations. *See* Atmospheric concentrations
Pope et al. (2002), 301
Pope et al. (2009), 300–301
Population Division, of Department of Economic and Social Affairs (U.N.), 386
Population exposure, in health benefits study, 291–292, 294, *295*
Population growth
  in base case for future policy scenarios (2013–2020), 109, 405
  in base case for past policy scenarios (2006–2010), 294

Population projections, in economic model, 385–386, *386*
Power dispatch model, 342, 370n11
Power plants
  coal-fired (*see* Coal-fired power plants)
  emission factors for, *236*
  energy consumption by
    in 2005, *232*
    in 2010, *235*
Power sector
  development of, 161, *162*, *163*
  in economic model, 337
Precalciner kilns, 204
  cement production by, 206, *207*
  estimates for 2010 of, 217, *217*
  emission factors for
    of particulate matter, *209*, 209–210, *210*
    of SO₂, NOₓ, and CO, 207–208, *208*
  and spatial distribution of emissions, 214
Premature deaths. *See* Mortality, premature
Price effect, 359, 371n17
Primary energy consumption, by sector, 234–235, *235*
Primary pollutants
  atmospheric modeling of, 275, 282–283
  defined, 3, 263
  under past policy scenarios, 64
  scientific research on, 23, 33–34
  sources of data on, 13
Private savings rate, in economic model, 332, 384
Process emissions, 18, 204–206, 214, 228–229, 231, 354
Production sectors, in economic model, *68*, *69*, *70*, 335, 375–378
Production technology, in economic model, 335
Productivity growth, in economic model, 335, 389
Propane (C₃H₈) emissions, *244*
Provincial emissions, in 2005, 238–242, *239*
Public Health and Air Pollution in Asia (PAPA) program (U.S.), 303
Pulverized combustion (PC) boiler, for coal-fired power plant, *170*, *180*, 180–181, *181*, 182

Qian et al. (2007), 294
Quality of labor input index, 386
Quantity index of government purchases, 356
Quantity of industry output, 375–377, 390n1
Quicklime (CaO), in clinker, 208

Regional emissions, in 2005, 238–242, *239*
Relative risk (RR), in health benefits study, 298–301, *299*, *300*, 303, *304–305*
Research Institute of Fiscal Science (China), 355
"Revealed preference" approach, to valuation of health damages, 394
Revenues
  from future carbon tax scenarios, 108, *112*, 114, *123–126*, 356
  lump sum transfer back to households in past carbon tax scenario of, 356
  economic impact of, 356–362, *357*, *358*, *361*
  vs. tax cut revenue case, 362–363
  used to cut enterprise income tax rate in future carbon tax scenario, *112*, 114
  agricultural benefits of, 143, *144–147*
  comparing results of, *148*, 150
  effects on economy and energy use of, *117–119*, 122–123
  effects on emissions of, 129–132, *130–131*
  effects on pollution concentration of, 133–135, *134–135*, 137
  in end year, 114, *116*
  in first year, 114, *115*
  health benefits of, 138–143, *139*, *141*, *142*
  used to cut existing distortionary taxes in past carbon tax scenario, 75, 356
  economic impact of, *357*, *359*
  vs. lump sum case, 362–363
  used to protect vulnerable industries in future carbon tax scenario, 108, *123–125*
  effects on economy and energy use of, *117–119*
  in end year, 114, *116*
  in first year, 114, *115*

Rice production
  annual yield, 311, *312*
  effects of future carbon tax scenarios on, 138, *144*, *146*
  effects of past carbon tax scenario on, *320*, *323*
  exposure–relative yield functions for, *319*
  spatial distribution of, 313, *314–315*
Richter et al. (2005), 30
Rotary kilns, 203–204
  cement production by, 206, *207*
  estimates for 2010 of, 217, *217*
  emission factors for
  of particulate matter, *209*, 209–210, *210*
  of SO_2, NO_x, and CO, 207–208, *208*
RR (relative risk), in health benefits study, 298–301, *299*, *300*, 303, *304–305*

Saikawa et al. (2009), 42, 46, 48
SAM (Social Accounting Matrix), 336, 383–384, *385*, 388, 390
"San Gong" power dispatch model, 342, 370n11
Satellite observations of atmospheric pollutants, 29–35, *36–37*, 40
Savings, in economic model, 332, 384
Scientific research, 13–14, 22–40
  on atmospheric concentrations, 28–40
  of CO_2, 38–40, *39*
  data sources for, 28–30
  of NO_x, 30–32, *31*
  of ozone, 35–38, *36–37*
  of particulate matter, 32–34, *33*
  of SO_2, 34–35
  on emissions, 22–28
  of CO_2, 26–28, *27*
  of conventional air pollutants, 23–26, *24–26*
  uncertainties in, 28
SCR. *See* Selective catalytic reduction (SCR)
Seasonality
  of ozone concentrations, 271–273, *272*, *282–285*
  of $PM_{2.5}$ concentrations, 273–274, *273–274*
  of SO_2 and sulfate concentrations, *279*, *279–280*

Secondary pollutants
  atmospheric modeling of
    under 2010 base case, 275, 278
    under 2010 carbon tax case, 280,
      282–283
    conclusions on, 287
    Eulerian gridded chemical transport
      models for, 265–266
  defined, 4–5, 263
  under past policy scenarios, 64
  scientific research on, 32–35
Sector emissions, in 2005, 238–242, *240*
Sector output, for economic model, 332,
  333, *334*
Selective catalytic reduction (SCR), in
  coal-fired power plant, 169, *170*, 171,
  192–193, 197
SEPA. *See* State Environmental Protection
  Administration (SEPA, China)
Shaft kilns, 203–204
  cement production by, 206, *207*
  estimates for 2010 of, 217, *217*
  emission factors for
    of particulate matter, 209, 209–210, *210*
    of SO₂, NOₓ, and CO, 207–208, *208*
  and spatial distribution of emissions, 214
Short-term measurement campaigns, 29
Sinton (2001), 12
Sinton and Fridley (2003), 12
Slant column density, 51n23
Small power plant shutdown policy, 61,
  163, 165–166
  economics of, 340–342, *341*, *342*
  impacts of, 344–350, *347–349*
SO₂. *See* Sulfur dioxide (SO₂)
Social Accounting Matrix (SAM), 336,
  383–384, *385*, 388, 390
Solow type model, 331, 369–370n2
Solvent consumption, 231
Sonde-based measurement campaigns, 30
"Soot," 50n9
Soybean production, 311–312
Space-based observations, of atmospheric
  pollutants, 29–35, *36–37*, 40
Spatial allocation, in economy-wide
  emission inventory, 247–248, *248*,
  250–251

SPECIATE database, 242
"Stated preference" approach, to valuation
  of health damages, 394
State Environmental Protection
  Administration (SEPA, China)
  and coal-fired power plant emissions, 169
  and emission inventory, 238
  and health benefits, 299
  integrated "Environmental Cost Model"
    of, *41*, 44–45, 48
Steel industry. *See* Iron and steel industry
Streets and Waldhoff (2000), *252*
Streets et al. (2000), *252*
Streets et al. (2001), *253*
Streets et al. (2003), 189, *191*, 225, 238,
  242, *252–253*
Subsidies, to energy-intensive, trade-
  exposed (EITE) industries
  effects on economy and energy use of,
    115–119
Subsidies, to energy-intensive, trade-
  exposed (EITE) industries under future
  carbon tax scenario, 108, *123–125*
Su et al. (2009), 105
Sulfate concentrations
  atmospheric modeling of, 268, 273–274
    in 2010 base case, 275, 276–277
    in 2010 11th FYP SO₂ control case,
      278–280, *279*
  health effects of, 293–294
  reductions of
    effect on nitrate concentrations of, 34–35
    in 11th FYP SO₂ control case, 64, 86,
      278–280, *279*
    in past carbon tax scenario, 88
  seasonality of, *279*, 279–280
Sulfur content
  for coal-fired power plants, *170*, 181, *182*
  and level of SO₂ emissions, 228
Sulfur dioxide (SO₂) concentrations
  atmospheric modeling of
    in 2010 base case, 275, 276–277
    in 2010 carbon tax case, *280–281*,
      280–282
    in 2010 11th FYP SO₂ control case,
      278–280, *279*
  effect of 11th FYP SO₂ controls on, 352

effects of future carbon tax scenarios on, 132–137, *134–137*
official statistics on, *20*, 21
scientific research on, 34–35
seasonality of, *279*, 279–280
Sulfur dioxide (SO₂) emissions
  from cement production
    from 1990 to 2008, *211*, 212, *213*
    emission factors of, 207–208, *208*
    estimates for 2010 of, 218, *218*, 219, *219*
    spatial distribution of, 213, *215–216*
  from coal-fired power plants
    from 2000 to 2005, 187–189, *188*
    in 2010, 192
    vs. AP-42, 183
    database of, 175–183, *180*
    field measurements of, 169–171, *170*
    vs. other studies, 189–191, *191*
  from combusting fuel, 228
  in economic model, *334*
  effects of future carbon tax scenarios on, 127–132
    vs. base case, 128–129, *129*
    with various carbon tax cases, *130–131*, 132, *148*
  effects of past policy scenarios on
    under base case, 77–79, *78*, *80*
    under carbon tax, 81–84, *83*, *357*
    under SO₂ controls, 79, 81, *82*, *348*, *352*
  in emission inventory
    emission factors of, 235–237, *236*
    gridded, 248, *250–251*
    national, regional, and sector distribution of in 2005, 238–240, *239*, *240*
    by scenario and sector in 2010, 243–247, *244–245*
    trends in, 249–255, *252–253*
  health effects of, 293–294
  official statistics on, 17–18
  by production sector, *70*
  scientific research on, 25, *25*
Sulfur dioxide (SO₂) emissions reduction
  under 11th FYP SO₂ controls, 59 (*see also* 11th Five-Year Plan (11th FYP, 2006–2010) SO₂ controls)
  vs. carbon tax and base case, 95–100, *97–99*

effects on economy and energy use of, 69–71, *72*
effects on emissions of, 79–81, *82*
effects on pollution concentrations of, 86, 87
health and agricultural benefits of, 94, *94*
integrated framework to assess, 64–67, *65*
overview of, 61
by flue gas desulfurization, 6
SUM06 index, 315, 316
"Supersite" monitoring stations, 19, 30
Surface measurement sites, 29

Tax rates, in economic model, 379–380, 388, *388*
Technology mandates, 8, 62, 67
Teragram (Tg), 220n3, 238, 256n4
Terms of trade
  defined, 155n14
  in economic model, *335*
  for oil imports, 117
TFP (total factor productivity) growth, 389
Thermal formation, in cement production, 207
Thermal power sector development, 161, *162*, *163*
Three-dimensional (3-D) atmospheric chemical transport models, 265–267. *See also* Global atmospheric transport and chemistry (GEOS-Chem) model
Tian (2003), 189, *191*, 252
"Top-down" methods, 22, 51n16
Total factor productivity (TFP) growth, 389
Total suspended particulate (TSP) concentrations, official statistics on, *20*, 21
Total suspended particulate (TSP) emissions
  from cement production, 213, *218*, *219*
  in economic model, 332, *334*
  in economy-wide emission inventory, *236*
  health effects from, 298
  official statistics on, 17, 18
  scientific research on, 23–25, *24*
Trade elasticities, in economic model, 390
Trade intensity, *123*

Transport and Chemical Evolution over the
Pacific (TRACE-P) experiment, 225
Transportation sector
activity levels for
in 2005, 231–233, *232, 233*
in 2010, *234*, 234–235, *235*
in economy-wide emission inventory, 229
effects of past carbon tax scenario on, 360
emission factors for, *236*, 238
new assumptions and methods in future
carbon tax scenarios for, 404
Troposphere
GEOS-Chem model for, 268–269
ozone concentrations in, 35
3-D atmospheric chemical transport
models for, 266
TSP. *See* Total suspended particulate (TSP)
12th Five-Year Plan (12th FYP), 4, 101,
104–105, 197, 254

Uncertainties in scientific research
on coal-fired power plants, 180, 182,
185–187
on economics of environmental policy,
368
on health benefits, 138–140, 300,
304–305
on national CO_2 emission estimates, 27,
28
and sources of data, 11–13
Unit-based methodology, for coal-fired
power plants, 163–165, *164*
Unit carbon tax, 354–355, 371n14
United Nations Framework Convention on
Climate Change (UNFCCC, 1992), 18,
62, 103
Urbanization rate, 387, *387*
Urban monitoring stations, 19, 50n11
Urban plumes, 29
U.S.-China Strategic Economic Dialogue,
45

Value added, for economic model, 332,
333, 334
Value of statistical life (VSL), 394, 405
for mortality valuation, *395*, 395–397,
397

van Aardenne et al. (2005), 189–191, *191,
252*
van Donkelaar et al. (2010), 51–52n27
VCD. *See* Vertical column density (VCD)
Vehicles
activity levels for
in 2005, 231–233, *232, 233*
in 2010, *234*, 234–235, *235*
emission factors for, *236*, 238
in emission inventory, 229
emission standards for, 4
fuel economy standards for, 4
Vennemo et al. (2009), 42, 46, 48, 363
Vertical column density (VCD), 30
of NO_X, 30–32
of SO_2, 34–35, *36–37*, 52n29
Viscusi and Aldy (2003), 393
Volatile organic compounds (VOCs)
and emissions estimation
under future carbon tax scenarios, 137,
155n21
under past policy scenarios, 76–77
non-methane (*see* Non-methane volatile
organic compounds (NMVOC))
in ozone formation, 283
scientific research on, 23
VSL (value of statistical life), 394, 405
for mortality valuation, *395*, 395–397,
397
Vulnerable industries, future carbon tax
revenues used to protect, 108,
123–125
effects on economy and energy use of,
115–119

W126 index, 316
Wages
in economic model, 383
and health risks, 394
Wall-fired boilers, for coal-fired power
plant, 181–182
Wang, Li, and Zhang (2011), 105–106
Wang, L. T. (2006), 242
Wang, S. X. (2001), 189, *191*, 237
Wang, T., et al. (2009), 38
Wang, Y. X., Munger, et al. (2010), 27–28,
39, *39*

Wang, Y. X. et al. (2004), 225–226
Wang, Y. X. et al. (2007), 226
Wang, Y. X. et al. (2009), 38
Wang and Mullahy (2006), 393, *395*, 396
Waxman-Markey mechanism, *124*, 355
Wei et al. (2008), 237, 242, 243
Wet deposition, in 3-D atmospheric
  chemical transport models, 266
Wet scrubber, for coal-fired power plants,
  170, 174, *181*, 185, 189
W-flame boilers, for coal-fired power plant,
  182
Wheat production
 annual yield, 311, *312*
 effects of future carbon tax scenarios on,
   138, *144*, *145*
 effects of past carbon tax scenario on,
   319–323, *320–323*
 exposure–relative yield functions for, *319*
 spatial distribution of, 313, *314–315*
WHO (World Health Organization), 295
"Willingness-to-pay" (WTP) approach, to
  valuation of health damages, 393–394,
  398
Wong, T. W., et al. (1999), 303
Work hours, in economic model, 335,
  370n4
World Bank (1997), 301
World Bank-SEPA (2007), *41*, 44–45, 48
World Health Organization (WHO), 295
WTP ("willingness-to-pay") approach, to
  valuation of health damages, 393–394,
  398

Xu, Li, and Huang (1995), 302

Yi et al. (2011), 106

Zhang, J. F., and Smith (2007), 47
Zhang, Q. (2005), 237
Zhang, Q. et al. (2007a), 189, *191*, 237,
  240, *252*
Zhang, Q. et al. (2007b), 189, 191, *191*,
  240, 241
Zhang, Q. et al. (2009), *252–253*, 269
Zhao, Nielsen, and McElroy (2012), 26–28,
  27

Zhao, Zhang, and Nielsen (2013), 35,
  36–37
Zhao et al. (2008), 176
Zhao et al. (2010), 176, 237
Zhao et al. (2011), *252–253*
Zhou and Hammitt (2007), 393, 394, *395*,
  396, 397, 398
Zhou et al. (2010), 300
Zhu, Wang, and Zheng (2004), 189, 191,
  191